EESES
Electric Energy Systems and Engineering Series

Editors: J. G. Kassakian · D. H. Naunin

Pei-bai Zhou

Numerical Analysis of Electromagnetic Fields

With 157 Figures

Springer-Verlag Berlin Heidelberg GmbH

Professor Pei-bai Zhou
Dept. of Electrical Engineering
Xi'an Jiao-tong University
Xi'an Shaanxi, 710049, P.R. China

Library of Congress Cataloging-in-Publication Data
Zhou, Pei-bai, 1936–
Numerical analysis of electromagnetic fields / Pei-bai Zhou.
p. cm. — (Electric energy systems and engineering series)
Includes bibliographical references and index.

1. Electromagnetic fields – Mathematics. 2. Field theory
(Physics) – Mathematics. 3. Numerical analysis. 4. Finite element
method. 5. Boundary element methods. I. Title. II. Series.
QC665.E4Z46 1993   92-39200
530.1'41 — dc20   CIP

This work is subject to copyright. All rights are reserved, whether the whole or part of the material is concerned, specifically the rights of translation, reprinting, re-use of illustrations, broadcasting, reproduction on microfilms or in other ways, and storage in data banks. Duplication of this publication or parts thereof is only permitted under the provisions of the German Copyright Law of September 9, 1965, in its current version and a copyright fee must always be paid. Violations fall under the prosecution act of the German Copyright Law.

ISBN 978-3-642-50321-4   ISBN 978-3-642-50319-1 (eBook)
DOI 10.1007/978-3-642-50319-1

© Springer-Verlag Berlin Heidelberg 1993

Originally published by Springer-Verlag Berlin Heidelberg New York in 1993.

The use of registered names, trademarks, etc. in this publication does not imply, even in the absence of a specific statement, that such names are exempt from the relevant protective laws and regulations and therefore free for general use.

Typesetting: Macmillan (India) Ltd., Bangalore, India
61/3020 – 5 4 3 2 1 0 – Printed on acid-free paper

Series Editors:

Prof. J. G. Kassakian
Massachusetts Institute of Technology,
77 Massachusetts Ave., Cambridge, MA 02139, USA

Prof. D. H. Naunin
Institut für Elektronik, Technische Universität Berlin,
Einsteinufer 19, W-1000 Berlin 10, FRG

Introduction to the Electric Energy Systems and Engineering Series

Concerns for continued supply and efficient use of energy have recently become important forces shaping our lives. Because of the influence which energy issues have on the economy, international relations, national security, and individual well-being, it is necessary that there exists a reliable, available and accurate source of information on energy in the broadest sense. Since a major form of energy is electrical, this new book series titled *Electric Energy Systems and Engineering* has been launched to provide such an information base in this important area.

The series coverage will include the following areas and their interaction and coordination: generation, transmission, distribution, conversion, storage, utilization, economics.

Although the series is to include introductory and background volumes, special emphasis will be placed on: new technologies, new adaptions of old technologies, materials and components, measurement techniques, control – including the application of microprocessors in control systems, analysis and planning methodologies, simulation, relationship to, and interaction with, other disciplines.

The aim of this series is to provide a comprehensive source of information for the developer, planner, or user of electrical energy. It will also serve as a visible and accessible forum for the publication of selected research results and monographs of timely interest. The series is expected to contain introductory level material of a tutorial nature, as well as advanced texts and references for graduate students, engineers and scientists.

The editors hope that this series will fill a gap and find interested readers.

John G. Kassakian · Dietrich H. Naunin

Preface

Numerical methods for solving boundary value problems have developed rapidly. Knowledge of these methods is important both for engineers and scientists. There are many books published that deal with various approximate methods such as the finite element method, the boundary element method and so on. However, there is no textbook that includes all of these methods. This book is intended to fill this gap. The book is designed to be suitable for graduate students in engineering science, for senior undergraduate students as well as for scientists and engineers who are interested in electromagnetic fields.

Objective

Numerical calculation is the combination of mathematical methods and field theory. A great number of mathematical concepts, principles and techniques are discussed and many computational techniques are considered in dealing with practical problems. The purpose of this book is to provide students with a solid background in numerical analysis of the field problems. The book emphasizes the basic theories and universal principles of different numerical methods and describes why and how different methods work. Readers will then understand any methods which have not been introduced and will be able to develop their own new methods.

Organization

Many of the most important numerical methods are covered in this book. All of these are discussed and compared with each other so that the reader has a clear picture of their particular advantage, disadvantage and the relation between each of them. The book is divided into four parts and twelve chapters.

The first part deals with the **universal concepts of numerical analysis for electromagnetic field problems** in Chapter 1 and 2. This is a review and extension of electromagnetic field theory, followed by a general outline of approximate methods. Green's theorem, Green's function, fundamental solutions and equivalent surface sources are the basic tools reviewed in this section. The concepts of discretization, error minimization of the approximation and the basic principles of numerical methods are introduced before any specific method is discussed.

The following two parts discuss specific methods. Part two is concerned with **domain methods**. It includes finite difference and finite element method. Chapter 3 explains the finite difference method. The finite element method is discussed in three chapters, designed step by step. Chapter 4 introduces the general idea and procedures of the finite element method. Here the matrix equation of the finite element method is derived by using the principle of Galerkin's weighted residuals. Chapter 5 descibes additional properties and applications of the finite element method. The discretization equation is derived using the variational principle. The equivalent functionals of different types of field problems are derived. The important techniques of domain discretizations are dealt with in Chapter 6, because they are used in both the finite element method and in other approximate methods.

Part three outlines **boundary methods**, as is the solution of integral equations. Four methods are described in the following four chapters. Chapter 7 considers the charge simulation method, being one of the simplest based on integral equations. The integral equation for problems containing several materials is in Chapter 8. The specific problem of integration around singularities of integral equations is introduced in this chapter. The boundary element method is given in Chapter 9. The charge simulation, surface charge simulation and the boundary element methods are special cases of the moment method, as is shown in the last chapter.

The last part of the book discusses the **optimization of electromagnetic field problems**. Optimum design is often the purpose of field analysis. Chapter 11 describes the mathematical tools used in optimization design. It introduces general methods to search for the extremum value of a given objective function. This chapter also demonstrates the solution methods for the solution of algebraic equations derived in parts two and three. The last chapter is the application of the combination of all these methods introduced in the text for purposes of optimized design.

Further applications, wider discussion and new techniques are listed in the references.

Summaries

Each chapter includes a brief summary of its content. Important statements are in italic.

Appendix

Some related mathematical formulations, approximate algorithms for calculating special functions, are given in the appendix belonging to each chapter.

Xi'an, March 1993   Zhou Pei-bai

Acknowledgement

The author wishes to acknowledge the co-operation and support of scientists and colleagues from many countries. The book was first recommended to Springer-Verlag by Professor A. J. Schwab, Technical University of Karlsruhe. The editor of this book, Mr. von Hagen, Mrs. Raufelder and the vice president of Springer-Verlag, New York, Mr. Grossmann, gave many helpful hints to the author during the preparation of the manuscript.

Encouragement and support for the publication of this book was given by Professor P. P. Biringer and Professor J. D. Lavers, University of Toronto. Both agreed to correct the manuscript and support the author while at the University of Toronto for a year to complete the manuscript. It was reviewed by Professor P. P. Biringer and J. D. Lavers, Dr. G. Bendzsak, Professors A. Konrad, P. E. Burke and S. Dmitrevsky who helped to improve the text and the contents. Special thanks are due to Professor Biringer for the substantial time he spent on the manuscript. The author expresses her warm thanks to all of them for their unselfish help.

Many friends who are working and studying at the University of Toronto also helped the author with the English language and gave valuable advice. Special thanks are given to Dr. Laszlo Kadar, Dr. K. V. Namjoshi, Mr. Chen Qian-ping and Miss Wang Jian-guo.

All my superiors and colleagues of Xi'an Jiao-tong University have supported the author to take leave for a year. The English language Professor Xie Jie-shu helped at the beginning of the manuscript. Many graduate students at Xi'an Jiao-tong University attended the course, read the manuscript, gave suggestions, corrected errors, and help the book to be completed. Dr. Chen Mei-juan and Miss Wang Ping are specially mentioned.

The author wants to thank her old colleagues, the group members studying electromagnetic field problems. Until 1985 she enjoyed the good cooperation concerning the topic of numerical analysis of electromagnetic fields.

Especially, the author has to express her thanks to her husband, Professor Ma Nai-xiang, and all members of her family who encouraged and supported the writing of this text.

Finally, I would like to thank Professor D. Naunin who oversaw the final review of the whole text.

Xi'an, March 1993   Zhou Pei-bai

Contents

Part 1 Universal Concepts for Numerical Analysis of Electromagnetic Field Problems

Chapter 1
Fundamental Concepts of Electromagnetic Field Theory 3

1.1	Maxwell's equations and boundary value problems	3
1.1.1	Potential equations in different frequency ranges	4
1.1.2	Boundary conditions of the interface	8
1.1.3	Boundary value problems	10
1.2	Green's theorem, Green's functions and fundamental solutions	11
1.2.1	Green's theorem	12
1.2.2	Vector analogue of Green's theorem	14
1.2.3	Green's function	15
1.2.3.1	Dirac-delta function	15
1.2.3.2	Green's function	16
1.2.4	Fundamental solutions	18
1.3	Equivalent sources	19
1.3.1	Single layer charge distribution	20
1.3.2	Double layer source distributions	22
1.3.3	Equivalent polarization charge and magnetization current ..	26
1.4	Integral equations of electromagnetic fields	28
1.4.1	Integral form of Poisson's equation	28
1.4.2	Integral equation for the exterior region	29
1.5	Summary	30
References	....	31
Appendix 1.1 The integral equation of 3-D magnetic fields		31

Chapter 2
General Outline of Numerical Methods 35

2.1	Introduction	35
2.2	Operator equations	37
2.2.1	Hilbert space	38

2.2.2	Definition and properties of operators	40
2.2.3	The relationship between the properties of the operators and the solution of operator equations	42
2.2.4	Operator equations of electromagnetic fields	43
2.3	Principles of error minimization	46
2.3.1	Principle of weighted residuals	47
2.3.2	Orthogonal projection principle	48
2.3.2.1	Projection operator	49
2.3.2.2	Orthogonal projection	49
2.3.2.3	Orthogonal projection methods	51
2.3.2.4	Non-orthogonal projection methods	51
2.3.3	Variational principle	51
2.4	Categories of various numerical methods	52
2.4.1	Methods of weighted residuals	53
2.4.1.1	Method of moments	53
2.4.1.2	Galerkin's finite element method	54
2.4.1.3	Collocation methods	55
2.4.1.4	Boundary element methods	55
2.4.2	Variational approach	56
2.5	Summary	58
References		59

Part 2 Domain Methods

Chapter 3
Finite Difference Method (FDM) 63

3.1	Introduction	63
3.2	Difference formulation of Poisson's equation	64
3.2.1	Discretization mode for 2-D problems	64
3.2.2	Difference equations in 2-D Cartesian coordinates	64
3.2.3	Discretization equation in polar coordinates	69
3.2.4	Discretization formula of axisymmetric fields	72
3.2.5	Discretization formula of the non-linear magnetic fields	73
3.2.6	Difference equations for time-dependent problems	73
3.3	Solution methods for difference equations	76
3.3.1	Properties of simultaneous equations	76
3.3.2	Successive over-relaxation (SOR) method	77
3.3.3	Convergence criterion	80
3.4.	Difference formulations of arbitrary boundaries and interfacial boundaries between different materials	80
3.4.1	Difference formulations on the lines of symmetry	81
3.4.2	Difference equation of a curved boundary	81
3.4.3	Difference formulations for the interface of different materials	82

3.5	Examples	84
3.6	Further discussions about the finite difference method	87
3.6.1	Physical explanation of the finite difference method	87
3.6.2	The error analysis of the finite difference method	88
3.6.3	Difference equation and the principle of weighted residuals	90
3.6.4	Difference equation and the variational principle	91
3.7	Summary	92
References		93

Chapter 4
Fundamentals of Finite Element Method (FFM) 95

4.1	Introduction	95
4.2	General procedures of the finite element method	96
4.2.1	Domain discretization and shape functions	97
4.2.2	Method using Galerkin residuals	100
4.2.2.1	Element matrix equations	101
4.2.2.2	System matrix equation	106
4.2.2.3	Storage of the system matrix	109
4.2.2.4	Treatment of the Dirichlet boundary condition	111
4.3	Solution methods of finite element equations	113
4.3.1	Direct methods	113
4.3.1.1	Gaussian elimination method	113
4.3.1.2	Cholesky's decomposition (triangular decomposition)	115
4.3.2	Iterative methods	117
4.3.2.1	Method of over-relaxation iteration	118
4.3.2.2	Conjugate-gradient method (CGM)	119
4.4	Mesh generation	120
4.4.1	Mesh generation of a triangular element	121
4.4.2	Automatic mesh generation	123
4.5	Examples	124
4.6	Summary	127
References		128

Chapter 5
Variational Finite Element Method 130

5.1	Introduction	130
5.2	Basic concepts of the functional and its variations	131
5.2.1	Definition of the functional and its variations	132
5.2.1.1	The functional	132
5.2.1.2	The differentiation and variation of a function	134
5.2.1.3	Variation of the functional	135

5.2.2	Calculus of variations and Euler's equation	136
5.2.2.1	Euler's equation	137
5.2.2.2	Euler's equation for multivariable functions	138
5.2.2.3	The shortest length of a curve	140
5.2.3	Relationship between the operator equation and the functional	141
5.3	Variational expressions for electromagnetic field problems	142
5.3.1	Variational expression for Poisson's equation	143
5.3.1.1	Mathematical manipulation	143
5.3.1.2	Physical manipulation	146
5.3.2	Variational expressions for Poisson's equations in piece-wise homogeneous materials	147
5.3.3	Variational expression for the scalar Helmholtz equation	148
5.3.4	Variational expression for the magnetic field in a non-linear medium	150
5.4	Variational finite element method	153
5.4.1	Ritz method	153
5.4.2	Finite element method (FEM)	155
5.4.2.1	Domain discretization	155
5.4.2.2	Finite element equation of a Laplacian problem	156
5.4.2.3	Finite element equation for 2-D magnetic fields	161
5.4.2.4	Finite element equation for non-linear magnetic fields	163
5.4.2.5	Finite element equation for Helmholtz's equation (2-D-case)	164
5.5	Special problems using the finite element method	166
5.5.1	Approaching floating electrodes by the variational finite element method	166
5.5.2	Open boundary problems	167
5.5.2.1	Introduction	167
5.5.2.2	Ballooning method	168
5.6	Summary	170
References		171

Chapter 6
Elements and Shape Functions 172

6.1	Introduction	172
6.2	Types and requirements of the approximating functions	173
6.2.1	Lagrange and Hermite shape functions	173
6.2.2	Requirements of the approximating functions	174
6.3	Global, natural, and local coordinates	176
6.3.1	Natural coordinates	176
6.3.2	Local coordinates	182
6.4	Lagrange shape function	183
6.4.1	Triangular elements	184
6.4.2	Quadrilateral elements	186
6.4.3	Tetrahedral and hexahedral elements	188

6.5	Parametric elements	189
6.6	Element matrix equation	191
6.6.1	Coordinate transformations, Jacobian matrix	192
6.6.2	Evaluation of the Lagrangian element matrix	193
6.6.3	Universal matrix	195
6.7	Hermite shape function	199
6.7.1	One dimensional Hermite shape function	199
6.7.2	Triangular Hermite shape functions	200
6.7.3	Evaluation of a Hermite element matrix	202
6.8	Application discussions	204
6.9	Summary	206
References		206
Appendix 6.1	Langrangian shape functions for 2-D cases	207
Appendix 6.2	Commonly used shape functions for 3-D cases	208
Appendix 6.3	The universal matrix of axisymmetric fields	209

Part 3 Boundary Methods

Chapter 7
Charge Simulation Method (CSM) ... 215

7.1	Introduction	215
7.2	Matrix equations of simulated charges	216
7.2.1	Matrix equation in homogeneous dielectrics	216
7.2.1.1	Governing equation subject to Dirichlet boundary conditions	216
7.2.1.2	Governing equation subject to Neumann boundary conditions	218
7.2.1.3	Mixed boundary conditions and free potential conductors	218
7.2.1.4	Matrix form of Poisson's equation	219
7.2.2	Matrix equation in piece-wise homogeneous dielectrics	220
7.3	Commonly used simulated charges	221
7.3.1	Point charge	222
7.3.2	Line charge	223
7.3.3	Ring charge	223
7.3.4	Charged elliptic cylinder	225
7.4	Applications of the charge simulation method	226
7.5	Coordinate transformations	232
7.5.1	Transformation matrix	233
7.5.2	Inverse transformation of the field strength	234
7.6	Optimized charge simulation method (OCSM)	235
7.6.1	Objective function	235
7.6.2	Transformation of constrained conditions	237
7.6.3	Examples	237
7.7	Error analysis in the charge simulation method	241
7.7.1	Properties of the errors	241

7.7.2	Error distribution pattern along the electrode contour	242
7.7.3	Factors influencing the errors	243
7.8	Summary	246
References		247
Appendix 7.1	Formulations for a point charge	248
Appendix 7.2	Formulations for a line charge	248
Appendix 7.3	Formulations for a ring charge	249
Appendix 7.4	Formulations for a charged elliptic cylinder	249
Appendix 7.5	Approximate formulations for calculating $K(k)$ and $E(k)$	250

Chapter 8
Surface Charge Simulation Method (SSM) 251

8.1	Introduction	251
8.1.1	Example	252
8.2	Surface integral equations	254
8.2.1	Single layer or double layer integral equations	254
8.2.2	Integral equations of the interfacial surface	255
8.3	Types of surface boundary elements and surface charge densities	257
8.3.1	Representations of boundary and charge density	257
8.3.2	Potential and field strength coefficients for 2-D and axisymmetrical problems	258
8.3.2.1	Planar element with constant or linear charge density	259
8.3.2.2	Arced element with constant or linear charge density	262
8.3.2.3	Ring element with linear charge density	264
8.3.3	Elements for 3-D problems	267
8.3.3.1	Planar triangular element	267
8.3.3.2	Cylindrical tetragonal bilinear element	267
8.3.3.3	Isoparametric high order element	268
8.3.3.4	Spline function element	269
8.4	Magnetic surface charge simulation method	270
8.5	Evaluation of singular integrals	272
8.5.1	The semi-analytical technique	273
8.5.2	Method using coordinate transformations	274
8.5.3	Numerical techniques	275
8.5.4	Combine the analytical integral and Gaussian quadrature	275
8.6	Applications	275
8.7	Summary	280
References		280
Appendix 8.1	Potential and field strength coefficients of 2-D planar elements with constant and linear charge density	281
Appendix 8.2	Potential and field strength coefficients of 2-D arced elements with constant and linear charge density	283
Appendix 8.3	Coefficients of ring elements with linear charge density	285

Chapter 9
Boundary Element Method (BEM) ... 287

9.1	Introduction	287
9.2	Boundary element equations	288
9.2.1	Method of weighted residuals	288
9.2.2	Green's theorem	291
9.2.3	Variational principle	291
9.2.4	Boundary integral equation	292
9.2.5	Indirect boundary integral equation	295
9.3	Matrix formulations of the boundary integral equation	295
9.3.1	Discretization and shape functions	296
9.3.2	Matrix equation of a 2-dimensional constant element	298
9.3.2.1	Evaluation of H_{ij} and G_{ij}	300
9.3.2.2	Evaluation of H_{ii} and G_{ii}	301
9.3.3	Matrix equation of 2-D linear elements	302
9.3.4	Matrix form of Poisson's equation	304
9.3.5	Matrix equation of a piecewise homogeneous domain	305
9.3.6	Matrix equation of axisymmetric problems	306
9.3.7	Discretization of 3-dimensional problems	307
9.3.8	Use of symmetry	309
9.4	Eddy current problems	310
9.4.1	Eddy current equations	310
9.4.1.1	A-φ formulations	311
9.4.1.2	T-Ω formulations	311
9.4.2	One-dimensional solution of an Eddy current problem	312
9.4.3	BEM for solving Eddy current problems	313
9.4.4	Surface impedance boundary conditions	316
9.5	Non-linear and time-dependent problems	317
9.5.1	BEM for non-linear problems	317
9.5.2	Time-dependent problems	320
9.6	Summary	321
References		322
Appendix 9.1 Bessel function		323

Chapter 10
Moment Methods ... 327

10.1	Introduction	327
10.2	Basis functions and weighting functions	330
10.2.1	Galerkin's methods	332
10.2.2	Point matching method	332
10.2.3	Sub-regions and sub-sectional basis	334
10.3	Interpretation using variations	335

10.4	Moment methods for solving static field problems	336
10.4.1	Charge distribution of an isolated plate	336
10.4.2	Charge distribution of a charged cylinder	338
10.5	Moment methods for solving eddy current problems	341
10.5.1	Integral equation of a 2-D eddy current problem	341
10.5.2	Sub-sectional basis method	342
10.6	Moment methods to solve the current distribution of a line antenna	343
10.6.1	Integral equation of a line antenna	343
10.6.2	Solution of Hallen's equation	345
10.7	Summary	347
References		348

Part 4 Optimization Methods of Electromagnetic Field Problems

Chapter 11
Methods of Applied Optimization 351

11.1	Introduction	351
11.2	Fundamental concepts	351
11.2.1	Necessary and sufficient conditions for the local minimum	352
11.2.2	Geometrical interpretation of the minimizer	354
11.2.3	Quadratic functions	355
11.2.4	Basic method for solving unconstrained non-linear optimization problems	356
11.2.5	Stability and convergence	357
11.3	Linear search and single variable optimization	358
11.3.1	Golden section method	358
11.3.2	Methods of polynomial interpolation	360
11.4	Analytic methods of unconstrained optimization problems	362
11.4.1	The method of steepest descent	363
11.4.2	Conjugate gradient method	364
11.4.2.1	Conjugate direction	364
11.4.2.2	Quadratic convergence	364
11.4.2.3	Selection of conjugate directions	365
11.4.3	Quasi-Newton's methods	367
11.4.3.1	Davidon-Fletcher-Powell (DFP) method	368
11.4.3.2	BFGS formulation	370
11.4.3.3	B matrix formulae	371
11.4.3.4	Cholesky factorization of the Hessian matrix	371
11.4.4	Method of non-linear least squares	372
11.4.4.1	Gauss-Newton method	372
11.4.4.2	Levenberg-Marquardt method	373
11.5	Function comparison methods	374

11.5.1	Polytype method	374
11.5.2	Powell's method of quadratic convergence	375
11.6	Constrained optimization methods	376
11.6.1	Basic concepts of constrained optimization	377
11.6.2	Kuhn-Tucker conditions	378
11.6.2.1	Lagrange multiplier method	378
11.6.2.2	Necessary condition of the first order	379
11.6.2.3	Necessary and sufficient conditions of the second order	380
11.6.3	Penalty and barrier function methods	381
11.6.4	Sequential unconstrained minimization technique	384
11.7	Summary	385
References		386

Chapter 12
Optimizing Electromagnetic Devices ... 387

12.1	Introduction	387
12.2	General concepts of optimum design	388
12.2.1	Objective function	388
12.2.2	Mathematical expressions of the boundary value problem	389
12.2.3	Optimization methods	390
12.2.4	Categories of optimization	391
12.3	Contour optimization	391
12.3.1	Method of curvature adjustment	391
12.3.2	Method of charge redistribution	393
12.3.3	Contour optimization by using non-linear programming	395
12.4	Problems of domain optimization	396
12.4.1	Field synthesis by using Fredholm's integral equation	396
12.4.2	Domain optimization by using non-linear programming	398
12.5	Summary	400
References		400

Subject Index ... 403

Part One
Universal Concepts for the Numerical Analysis of Electromagnetic Field Problems

Many of the basic concepts that are used when undertaking numerical analysis of electromagnetic field problems are reviewed in this section. The first chapter provides the theoretical basis for numerical analysis of electromagnetic field problems. The second chapter provides a general outline of the various numerical methods introduced in this book.

Chapter 1

Fundamental Concepts of Electromagnetic Field Theory

The solution of many practical electromagnetic field problems can only be undertaken by applying numerical methods. Before such a solution can be undertaken, it is important that a correct mathematical model be established for the problem considered. Maxwell's equations and the associated boundary conditions provide the necessary basis for the modelling of practical electromagnetic problems which are reviewed in this chapter. Further, as Green's theorem, fundamental solutions, and equivalent sources are the basic tools used in some numerical techniques, they are also presented here.

1.1 Maxwell's equations and boundary value problems

In free space, Maxwell's equations and the constitutive equations are:

$$\nabla \times \mathbf{H} = \frac{\partial \mathbf{D}}{\partial t} + \mathbf{J} \tag{1.1.1}$$

$$\nabla \times \mathbf{E} = -\frac{\partial \mathbf{B}}{\partial t} \tag{1.1.2}$$

$$\nabla \cdot \mathbf{B} = 0 \tag{1.1.3}$$

$$\nabla \cdot \mathbf{D} = \rho \tag{1.1.4}$$

$$\mathbf{B} = \mu_0 \mathbf{H} \tag{1.1.5}$$

$$\mathbf{D} = \varepsilon_0 \mathbf{E} . \tag{1.1.6}$$

In the presence of conducting materials, the principle of charge conservation is expressed by the relation:

$$\nabla \cdot \mathbf{J} + \frac{\partial \rho}{\partial t} = 0 . \tag{1.1.7}$$

The current density $\mathbf{J}$ and the electric field $\mathbf{E}$ are related by Ohm's law:

$$\mathbf{J} = \gamma \mathbf{E} . \tag{1.1.8}$$

Finally, if the problem involves conducting bodies moving through magnetic fields, the total electric field must include a component $\mathbf{E}_v$ that is due to the velocity effects:

$$\mathbf{E}_v = \mathbf{v} \times \mathbf{B} . \tag{1.1.9}$$

In these equations, $\mathbf{H}$, $\mathbf{E}$ are vectors of magnetic and electric field strength, $\mathbf{B}$, $\mathbf{D}$ are the vectors of magnetic and electric flux density, $\mathbf{J}$ is the conduction current density and ρ is the electric charge density. Finally, ε_0, μ_0 are the permittivity and permeability of the free space while γ is the electrical conductivity of the material.

In dielectric and magnetic materials, the polarization vector $\mathbf{P}$ and the magnetization vector $\mathbf{M}$ are defined as folows:

$$\mathbf{P} = \mathbf{D} - \varepsilon_0 \mathbf{E} \tag{1.1.10}$$

$$\mathbf{M} = \frac{1}{\mu_0} \mathbf{B} - \mathbf{H} . \tag{1.1.11}$$

Maxwell's equations can then be written as:

$$\nabla \times \mathbf{E} = -\frac{\partial \mathbf{B}}{\partial t} \tag{1.1.12}$$

$$\nabla \times \mathbf{B} = \mu_0 \varepsilon_0 \frac{\partial \mathbf{E}}{\partial t} + \mu_0 \left(\mathbf{J} + \frac{\partial \mathbf{P}}{\partial t} + \nabla \times \mathbf{M} \right) \tag{1.1.13}$$

$$\nabla \cdot \mathbf{B} = 0 \tag{1.1.14}$$

$$\nabla \cdot \mathbf{E} = \frac{1}{\varepsilon_0} (\rho - \nabla \cdot \mathbf{P}) . \tag{1.1.15}$$

In Eq. (1.1.13) it will be noted that the dielectric and magnetic materials give rise to an equivalent current density of the form $\partial \mathbf{P}/\partial t + \nabla \times \mathbf{M}$. Similarly, in Eq. (1.1.15), the dielectric material gives rise to an equivalent volume charge density of the form $-\nabla \cdot \mathbf{P}$.

1.1.1 Potential equations in different frequency ranges [1]

Rapidly varying time-dependent fields

When the time variation of fields is rapid, the electric field and magnetic field are coupled to each other. The field distributions are dependent on both time and position, $\mathbf{E}(\mathbf{r}, t)$, $\mathbf{B}(\mathbf{r}, t)$. The time varying magnetic field induces the rotational electric field and the time varying electric field produces the rotational magnetic field. All the field quantities are fully dynamic.

In lossless media and source-free regions it is a simple matter to show that $\mathbf{E}$ and $\mathbf{H}$ satisfy a wave equation. In the case of $\mathbf{E}$, for example, by taking the curl

1.1 Maxwell's equations and boundary value problems

of Eq. (1.1.2) and then substituting Eq. (1.1.1) for H, it follows that:

$$\nabla^2 \mathbf{E} - \mu\varepsilon \frac{\partial^2 \mathbf{E}}{\partial t^2} = 0 \tag{1.1.16}$$

which is a wave equation for **E**. Similarly, it can be shown that **H** satisfies the identical equation:

$$\nabla^2 \mathbf{H} - \mu\varepsilon \frac{\partial^2 \mathbf{H}}{\partial t^2} = 0 . \tag{1.1.17}$$

When solving Maxwell's equation it is often convenient to introduce the scalar and vector potential functions φ and **A** as indicated below:

$$\mathbf{E} = -\frac{\partial \mathbf{A}}{\partial t} - \nabla\varphi \tag{1.1.18}$$

$$\mathbf{B} = \nabla \times \mathbf{A} . \tag{1.1.19}$$

It can then be shown that if the problem involves sources ρ and **J** the potentials satisfy inhomogeneous wave equations:

$$\nabla^2 \varphi - \mu\varepsilon \frac{\partial^2 \varphi}{\partial t^2} = -\rho/\varepsilon \tag{1.1.20}$$

$$\nabla^2 \mathbf{A} - \mu\varepsilon \frac{\partial^2 \mathbf{A}}{\partial t^2} = -\mu \mathbf{J} . \tag{1.1.21}$$

In deriving Eqs. (1.1.20) and (1.1.21) Lorentz's gauge has been assumed:

$$\nabla \cdot \mathbf{A} = -\mu\varepsilon \frac{\partial \varphi}{\partial t} . \tag{1.1.22}$$

In the case of lossy media, the wave equation is obtained by using the gauge of:

$$\nabla \cdot \mathbf{A} = -\mu\varepsilon \frac{\partial \varphi}{\partial t} - \mu\gamma\varphi . \tag{1.1.23}$$

The resulting equations have the form:

$$\nabla^2 \varphi - \mu\varepsilon \frac{\partial^2 \varphi}{\partial t^2} - \mu\gamma \frac{\partial \varphi}{\partial t} = -\rho/\varepsilon \tag{1.1.24}$$

$$\nabla^2 \mathbf{A} - \mu\varepsilon \frac{\partial^2 \mathbf{A}}{\partial t^2} - \mu\gamma \frac{\partial \mathbf{A}}{\partial t} = -\mu \mathbf{J} . \tag{1.1.25}$$

The above equations are used for computing waves radiating from antennas, waves scattered by material bodies, and waves propagating in wave-guides or other electronic devices.

Steady-state fields

When the time variation in a problem is relatively slow, the steady-state approximation may be used. The criterion for 'slow' as opposed to rapid variation is:

$$\gamma \gg \omega\varepsilon ,\qquad (1.1.26)$$

where ω is the frequency of sinusoid.

This criterion means that the conducting currents dominate the problem and the displacement currents can be neglected. Hence the rotational magnetic field induced by the electric field no longer exists. There is no relation between the position changing and the time varying field. *Hence there is no wave transmission.*

Usually, in steady-state field problems, the quantities $\mathbf{E}(\mathbf{r}, t)$, $\mathbf{H}(\mathbf{r}, t)$, $\mathbf{J}(\mathbf{r}, t)$, and $\rho(\mathbf{r}, t)$ are time harmonic functions. The field distribution therefore depends only on the position and phase delay at each point in space. In this case Maxwell's equations are reduced to:

$$\nabla \times \mathbf{H} = \mathbf{J} \qquad (1.1.27)$$

$$\nabla \times \mathbf{E} = -j\omega \mathbf{B} \qquad (1.1.28)$$

$$\nabla \cdot \mathbf{B} = 0 \qquad (1.1.29)$$

$$\nabla \cdot \mathbf{D} = 0 . \qquad (1.1.30)$$

When μ, γ are constants, $\mathbf{E}$ and $\mathbf{H}$ obey a parabolic diffusion equation:

$$\nabla^2 \mathbf{H} = j\omega\mu\gamma \mathbf{H}$$

$$\nabla^2 \mathbf{E} = j\omega\mu\gamma \mathbf{E} . \qquad (1.1.31)$$

For such problems it is convenient to employ a magnetic vector potential $\mathbf{A}$ or an electric vector potential $\mathbf{T}$ [2]. The definition of $\mathbf{A}$ and $\mathbf{T}$ follows directly from the Maxwell's equations of $\nabla \cdot \mathbf{B} = 0$ and $\nabla \cdot \mathbf{J} = 0$:

$$\begin{cases} \mathbf{B} = \nabla \times \mathbf{A} \\ \mathbf{J} = \nabla \times \mathbf{T} . \end{cases} \qquad (1.1.32)$$

Because of Ampere's law the relationship between the magnetic field strength $\mathbf{H}$ and the electric vector potential $\mathbf{T}$ is given by:

$$\mathbf{H} = \mathbf{T} - \nabla\Omega \qquad (1.1.33)$$

where Ω is a scalar magnetic potential. Equation (1.1.33) is derived from Eqs. (1.1.27), (1.1.32) and $\nabla \times \nabla\Omega = 0$. Similarity between the two vector potentials $\mathbf{A}$ and $\mathbf{T}$ is illustrated in Fig. 1.1.1.

Differential equations for the two vector potentials can be obtained by substituting Eqs. (1.1.18), (1.1.19) and (1.1.33) into Maxwell's equations. After

1.1 Maxwell's equations and boundary value problems

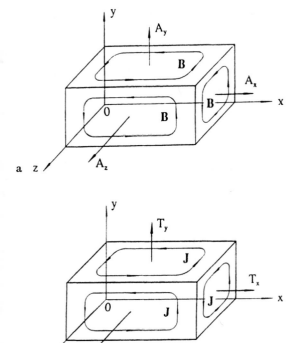

Fig. 1.1.1a, b. The similarity between the vector potentials **A** and **T**

a simple manipulation the following two equations are obtained:

$$\nabla \times \left(\frac{1}{\mu} \nabla \times \mathbf{A}\right) + j\omega\gamma \mathbf{A} + \gamma \nabla \varphi = \mathbf{J}_s \qquad (1.1.34)$$

$$\nabla \times \left(\frac{1}{\gamma} \nabla \times \mathbf{T}\right) + j\omega\mu \mathbf{T} - j\omega\mu \nabla \Omega = 0 \qquad (1.1.35)$$

where $\mathbf{J}_s$ is an imposed current density.

The most important application of a steady-state approximation is in determining the eddy current distribution in conducting regions and in iron cores. Depending on the material constants, the approximation may be valid up to the X-ray frequency range.

Static and quasi-static fields

*In static fields quantities ρ, **J**, **E**, **H** are time independent, i.e. $\partial/\partial t = 0$ and the field distributions are functions of positions only.* If the frequency is sufficiently low, the rotational electric field induced by the magnetic field of the displacement current is infinitely small. The field distribution of this case is then in fact a static distribution, it is called 'quasi-static' field. The criterion of quasi-static is $L \ll \lambda$. Where λ is the wave length, L is the dimension of the field region.

In the case of static and quasi-static fields Maxwell's equations reduce to the following form:

$$\nabla \times \mathbf{E} = 0, \qquad \nabla \cdot \mathbf{D} = \rho, \qquad \nabla \cdot \mathbf{J} = -\frac{\partial \rho}{\partial t} \tag{1.1.36}$$

$$\nabla \times \mathbf{H} = \mathbf{J}, \qquad \nabla \cdot \mathbf{B} = 0, \qquad \nabla \cdot \mathbf{J} = 0 \tag{1.1.37}$$

Based on $\nabla \times \mathbf{E} = 0$, $\nabla \cdot \mathbf{B} = 0$, both the scalar electric and magnetic potentials φ, φ_m, and the vector magnetic potential $\mathbf{A}$ are introduced in following forms:

$$\mathbf{E} = -\nabla \varphi \tag{1.1.38}$$

$$\mathbf{B} = \nabla \times \mathbf{A}$$

$$\mathbf{H} = -\nabla \varphi_m \quad \text{(in a current free region)} . \tag{1.1.39}$$

Based on Eqs. (1.1.36), (1.1.37) and (1.1.39), Poisson's and Laplace's equations are derived:

$$\nabla^2 \varphi = -\rho/\varepsilon \tag{1.1.40}$$

$$\nabla^2 \mathbf{A} = -\mu \mathbf{J} \tag{1.1.41}$$

$$\nabla^2 \varphi_m = 0 . \tag{1.1.42}$$

Usually, in static and quasi-static cases, Coulomb's gauge is satisfied as

$$\nabla \cdot \mathbf{A} = 0 . \tag{1.1.43}$$

1.1.2 Boundary conditions of the interface

At the interface of different materials the integral form of Maxwell's equations is reduced to:

$$\mathbf{n} \cdot (\mathbf{B}_1 - \mathbf{B}_2) = 0 \tag{1.1.44}$$

$$\mathbf{n} \cdot (\mathbf{D}_1 - \mathbf{D}_2) = \sigma \tag{1.1.45}$$

$$\mathbf{n} \times (\mathbf{E}_1 - \mathbf{E}_2) = 0 \tag{1.1.46}$$

$$\mathbf{n} \times (\mathbf{H}_1 - \mathbf{H}_2) = \mathbf{K} \tag{1.1.47}$$

$$\mathbf{n} \cdot (\mathbf{J}_1 - \mathbf{J}_2) = -\frac{d\sigma}{dt} , \tag{1.1.48}$$

where $\mathbf{n}$ is the unit normal vector to the interface as shown in Fig. 1.1.2. $\mathbf{E}_1$, $\mathbf{D}_1$, $\mathbf{B}_1$, $\mathbf{H}_1$, $\mathbf{J}_1$ and $\mathbf{E}_2$, $\mathbf{D}_2$, $\mathbf{B}_2$, $\mathbf{H}_2$, $\mathbf{J}_2$ are field vectors on both sides of the interface, respectively. $\mathbf{K}$ and σ are densities of surface current and charges.

1.1 Maxwell's equations and boundary value problems

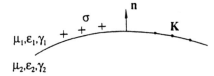

Fig. 1.1.2. Interface boundary conditions of E and B

If the scalar electric potential is chosen as variable, the interfacial boundary conditions are:

$$\begin{cases} \varphi_1 = \varphi_2 \\ \varepsilon_1 \dfrac{\partial \varphi_1}{\partial n} - \varepsilon_2 \dfrac{\partial \varphi_2}{\partial n} = \sigma \end{cases} \quad (1.1.49)$$

For a translational symmetric magnetic field the interface boundary conditions are:

$$\begin{cases} A_1 = A_2 \\ \dfrac{1}{\mu_1} \dfrac{\partial A_1}{\partial n} = \dfrac{1}{\mu_2} \dfrac{\partial A_2}{\partial n} \end{cases} \quad (1.1.50)$$

If the magnetic field problems of 3-D cases are considered, the vector magnetic potential A is composed of three components, e.g.

$$\mathbf{A} = A_n \mathbf{n} + A_t \mathbf{t} + A_s \mathbf{s} \quad (1.1.51)$$

where A_n, A_t, A_s are three components of A. t and s are two unit vectors orthogonal to the normal direction n. Under Coulomb's gauge, the normal component A_n satisfies:

$$A_{1n} = A_{2n} . \quad (1.1.52)$$

Continuity of the tangential components of the magnetic field strength H is expressed by the equation:

$$\mathbf{n} \times \left(\dfrac{1}{\mu_1} \nabla \times \mathbf{A}_1 \right) = \mathbf{n} \times \left(\dfrac{1}{\mu_2} \nabla \times \mathbf{A}_2 \right) . \quad (1.1.53)$$

The above equation can be decomposed into two equations [3]:

$$\left(\dfrac{\partial A_t}{\partial n} \right)_1 = \dfrac{\mu_1}{\mu_2} \left(\dfrac{\partial A_t}{\partial n} \right)_2 - \left(\dfrac{\mu_1}{\mu_2} - 1 \right) \dfrac{\partial A_n}{\partial t} \quad (1.1.54)$$

$$\left(\dfrac{\partial A_s}{\partial n} \right)_1 = \dfrac{\mu_1}{\mu_2} \left(\dfrac{\partial A_s}{\partial n} \right)_2 - \left(\dfrac{\mu_1}{\mu_2} - 1 \right) \dfrac{\partial A_n}{\partial s} . \quad (1.1.55)$$

Equations (1.1.54) and (1.1.55) show that the boundary conditions for a 3-D magnetic field are more complicated than for a scalar field. Thus the appropriate choice of the mathematical model, the unknown variables, and the gauge conditions is significant for the solution of 3-D magnetic field problems.

Methods for solving 3-D magnetic field problems are of great interest to scientists and engineers.

1.1.3 Boundary value problems

The task of determining a solution of a differential equation which is subject to certain boundary conditions is called a boundary value problem. Usually boundary conditions can be generalized into three kinds, viz.:

Dirichlet condition $\quad \varphi|_\Gamma = f(\Gamma)$ (1.1.56)

Neumann condition $\quad \dfrac{\partial \varphi}{\partial n}\bigg|_\Gamma = f(\Gamma)$ (1.1.57)

Robin condition $\quad f_1(\Gamma)\dfrac{\partial \varphi}{\partial n} + f_2(\Gamma)\varphi = f_3(\Gamma)$. (1.1.58)

The governing equations corresponding to the above three boundary conditions are called Dirichlet, Neumann and Robin, respectively. In these equations, $f(\Gamma)$, $f_1(\Gamma)$, $f_2(\Gamma)$, $f_3(\Gamma)$ are the known functions. Usually the Dirichlet boundary condition is called the essential boundary condition or the boundary condition of the first kind. Similarly the Neumann condition is designated the boundary condition of the second kind, the Robin boundary condition is called the boundary condition of the third kind. In a particular problem the boundary condition may be the combination of the first and the second kind. In solving a practical problem, the mathematical description of the boundary conditions is significant. If the vector magnetic potential is chosen as unknown and considered in the 2-D case, the following boundary conditions are commonly valid:

$A = 0 \quad$ along the line of flux density (1.1.59)

$\dfrac{\partial A}{\partial n} = 0 \quad$ along the surface of ferromagnetic material (1.1.60)

$\dfrac{\partial A}{\partial n} = 0 \quad$ along the line of geometric symmetry. (1.1.61)

For example, in Fig. 1.1.3(a) and (b), there are two pairs of current-carrying conductors, extending the field to infinity. Due to the geometric symmetry, only a quarter of the domain needs to be considered. Figure 1.1.3(a) shows that the current in the two conductors is in opposite directions, the y-axis is the B-line, hence it is $A = $ const. It can be assumed that $A = 0$ is on the y-axis. The x-axis is a line of geometric symmetry, with B-lines orthogonal to the x-axis. Hence the x-axis satisfies the condition of $\partial A / \partial n = 0$. At infinity $\lim\limits_{r \to \infty} rA =$ limited. In Fig. 1.1.3(b) the currents in the two conductors are in the same direction, both x and y-axes are lines of symmetry, hence these two axes satisfy $\partial A / \partial n = 0$. Figure 1.1.3(c) shows a model of an electric machine; the stator and the rotator

1.2 Green's theorem, Green's function, and fundamental solutions

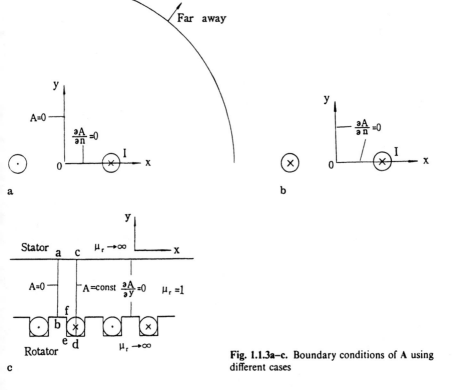

Fig. 1.1.3a–c. Boundary conditions of A using different cases

are made of ferromagnetic material. All the B lines are orthogonal to the surfaces of the ferromagnetic material, hence these surfaces satisfy $\partial A/\partial n = 0$. When considering the symmetry, only the region of *abfedca* needs to be classified. The boundary conditions of this area are shown in the figure. All the surfaces of the magnetic material satisfy $\partial A/\partial n = 0$.

The uniqueness theorem proves that the solution of the governing equation which is subject to specific boundary conditions is unique no matter which method is used.

1.2 Green's theorem, Green's function, and fundamental solutions

The boundary integral equation methods are suitable for numerical analysis of field problems, the reason being is that they depend solely on the modelling of the boundaries and the interfacial surfaces. Green's function and fundamental solutions are the basic functions used in integral equation methods. The integral equations of any problems are usually derived from differential equations using Green's theorem.

1.2.1 Green's theorem

Green's theorem is one of the most useful theorems in solving electromagnetic problems. Many of the solution methods, including classical and numerical methods, are based on Green's theorem. It is derived directly from the divergence theorem.

It is well known that

$$\int_\Omega \nabla \cdot \mathbf{A}\, d\Omega = \oint_s \mathbf{A} \cdot d\mathbf{S} \tag{1.2.1}$$

where Ω is a region bounded by a closed surface S. $\mathbf{A}$ is a vector function of the position. Assume that u and v are two arbitrary scalar functions of the position. If u, v and their first and second derivatives are continuous in the space of Ω and on the surface S, the application of the divergence theorem to the vector $u\nabla v$ yields:

$$\int_\Omega \nabla \cdot (u\nabla v)\, d\Omega = \oint_s (u\nabla v)\, dS \cdot \mathbf{n} . \tag{1.2.2}$$

$\mathbf{n}$ is the external normal unit vector of the surface S. With $\mathbf{S} = S\mathbf{n}$, using the vector identity

$$\nabla \cdot (u\nabla v) = u\nabla^2 v + \nabla u \cdot \nabla v \tag{1.2.3}$$

and noting that

$$u\nabla v \cdot \mathbf{n} = u\frac{\partial v}{\partial n} \tag{1.2.4}$$

the divergence theorem of Eq. (1.2.2) is transformed into:

$$\int_\Omega (u\nabla^2 v + \nabla u \cdot \nabla v)\, d\Omega = \oint_s u\frac{\partial v}{\partial n}\, dS . \tag{1.2.5}$$

This is known as Green's first identity.

If the roles of the function u and v in Eq. (1.2.5) are exchanged, the result is:

$$\int_\Omega (v\nabla^2 u + \nabla v \cdot \nabla u)\, d\Omega = \oint_s v\frac{\partial u}{\partial n}\, dS . \tag{1.2.6}$$

This equation is the symmetric form of Green's first identity. Subtracting Eq. (1.2.6) from Eq. (1.2.5) yields:

$$\int_\Omega (u\nabla^2 v - v\nabla^2 u)\, d\Omega = \oint_s \left(u\frac{\partial v}{\partial n} - v\frac{\partial u}{\partial n} \right) dS . \tag{1.2.7}$$

1.2 Green's theorem, Green's function, and fundamental solutions

This is Green's second identity, frequently referred to as Green's theorem. It is an integral theorem involving the gradient of the integrand. *This theorem transfers a volume integration to a surface integration.*

In a particular case, let $u = v$ and u be a solution of Laplace's equation, then Eq. (1.2.5) reduces to:

$$\int_\Omega (\nabla u)^2 \, d\Omega = \oint_s u \frac{\partial u}{\partial n} \, dS . \tag{1.2.8}$$

By means of Green's theorem the potential at a fixed point $P(\mathbf{r})$ within the volume Ω can be expressed in terms of a volume integral plus a surface integral over S, as:

$$\varphi(\mathbf{r}) = \frac{1}{4\pi\varepsilon} \int_\Omega \frac{\rho(\mathbf{r}')}{R} \, d\Omega + \frac{1}{4\pi} \oint_s \left[\frac{1}{R} \frac{\partial \varphi}{\partial n} - \varphi \frac{\partial}{\partial n}\left(\frac{1}{R}\right) \right] dS . \tag{1.2.9}$$

This is an integral equation of potential $\varphi(\mathbf{r})$. It does not represent the solution of the potential. In this equation $\rho(\mathbf{r}')$ is the density of a volume charge, $R = |\mathbf{r} - \mathbf{r}'|$, Ω is a volume enclosed by the closed surface S, as shown in Fig. 1.2.1. The process of deriving Eq. (1.2.9) will be given in Sect. 1.4.1.

Equation (1.2.9) *demonstrates that the potential* $\varphi(\mathbf{r})$ *in volume* Ω *is determined by the volume source density* $\rho(\mathbf{r}')$ *inside the surface S and the potential* φ *and its normal derivatives of the first order* $\partial\varphi/\partial n$ *on the surface S.* If there is no charge in volume Ω, then the potential within the volume is determined by the potential φ and its normal derivatives over the surface, i.e.

$$\varphi(\mathbf{r}) = \frac{1}{4\pi} \oint_s \left[\frac{1}{R} \frac{\partial \varphi}{\partial n} - \varphi \frac{\partial}{\partial n}\left(\frac{1}{R}\right) \right] dS . \tag{1.2.10}$$

Thus the surface integral term of Eq. (1.2.9) *represents the contributions of the sources outside the surface S. In other words, the boundary conditions represent*

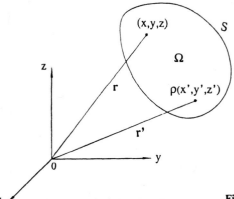

Fig. 1.2.1. The region of the interior problem

the contributions of the sources outside the surface S. This conclusion implies that the boundary conditions can also be represented by equivalent exterior sources. This is the theoretical foundation of the Charge Simulation Method – where the boundary condition is simulated by an equivalent source outside the region of interest. If there are no exterior sources the surface integral must vanish. Examination of Eq. (1.2.8), indicates that if the function u in Eq. (1.2.8) is a harmonic function (e.g. it is the solution of Laplace's equation) and $v = 1$, then Green's second identity is reduced to:

$$\int_s \frac{\partial u}{\partial n} \, dS = 0 \,. \tag{1.2.11}$$

This means that if u is the potential of an electrostatic field and subject to the Neumann boundary condition $|\partial u/\partial n|_s = g$, then the function g has to satisfy the following condition:

$$\int_s g \, dS = 0 \,. \tag{1.2.12}$$

1.2.2 Vector analogue of Green's theorem

The purpose of this section is to introduce an integral equation in which the vector potential **A** is considered as unknown. The method is based on the introduction of a vector identity in Gauss's theorem. Suppose **P** and **Q** are continuous vector functions of position in volume Ω closed by the regular surface S, and both **P** and **Q** have partial derivatives of first and second order over the surface S and in volume Ω. Using the divergence theorem

$$\int_\Omega \nabla \cdot (\mathbf{P} \times \nabla \times \mathbf{Q}) \, d\Omega = \oint_s (\mathbf{P} \times \nabla \times \mathbf{Q}) \cdot \mathbf{dS} \tag{1.2.13}$$

and expanding the integrand of the volume integral by using the vector identity

$$\nabla \cdot (\mathbf{A} \times \mathbf{B}) = \mathbf{B} \cdot \nabla \times \mathbf{A} - \mathbf{A} \cdot \nabla \times \mathbf{B} \tag{1.2.14}$$

one obtains

$$\int_\Omega (\nabla \times \mathbf{P} \cdot \nabla \times \mathbf{Q} - \mathbf{P} \cdot \nabla \times \nabla \times \mathbf{Q}) \, d\Omega = \oint_s (\mathbf{P} \times \nabla \times \mathbf{Q}) \, dS \cdot \mathbf{n} \,. \tag{1.2.15}$$

This is the vector analogue of the scalar form of Green's first identity.

By using the same process as before the vector analogue of a scalar form of Green's second identity is obtained:

$$\int_\Omega (\mathbf{Q} \cdot \nabla \times \nabla \times \mathbf{P} - \mathbf{P} \cdot \nabla \times \nabla \times \mathbf{Q}) \, d\Omega = \oint_s (\mathbf{P} \times \nabla \times \mathbf{Q} - \mathbf{Q} \times \nabla \times \mathbf{P}) \cdot \mathbf{n} \, dS \,. \tag{1.2.16}$$

Applying the vector form of Green's theorem the integral equation for the vector potential **A** is

$$\mathbf{A}(\mathbf{r}) = \frac{\mu_0}{4\pi} \int_\Omega \frac{\mathbf{J}(\mathbf{r}')}{R} d\Omega + \frac{1}{4\pi} \oint_s \left[\frac{\mathbf{B} \times \mathbf{n}}{R} + (\mathbf{A} \times \mathbf{n}) \times \nabla\left(\frac{1}{R}\right) \right.$$
$$\left. - (\mathbf{n} \cdot \mathbf{A}) \nabla\left(\frac{1}{R}\right) \right] dS \ . \qquad (1.2.17)$$

The process of deriving this equation is given in Appendix 1.1.

1.2.3 Green's function [4, 5]

In a linear, uniform, and isotropic medium Green's function is a response function relating the field point (**r**) *and the source point* (**r**′). Hence Green's function is extremely important in field analysis. In this section the definition and basic application of Green's function are introduced.

1.2.3.1 Dirac-delta Function [6]

Recall the definition of the Dirac-delta function:

$$\begin{cases} \delta(\mathbf{r} - \mathbf{r}') = 0 & \mathbf{r} \neq \mathbf{r}' \\ \int_{-\infty}^{\infty} \delta(\mathbf{r} - \mathbf{r}') d\mathbf{r} = 1 & \mathbf{r} = \mathbf{r}' \ . \end{cases} \qquad (1.2.18)$$

It follows that

$$f(\mathbf{r}') = \int_\Omega f(\mathbf{r}) \delta(\mathbf{r}, \mathbf{r}') d\Omega \qquad (1.2.19)$$

where

$$\delta(\mathbf{r}, \mathbf{r}') = \delta(x - x')\delta(y - y')\delta(z - z') \ . \qquad (1.2.20)$$

For an arbitrary operator equation

$$\mathscr{L} u(\mathbf{r}) = -f(\mathbf{r}') \quad u, f \in \Omega \qquad (1.2.21)$$

the function u can be expressed as

$$u(\mathbf{r}) = -\int_\Omega f(\mathbf{r}') \mathscr{L}^{-1} \delta(\mathbf{r}, \mathbf{r}') d\Omega \qquad (1.2.22)$$

where $\mathscr{L}$ represents any operator[1] (e.g. for Poisson's equation $\mathscr{L} = \nabla^2$) and $\mathscr{L}^{-1}$ the inverse operation of the operator $\mathscr{L}$. The sequence of $\int_\Omega$ and $\mathscr{L}$ can be changed since the operator $\mathscr{L}^{-1}$ has no effect on the variable $\mathbf{r}'$. Eq. (1.2.22) shows that if $\mathscr{L}^{-1} \delta(\mathbf{r}, \mathbf{r}')$ is known, then the function $u(\mathbf{r})$ is obtained.

---

[1] An operator represents a specific operation, it maps a function u into another function f.

1.2.3.2 Green's function

It is supposed that Green's function $G(\mathbf{r}, \mathbf{r}')$ satisfies the equation

$$G(\mathbf{r}, \mathbf{r}') = \mathscr{L}^{-1} - \delta(\mathbf{r}, \mathbf{r}') . \tag{1.2.23}$$

This means that Green's function is the solution of an operator equation subject to an impulse source. Multiplying the operator $\mathscr{L}$ on both sides of Eq. (1.2.23) from the left-hand side yields

$$\mathscr{L} G(\mathbf{r}, \mathbf{r}') = -\delta(\mathbf{r}, \mathbf{r}') . \tag{1.2.24}$$

If $\mathscr{L} \equiv \nabla^2$, then

$$\nabla^2 G(\mathbf{r}, \mathbf{r}') = -\delta(\mathbf{r}, \mathbf{r}') . \tag{1.2.25}$$

Taking the integral operation on both sides of Eq. (1.2.23) one obtains the expression of the inverse operator $\mathscr{L}^{-1}$, i.e.

$$\mathscr{L}^{-1} = \int_\Omega \mathscr{L}^{-1} \delta(\mathbf{r}, \mathbf{r}') d\Omega = -\int_\Omega G(\mathbf{r}, \mathbf{r}') d\Omega . \tag{1.2.26}$$

Thus, the inverse operator of the differential operator is an integral operator in which Green's function is the kernel. However, the function $G(\mathbf{r}, \mathbf{r}')$ is undefined in accordance with Eq. (1.2.24). If there is any function $g(\mathbf{r})$, only if $\mathscr{L} g = 0$, plus $G(\mathbf{r}, \mathbf{r}')$, e.g. $\tilde{G}(\mathbf{r}, \mathbf{r}') = G(\mathbf{r}, \mathbf{r}') + g(\mathbf{r})$, then Eq. (1.2.24), is still satisfied. Hence specific boundary conditions are necessary for the unique determination of Green's function.

Furthermore, if the solution of equation $\mathscr{L} G(\mathbf{r}, \mathbf{r}') = -\delta(\mathbf{r}, \mathbf{r}')$ under homogeneous boundary conditions is known, i.e. if $G(\mathbf{r}, \mathbf{r}')$, then the solution of equation $\mathscr{L} u = f$ under inhomogeneous boundary conditions can be obtained. The reason is that in Green's theorem

$$\int_\Omega (u \mathscr{L} v - v \mathscr{L} u) d\Omega = \oint_s \left(u \frac{\partial v}{\partial n} - v \frac{\partial u}{\partial n} \right) dS$$

if $v = G$, then Eq. (1.2.7) is transformed to

$$\int_\Omega [-u(\mathbf{r}) \delta(\mathbf{r}, \mathbf{r}') + G(\mathbf{r}, \mathbf{r}') f(\mathbf{r}')] d\Omega = \oint_s \left[u(\mathbf{r}) \frac{\partial G(\mathbf{r}, \mathbf{r}')}{\partial n} - G(\mathbf{r}, \mathbf{r}') \frac{\partial u(\mathbf{r})}{\partial n} \right] dS . \tag{1.2.27}$$

Combining Eqs. (1.2.19) and (1.2.27) yields

$$u(\mathbf{r}') = \int_\Omega G(\mathbf{r}, \mathbf{r}') f(\mathbf{r}) d\Omega + \oint_s \left[-u(\mathbf{r}) \frac{\partial G(\mathbf{r}, \mathbf{r}')}{\partial n} + G(\mathbf{r}, \mathbf{r}') \frac{\partial u(\mathbf{r})}{\partial n} \right] dS . \tag{1.2.28}$$

1.2 Green's theorem, Green's function, and fundamental solutions

Due to the symmetry of Green's function of Laplacian, i.e.

$$G(\mathbf{r}, \mathbf{r}') = G(\mathbf{r}', \mathbf{r}) \tag{1.2.29}$$

Eq. (1.2.28) becomes:

$$u(\mathbf{r}) = \int_\Omega G(\mathbf{r}, \mathbf{r}') f(\mathbf{r}') d\Omega + \oint_s \left[-u(\mathbf{r}) \frac{\partial G(\mathbf{r}, \mathbf{r}')}{\partial n} + G(\mathbf{r}, \mathbf{r}') \frac{\partial u}{\partial n} \right] dS . \tag{1.2.30}$$

This equation is Green's third identity. In Eq. (1.2.30) the boundary values of u and $\partial u/\partial n$ on the surface S are involved. For Dirichlet problems, let $G(\mathbf{r}, \mathbf{r}')|_s = 0$, Eq. (1.2.30) then reduces to

$$u(\mathbf{r}) = \int_\Omega G(\mathbf{r}, \mathbf{r}') f(\mathbf{r}') d\Omega - \oint_s u(\mathbf{r}) \frac{\partial}{\partial n} G(\mathbf{r}, \mathbf{r}') dS . \tag{1.2.31}$$

For Neumann problems, $\partial G(\mathbf{r}, \mathbf{r}')/\partial n|_s = 0$, Eq. (1.2.30) reduces to

$$u(\mathbf{r}) = \int_\Omega G(\mathbf{r}, \mathbf{r}') f(\mathbf{r}') d\Omega + \oint_s G(\mathbf{r}, \mathbf{r}') \frac{\partial u}{\partial n} dS . \tag{1.2.32}$$

For Robin problems the boundary condition is

$$\left[f_1(\mathbf{r}) \frac{\partial u(\mathbf{r})}{\partial n} + f_2(\mathbf{r}) u(\mathbf{r}) \right]\bigg|_s = f_3(\mathbf{r}) . \tag{1.2.33}$$

Let $G(\mathbf{r}, \mathbf{r}')$ satisfy the condition

$$\left[f_1(\mathbf{r}) \frac{\partial G(\mathbf{r}, \mathbf{r}')}{\partial n} + f_2(\mathbf{r}) G(\mathbf{r}, \mathbf{r}') \right]\bigg|_s = 0 .$$

Substituting Eq. (1.2.33) into Eq. (1.2.31) yields

$$u(\mathbf{r}) = \int_\Omega G(\mathbf{r}, \mathbf{r}') f(\mathbf{r}') d\Omega + \oint_s \frac{1}{f_1} f_3(\mathbf{r}) G(\mathbf{r}, \mathbf{r}') dS \tag{1.2.34}$$

or

$$u(\mathbf{r}) = \int_\Omega G(\mathbf{r}, \mathbf{r}') f(\mathbf{r}') d\Omega - \oint_s \frac{1}{f_2} f_3(\mathbf{r}) \frac{\partial G(\mathbf{r}, \mathbf{r}')}{\partial n} dS . \tag{1.2.35}$$

Conclusion: *If Green's function for any operator equation with homogeneous boundary conditions is known, then the field distribution produced by any continuously distributed sources under inhomogeneous boundary conditions will be given by the above integral equations.* For example, for Poisson's equation and homogeneous boundary conditions, Eq. (1.2.9) is reduced to

$$\varphi(\mathbf{r}) = \int_\Omega \frac{\rho(\mathbf{r}')}{\varepsilon_0} G(\mathbf{r}, \mathbf{r}') d\Omega . \tag{1.2.36}$$

This equation shows that if Green's function of the given operator equation is known then the solution under any kind of source distribution can be calculated by using Eq. (1.2.36). Therefore Green's function is among the basic tools for analysing various mathematical-physical problems.

Green's function of Poisson's equation in a 3-dimensional case in free space is

$$G(\mathbf{r}, \mathbf{r}') = \frac{1}{4\pi\varepsilon_0 |\mathbf{r} - \mathbf{r}'|} . \qquad (1.2.37)$$

This is the solution of Poisson's equation for a unit impulse source. If the influence of the ground is considered, then

$$G(\mathbf{r}, \mathbf{r}') = \frac{1}{4\pi\varepsilon} \left(\frac{1}{R} - \frac{1}{R_1} \right) \qquad (1.2.38)$$

where $R = |\mathbf{r} - \mathbf{r}'|$ and R_1 is the distance from the image source to the observation point.

1.2.4 Fundamental solutions

Regardless of the boundary conditions, the solution of an operator equation produced by a unit source in an infinite space is called the fundamental solution and it fulfils the following equation:

$$\mathscr{L} F(\mathbf{r}, \mathbf{r}') = -\delta(\mathbf{r} - \mathbf{r}') \qquad (1.2.39)$$

where $\mathscr{L}$ is an arbitrary operator. Note that the difference between Green's function and the fundamental solution is that Green's function is related to the boundary conditions but the fundamental solution is defined in a boundless free space. Alternatively, Green's function in a free space is the fundamental solution of the same operator equation. The fundamental solution of Laplace's equation in 2-D and 3-D cases are derived as follows.

In 2-D polar coordinates Laplace's equation is expanded to

$$\frac{1}{r} \frac{d}{dr} \left(r \frac{du}{dr} \right) = 0 .$$

Then the solution is

$$u = C_1 \ln r + C_2 . \qquad (1.2.40)$$

In a 3-D case the solution of $\frac{d}{dr}\left(r^2 \frac{du}{dr}\right) = 0$ is

$$u = \frac{C_1}{r} + C_2 . \qquad (1.2.41)$$

1.3 Equivalent sources

Table 1.2.1. Fundamental solutions of different equations in electromagnetic fields [7]

Governing equation	Fundamental solution	
	2-D case	3-D case
Laplace's equation $\nabla^2 F + \delta = 0$	$F = \dfrac{1}{2\pi} \ln \dfrac{1}{\|\mathbf{r}-\mathbf{r}'\|}$	$F = \dfrac{1}{4\pi} \dfrac{1}{\|\mathbf{r}-\mathbf{r}'\|}$
Helmholtz equation $(\nabla^2 + k^2) F + \delta = 0$ $k^2 = (\omega^2 \mu \varepsilon - j\omega\mu\gamma)$	$F = \dfrac{1}{4j} H_0^{(2)}(k\|\mathbf{r}-\mathbf{r}'\|)$ $H^{(2)}$ – Hankel function	$F = \dfrac{1}{4\pi\|\mathbf{r}-\mathbf{r}'\|} \exp(-jk\|\mathbf{r}-\mathbf{r}'\|)$
Diffusion equation $\nabla^2 F - \dfrac{1}{k}\dfrac{\partial F}{\partial t} + \delta(r)\delta(t) = 0$ $k = 1/\mu\gamma$	$F = \dfrac{-1}{4\pi kt} \exp\left(\dfrac{-\|\mathbf{r}-\mathbf{r}'\|^2}{4kt}\right)$	$F = \dfrac{-1}{(4\pi kt)^{3/2}} \exp\left(\dfrac{-\|\mathbf{r}-\mathbf{r}'\|^2}{4kt}\right)$
Wave equation $v^2\nabla^2 F - \dfrac{\partial^2 F}{\partial t^2} + \delta(\mathbf{r})\delta(t) = 0$ $v^2 = 1/\mu\varepsilon$	$F = \dfrac{-H(vt - \|\mathbf{r}-\mathbf{r}'\|)}{2\pi v(v^{3/2} - \|\mathbf{r}-\mathbf{r}'\|^2)}$	$F = \dfrac{\delta\left(t \pm \dfrac{\|\mathbf{r}-\mathbf{r}'\|}{v}\right)}{4\pi\|\mathbf{r}-\mathbf{r}'\|}$

With $C_1 = -1$, $C_2 = 0$; $C_1 = 1$, $C_2 = 0$ in Eq. (1.2.40) and Eq. (1.2.41), respectively, the fundamental solutions of the 2-D and 3-D Laplace's equations are

$$F(\mathbf{r},\mathbf{r}') = \ln \frac{1}{R}. \tag{1.2.42}$$

$$F(\mathbf{r},\mathbf{r}') = \frac{1}{R} \tag{1.2.43}$$

The fundamental solutions of commonly used differential equations in electromagnetic fields are listed in Table 1.2.1.

1.3 Equivalent sources

In numerical analysis the method of equivalent sources is commonly used in integral equation methods [8, 9]. In terms of a potential boundary value problem as shown in Fig. 1.3.1(a) the problem is replaced by an equivalent problem in which the potential boundary condition is replaced by a distributed single layer or double layer source as shown in Fig. 1.3.1(b) in free space. The

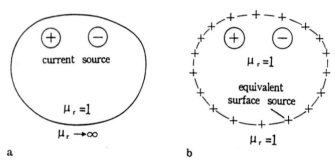

Fig. 1.3.1a, b. Equivalent single layer source

equivalent source may be electric or magnetic charges or currents. The characteristics of the layer sources are illustrated in this section.

1.3.1 Single layer charge distribution

It is known that if the charge is distributed on a surface S with density σ, which is a bounded and piecewise continuous function of position S, the potential at any point not on S is

$$\varphi(\mathbf{r}) = \frac{1}{4\pi\varepsilon_0} \int_s \frac{\sigma(\mathbf{r}')}{R} \, dS = \int_s \frac{\sigma(\mathbf{r}')}{\varepsilon_0} F(\mathbf{r}, \mathbf{r}') \, dS \qquad (1.3.1)$$

where $\varphi(\mathbf{r})$ is a continuous function of $\sigma(\mathbf{r}')$. $F(\mathbf{r}, \mathbf{r}')$ is the fundamental solution. If the point $(\mathbf{r})$ lies on S, the singularity of Eq. (1.3.1) has to be considered. A circle with a sufficiently small radius r_0 circumscribes this point as shown in Fig. 1.3.2.

The potential at any point can be expressed as $\varphi_s = \varphi_1 + \varphi_2$, where φ_1 is the contribution of the charge on the small disk shown in Fig. 1.3.2 and φ_2 is the contribution of the charge outside the disk. The component φ_1 is

$$\varphi_1 = \frac{1}{4\pi\varepsilon_0} \int_{S_0} \frac{\sigma}{r} \, dS = \frac{\sigma_0}{2\varepsilon_0} \int_0^{r_0} \frac{1}{r} r \, dr = \frac{\sigma_0 r_0}{2\varepsilon_0}. \qquad (1.3.2)$$

It is a definite value. In Eq. (1.3.2) σ_0 is the charge density of the small disk; it is considered as a constant, φ_1 is zero while r_0 tends to zero. Thus $\varphi_s = \varphi_1 + \varphi_2$ is still bounded and continuous. *The potential produced by a surface charge distribution is a bounded, continuous function of position of all points both on and off the surface, thus it is continuous across the surface*, i.e.

$$\varphi_+ = \varphi_- = \varphi_s. \qquad (1.3.3)$$

1.3 Equivalent sources

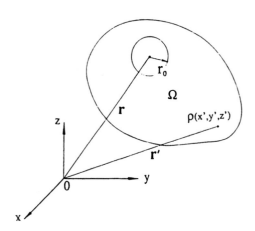

Fig. 1.3.2. Treatment of a singularity

The subscripts '+' and '−' denote the potential just outside and inside of the surface, respectively.

On the other hand the field intensity

$$E(\mathbf{r}) = -\nabla \varphi(\mathbf{r}) = \frac{1}{4\pi\varepsilon_0} \int_s \sigma(\mathbf{r}') \nabla\left(\frac{1}{R}\right) dS \qquad (1.3.4)$$

is continuous and has continuous derivatives of all orders at any points not on the surface. *However, the field intensity undergoes an abrupt change across the surface S.* According to Gauss's law

$$(\mathbf{D}_+ - \mathbf{D}_-) \cdot \mathbf{n} = \sigma$$

$$\mathbf{E}_+ - \mathbf{E}_- = \frac{\sigma}{\varepsilon} \mathbf{n} \qquad (1.3.5)$$

where $\mathbf{D}_+$, $\mathbf{D}_-$, $\mathbf{E}_+$, $\mathbf{E}_-$ are field vectors outside and inside the surface, $\mathbf{n}$ is the unit vector outward of the normal direction of S as shown in Fig. 1.3.3(a). If the surface charge density $\sigma(s)$ is known as a single layer source, it is coincident with an inhomogeneous Neumann boundary condition. In other words, *the boundary value problems with inhomogeneous boundary conditions of the second kind are identical to those of a single layer source on the boundary surface.*

For a distributed surface current density similar equations expressed by the vector potential **A** and the magnetic flux density are

$$\mathbf{A}(\mathbf{r}) = \frac{\mu_0}{4\pi} \int_s \frac{\mathbf{K}(\mathbf{r}')}{R} dS = \mu_0 \int_s \mathbf{K}(\mathbf{r}') G(\mathbf{r}, \mathbf{r}') ds'. \qquad (1.3.6)$$

$$\mathbf{B}(\mathbf{r}) = \frac{\mu_0}{4\pi} \int_s \mathbf{K}(\mathbf{r}') \times \nabla\left(\frac{1}{R}\right) dS = \mu_0 \int_s \mathbf{K}(\mathbf{r}') \nabla \times G(\mathbf{r}, \mathbf{r}') ds'. \qquad (1.3.7)$$

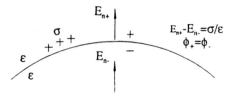

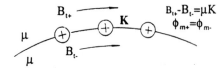

Fig. 1.3.3a, b. Field discontinuity along the single layer source

K is the surface current density. The interfacial boundary conditions of **B** are

$$\mathbf{B}_+ - \mathbf{B}_- = \mu_0 \mathbf{K} \times \mathbf{n} \tag{1.3.8}$$

$$\mathbf{n} \cdot (\mathbf{B}_+ - \mathbf{B}_-) = 0 \tag{1.3.9}$$

$$\mathbf{n} \times (\mathbf{B}_+ - \mathbf{B}_-) = \mu_0 \mathbf{n} \times (\mathbf{K} \times \mathbf{n}) = \mu_0 [(\mathbf{n} \cdot \mathbf{n})\mathbf{K} - (\mathbf{n} \cdot \mathbf{K})\mathbf{n}] = \mu_0 \mathbf{K}. \tag{1.3.10}$$

Hence, *there is an abrupt change of the tangential component of* **B** *as shown in Fig. 1.3.3(b)*.

1.3.2 Double layer source distributions [10]

The potential φ induced by an electric dipole $\mathbf{p} = q\mathbf{d}$ as shown in Fig. 1.3.4(a) is

$$\varphi(\mathbf{r}) = \frac{q}{4\pi\varepsilon_0}\left(\frac{1}{r_2} - \frac{1}{r_1}\right) = \frac{\mathbf{p}}{4\pi\varepsilon_0} \frac{\partial}{\partial d}\left(\frac{1}{r}\right) = \frac{p \cos(\mathbf{r}^0, \mathbf{d})}{4\pi\varepsilon_0 r^2}$$

$$= -\frac{\mathbf{p}}{4\pi\varepsilon_0}\nabla\left(\frac{1}{r}\right) = \frac{\mathbf{p}}{4\pi\varepsilon_0}\nabla'\left(\frac{1}{r}\right) \tag{1.3.11}$$

where $\mathbf{r}^0$ is an unit vector along the **r** direction, ∇' represents the spatial derivative of the source point.

Consider a double layer charge distribution, where the positive charges are distributed on the positive side of a closed surface S and the negative charges are distributed with a density of $-\sigma$ on the opposite side as shown in Fig. 1.3.4(b).

1.3 Equivalent sources

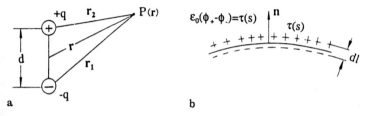

Fig. 1.3.4a, b. A double layer charge distribution

The double layer charge distribution is separated by the infinitesimal distance dl. The definition of the dipole moment per unit area is

$$\tau = \mathbf{n} \lim_{\substack{\sigma \to \infty \\ dl \to 0}} (\sigma \, dl) \, . \tag{1.3.12}$$

The potential induced by the double layer source at point $P(\mathbf{r})$ not on the surface S is

$$d\varphi(\mathbf{r}) = \frac{1}{4\pi\varepsilon_0} \frac{\tau \cos \theta}{R^2} dS = \frac{1}{4\pi\varepsilon_0} \tau \cdot \nabla \left(\frac{1}{R} \right) dS \, . \tag{1.3.13}$$

$\frac{\cos \theta}{r^2} ds$ is proportional to the solid angle $d\omega$ at point $P(\mathbf{r})$, then

$$\varphi(\mathbf{r}) = \frac{1}{4\pi\varepsilon_0} \int_s \tau(\mathbf{r}') \cdot \nabla \left(\frac{1}{r} \right) dS = \pm \frac{1}{4\pi\varepsilon_0} \int_s \tau \, d\omega \tag{1.3.14}$$

where $d\omega$ is the solid angle subtended at point $P(\mathbf{r})$ by surface ds, as shown in Fig. 1.3.5. The sign $\pm$ depends on which side of the surface S the observation point lies. The solid angle is positive, if the radius vector drawn from point $(\mathbf{r})$ to the element ds makes an acute angle with the positive normal $\mathbf{n}$ of the surface.

The main characteristic of the double layer distribution is that the potential is discontinuous on both sides of the layer. Suppose that the surface S is closed and the charge density is uniform. Then τ can be taken out from the integral. The positive charge lies on the outer side of S so τ has the same direction as the positive normal of the surface. As the property of the solid angle is

$$\oint d\omega = \begin{cases} 4\pi & P \text{ inside the surface} \\ 0 & P \text{ outside the surface} \end{cases} \tag{1.3.15}$$

thus

$$\varphi_+ = 0 \qquad \varphi_- = -\tau/\varepsilon_0 \tag{1.3.16}$$

and

$$\varphi_+ - \varphi_- = \tau/\varepsilon_0 \, . \tag{1.3.17}$$

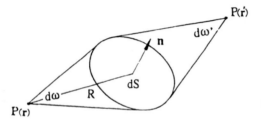

Fig. 1.3.5. Solid angle

Therefore, the potential undergoes an abrupt change of τ/ε_0 while the observation point is moving from the inner side to the outer side of a double layer source.

If surface S is not closed, Eq. (1.3.17) is still correct. The reason is that a surface S' can be added to close the surface S. The potential on both sides of the surface are superimposed by two parts, the contributions of surface charges on S and S'. The potential produced by the charge on S' is continuous but the potential $\varphi + \varphi'$ is discontinuous while the observation point passing through the surface S. Thus all the discontinuity is caused by the surface S. Based on the properties of a double layer, the boundary value problem with inhomogeneous boundary conditions of the first kind (φ = const.) can be represented by a certain distribution of the dipole layer source.

The continuity of the field strength due to a dipole layer is discussed below.

According to Gauss's law, on each side of the surface of the double layer the normal derivative of the potential has an abrupt change twice (one is $+\sigma$ and the other jump is $-\sigma$). Hence, E_n is continuous from one side to the other, i.e.

$$(\mathbf{E}_+ - \mathbf{E}_-) \cdot \mathbf{n} = 0 \tag{1.3.18}$$

or

$$\frac{\partial \varphi_+}{\partial n} = \frac{\partial \varphi_-}{\partial n} = \frac{\partial \varphi}{\partial n}\bigg|_s. \tag{1.3.19}$$

However, the tangential component of $\mathbf{E}$ may be discontinuous, because the potentials undergo an abrupt change on both sides, as shown in Fig. 1.3.6, where

$$\varphi_2 - \varphi_3 = \tau/\varepsilon_0$$

$$\varphi_4 - \varphi_1 = -\frac{1}{\varepsilon_0}(\tau + \nabla\tau \cdot \mathbf{db})$$

and

$$\varphi_2 - \varphi_1 = \Delta\varphi_+$$

$$\varphi_3 - \varphi_4 = \Delta\varphi_-$$

db is a small length along the surface, $\varphi_1, \ldots, \varphi_4$ are potentials very close to the interface but on the opposite side of the interface as shown with the points 1, 2, 3, 4 in Fig. 1.3.6. Due to

$$\sum \varphi_i = 0. \tag{1.3.20}$$

1.3 Equivalent sources

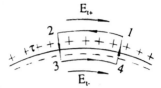

Fig. 1.3.6. Discontinuity of E_t around a double layer source

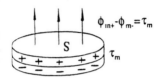

Fig. 1.3.7. Equivalent magnetic sheet

Substituting φ_1 to φ_4 into Eq. (1.3.20) leads to

$$\tau/\varepsilon_0 - (\tau + \nabla\tau \cdot \mathbf{db})/\varepsilon_0 + \Delta\varphi_- - \Delta\varphi_+ = 0 .$$

Then

$$\frac{\Delta\varphi_+}{db} - \frac{\Delta\varphi_-}{db} = -\frac{\nabla\tau}{\varepsilon_0} \cdot \mathbf{t}$$

and

$$\mathbf{E}_{t_+} - \mathbf{E}_{t_-} = -\nabla\tau/\varepsilon_0 \qquad (1.3.21)$$

where $\mathbf{t}$ is an unit vector tangential to the surface S. The abrupt change of $\mathbf{E}_t$ is the abrupt change of $\mathbf{E}$, as the normal component of $\mathbf{E}$ is continuous. I.e.

$$\mathbf{E}_+ - \mathbf{E}_- = -\nabla\tau/\varepsilon_0 . \qquad (1.3.22)$$

$\nabla\tau$ is the gradient of τ along the surface. If τ is uniform then $\mathbf{E}$ is continuous.

For the magnetic scalar potential φ_m the corresponding equation is

$$\varphi_{m_+} - \varphi_{m_-} = \tau_m = \sigma_m\, dl \qquad (1.3.23)$$

where σ_m is the magnetic surface charge density. τ_m is the surface density of the magnetic moment, and is known as the intensity of the magnetic sheet. The positive direction of τ_m is coincident with the current according to the right hand rule as shown in Fig. 1.3.7. This means that the effect of the double layer magnetic source may be substituted by an equivalent magnetic dipole with a magnetic moment IS ($IS = \tau_m S$).

For the magnetic vector potential

$$\mathbf{A}(\mathbf{r}) = \frac{\mu_0}{4\pi} \int_s \mathbf{M}(\mathbf{r}') \times \nabla\left(\frac{1}{R}\right) dS \qquad (1.3.24)$$

where $\mathbf{M}$ is the intensity of magnetization. $\mathbf{A}$ is discontinuous while it is passing through a surface with a magnetic moment, i.e.

$$\mathbf{A}_+ - \mathbf{A}_- = \mu_0 \mathbf{M} \times \mathbf{n} \qquad (1.3.25)$$

due to

$$(\mathbf{A}_+ - \mathbf{A}_-) \cdot \mathbf{n} = 0 \tag{1.3.26}$$

and

$$\mathbf{n} \times (\mathbf{A}_+ - \mathbf{A}_-) = \mathbf{n} \times (\mu_0 \mathbf{M} \times \mathbf{n}) = \mu_0 \mathbf{M} - \mu_0 (\mathbf{n} \cdot \mathbf{M}) \cdot \mathbf{n}. \tag{1.3.27}$$

Equation (1.3.27) indicates that only if the direction of magnetization is normal to the surface the tangential component of $\mathbf{A}$ is continuous.

1.3.3 Equivalent polarization charge and magnetization current

In dielectrics the potential produced by the dielectric polarization is due to the polarization dipoles. According to Eq. (1.3.14) it follows

$$\varphi(\mathbf{r}) = \int_\Omega \frac{\mathbf{P}(\mathbf{r}')}{4\pi\varepsilon_0} \cdot \nabla'\left(\frac{1}{R}\right) d\Omega \tag{1.3.28}$$

where $\mathbf{P} = N\mathbf{p} = Nq\mathbf{d}$ is the polarization vector. By using the vector identity

$$\nabla \cdot (f\mathbf{F}) = \nabla f \cdot \mathbf{F} + f\nabla \cdot \mathbf{F} \tag{1.3.29}$$

Eq. (1.3.28) is rearranged to

$$\varphi(\mathbf{r}) = \frac{1}{4\pi\varepsilon_0} \int_\Omega \nabla' \cdot \left(\frac{\mathbf{P}}{R}\right) d\Omega - \frac{1}{4\pi\varepsilon_0} \int_\Omega \frac{\nabla' \cdot \mathbf{P}}{R} d\Omega$$

$$= \frac{1}{4\pi\varepsilon_0} \oint_s \frac{1}{R} \mathbf{P} \cdot \mathbf{n}\, dS - \frac{1}{4\pi\varepsilon_0} \int_\Omega \frac{\nabla' \cdot \mathbf{P}}{R} d\Omega$$

$$= \frac{1}{4\pi\varepsilon_0} \int_\Omega \frac{\rho_b}{R} d\Omega + \frac{1}{4\pi\varepsilon_0} \oint_s \frac{\sigma_b}{R} dS \tag{1.3.30}$$

where

$$\rho_b = -\nabla' \cdot \mathbf{P} \tag{1.3.31}$$

and

$$\sigma_b = \mathbf{P} \cdot \mathbf{n} = P_n. \tag{1.3.32}$$

ρ_b and σ_b are the volume and surface density of the polarization. P_n is the normal component of the polarization vector. If the polarization is uniform, then $\rho_b = 0$.

Similar to Eq. (1.3.13) the magnetic scalar potential produced by the magnetic dipole moment is

$$d\varphi_m(\mathbf{r}) = \frac{1}{4\pi} \mathbf{M} \cdot \nabla'\left(\frac{1}{R}\right) = \frac{\mathbf{M} \cdot \mathbf{R}}{4\pi R^3} d\Omega. \tag{1.3.33}$$

1.3 Equivalent sources

Thus the total scalar potential produced by the magnetization is

$$\varphi_m(\mathbf{r}) = \frac{1}{4\pi} \int_\Omega \mathbf{M}(\mathbf{r}') \cdot \nabla'\left(\frac{1}{R}\right) d\Omega. \tag{1.3.34}$$

Using the same procedure as in Eq. (1.3.30) one obtains:

$$\varphi_m(\mathbf{r}) = \frac{1}{4\pi} \int_\Omega \frac{-\nabla' \cdot \mathbf{M}}{R} d\Omega + \frac{1}{4\pi} \oint_S \frac{\mathbf{M} \cdot \mathbf{n}}{R} dS$$

$$= \frac{1}{4\pi} \int_\Omega \frac{\rho_m}{R} d\Omega + \oint_S \frac{\mathbf{M} \cdot \mathbf{n}}{R} dS \tag{1.3.35}$$

$$\mathbf{H}(\mathbf{r}) = \frac{1}{4\pi} \int_\Omega \rho_m \nabla\left(\frac{1}{R}\right) d\Omega + \frac{1}{4\pi} \oint_S \sigma_m \nabla\left(\frac{1}{R}\right) dS \tag{1.3.36}$$

where

$$\mathbf{M} = \chi_m \mathbf{H} \tag{1.3.37}$$

$$-\nabla' \cdot \mathbf{M} = \rho_m \qquad \mathbf{M} \cdot \mathbf{n} = \sigma_m \tag{1.3.38}$$

and χ_m is the magnetic susceptability.

The magnetic vector potential produced by the magnetization is

$$\mathbf{A}(\mathbf{r}) = \frac{\mu_0}{4\pi} \int_\Omega \mathbf{M} \times \nabla'\left(\frac{1}{R}\right) d\Omega. \tag{1.3.39}$$

Using the vector identity

$$\nabla \times (f\mathbf{F}) = \nabla f \times \mathbf{F} + (\nabla \times \mathbf{F})f \tag{1.3.40}$$

and the divergence theorem in vector form

$$\int_\Omega \nabla \times \mathbf{F} \, d\Omega = \oint_S d\mathbf{S} \times \mathbf{F} = \oint_S \mathbf{n} \times \mathbf{F} \, dS \tag{1.3.41}$$

Equation (1.3.39) reduces to

$$\mathbf{A}(\mathbf{r}) = \frac{\mu_0}{4\pi} \int_\Omega \frac{\nabla' \times \mathbf{M}}{R} d\Omega - \frac{\mu_0}{4\pi} \int_\Omega \nabla' \times \left(\frac{\mathbf{M}}{R}\right) d\Omega$$

$$= \frac{\mu_0}{4\pi} \int_\Omega \frac{\nabla' \times \mathbf{M}}{R} d\Omega + \frac{\mu_0}{4\pi} \oint_S \frac{\mathbf{M} \times \mathbf{n}}{R} dS$$

$$= \frac{\mu_0}{4\pi} \int_\Omega \frac{\mathbf{J}_m}{R} d\Omega + \frac{\mu_0}{4\pi} \oint_S \frac{\mathbf{K}_m}{R} dS. \tag{1.3.42}$$

Then

$$\mathbf{B(r)} = \nabla \times \mathbf{A(r)} = \frac{\mu_0}{4\pi}\int_\Omega \frac{\mathbf{J}_m(\mathbf{r'}) \times \mathbf{R}}{R^3}\,d\Omega + \frac{\mu_0}{4\pi}\oint_s \frac{\mathbf{K}_m(\mathbf{r'}) \times \mathbf{R}}{R^3}\,dS \quad (1.3.43)$$

$$\mathbf{J}_m = \nabla' \times \mathbf{M} \quad (1.3.44)$$

$$\mathbf{K}_m = \mathbf{M} \times \mathbf{n} \quad (1.3.45)$$

where $\mathbf{J}_m$, $\mathbf{K}_m$ are the volume and surface current densities of magnetization. Equation (1.3.43) has the same form as Biotsavart's law. Hence, at the interface of different materials the discontinuity of **B** is caused by the surface magnetization current density, i.e.

$$\mathbf{K} = \mathbf{n} \times (\mathbf{M}_+ - \mathbf{M}_-) \quad (1.3.46)$$

where $\mathbf{M}_+$ and $\mathbf{M}_-$ are the intensities of the magnetization on both sides.

In reality the magnetization current is physical, but the magnetic charge is simulated. The equivalent current and the equivalent magnetization charge cannot be used simultaneously because they represent the rotational and point sources, respectively.

1.4 Integral equations of electromagnetic fields

Numerical methods for solving electromagnetic fields can be classified into two types: those based on the differential equation or those on the integral equation. *Applying Green's theorem and Green's function, the differential equations of electromagnetic fields can be expressed by the corresponding integral equations.* In this section only the integral equation of Poisson's equation is derived as an example. Other cases such as the integral equations of the interfacial surface of different materials and the surface integral equation will be outlined in Chaps. 8 and 9.

1.4.1 Integral form of Poisson's equation

In Sect. 1.2.3 Green's third identity was shown, viz.

$$\varphi(\mathbf{r}) = \int_\Omega f(\mathbf{r'})G(\mathbf{r},\mathbf{r'})\,d\Omega + \oint_s \left[G(\mathbf{r},\mathbf{r'})\frac{\partial\varphi(\mathbf{r})}{\partial n} - \varphi(\mathbf{r})\frac{\partial G(\mathbf{r},\mathbf{r'})}{\partial n}\right]dS \quad (1.4.1)$$

it is the integral form of Poisson's equation $\nabla^2\varphi = -f(\mathbf{r'})$. If $G(\mathbf{r},\mathbf{r'})$ is known, the solution for any source $f(\mathbf{r'})$ may be evaluated by Eq. (1.4.1). The two surface integrations in Eq. (1.4.1) represent the boundary conditions of the second and first kind: they can be replaced by single and double layer sources.

1.4 Integral equations of electromagnetic fields

For a 3-D static electric field with $f(\mathbf{r}') = \rho(\mathbf{r}')/\varepsilon$ and $G(\mathbf{r},\mathbf{r}') = 1/4\pi R$ Eq. (1.4.1) is equal to Eq. (1.2.9), i.e.

$$\varphi(\mathbf{r}) = \frac{1}{4\pi\varepsilon}\int_\Omega \frac{\rho(\mathbf{r}')}{R}d\Omega + \frac{1}{4\pi}\oint_S \left[\frac{1}{R}\frac{\partial\varphi(\mathbf{r})}{\partial n} - \varphi(\mathbf{r})\frac{\partial}{\partial n}\left(\frac{1}{R}\right)\right]dS.$$

A significant and important matter is that the surface integration of Eq. (1.2.9) has no effect on the outside region of S. The reason is that a single layer source produces the discontinuity of $\partial\varphi/\partial n$, i.e.

$$\left(\frac{\partial\varphi}{\partial n}\right)_+ - \left(\frac{\partial\varphi}{\partial n}\right)_- = \sigma/\varepsilon. \tag{1.4.2}$$

A double layer source produces discontinuity of the potential φ, i.e.

$$\varphi_+ - \varphi_- = \tau/\varepsilon.$$

On the surface the following conditions exist:

$$\varepsilon\frac{\partial\varphi}{\partial n} = \sigma \tag{1.4.3}$$

$$\varepsilon\varphi = \tau \tag{1.4.4}$$

Substitution of Eq. (1.4.3) into Eq. (1.4.2) yields:

$$\left(\frac{\partial\varphi}{\partial n}\right)_+ = 0. \tag{1.4.5}$$

Comparing Eq. (1.3.17) with Eq. (1.4.4) results in:

$$\varphi_+ = 0. \tag{1.4.6}$$

Thus Eq. (1.2.9) is only valid for the region inside the surface S.

It is concluded that one can always close off any portion of the field by a surface which reduces the field outside the surface to zero. Then the effect of the exterior sources on the interior is replaced by the surface source of a single or a double layer charge distributed on the boundary surface.

If the boundary surface is an equipotential surface, i.e. $\mathbf{n} \times \mathbf{E} = 0$, so that the dipole moment is zero, the potential inside the surface produced by the outer charges is equivalent to a single layer source with the density $\varepsilon(\partial\varphi/\partial n)$

1.4.2 Integral equation for the exterior region

Handling an exterior problem assume that there is a surface Γ_2 enclosing Ω_2. In the volume Ω_2 Green's second identity is

$$\int_{\Omega_2}(G\nabla^2\varphi - \varphi\nabla^2 G)d\Omega = \oint_{\Gamma_1+\Gamma_2}\left(G\frac{\partial\varphi}{\partial n} - \varphi\frac{\partial G}{\partial n}\right)d\Gamma. \tag{1.4.7}$$

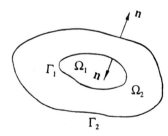

Fig. 1.4.1. The exterior problem

If Γ_2 tends to infinity, the surface integral $\oint_{\Gamma_2}\left(G\dfrac{\partial\varphi}{\partial n}-\varphi\dfrac{\partial G}{\partial n}\right)dS$ tends to zero. Equation (1.4.7) then becomes

$$\int_{\Omega_2}(G\nabla^2\varphi - \varphi\nabla^2 G)\,d\Omega = -\oint_{\Gamma_1}\left(G\dfrac{\partial\varphi}{\partial n}-\varphi\dfrac{\partial G}{\partial n}\right)d\Gamma. \tag{1.4.8}$$

Considering the definition of Green's function and Eq. (1.2.19) results in

$$\varphi(\mathbf{r}) = \int_{\Omega_2} f(\mathbf{r'})G(\mathbf{r},\mathbf{r'})\,d\Omega - \oint_{\Gamma_1}\left(G\dfrac{\partial\varphi}{\partial n}-\varphi\dfrac{\partial G}{\partial n}\right)d\Gamma \tag{1.4.9}$$

The difference between this equation and that of the interior is the sign in front of the term of the boundary integral.

1.5 Summary

In this chapter Maxwell's equations are summarized for the different ranges of frequencies. The field problems fall into three categories:

(1) Dynamic electromagnetic field: In this case the field distribution is dependent on both position and time.

(2) Steady-state field. In the case of $\gamma \gg \omega\varepsilon$ the displacement current is neglected. The field distribution is fixed into position and the phase is a function of the position. The eddy current problem is the main concern in this range of frequencies.

(3) Static and quasi-static fields: In these two cases the field distribution is solely a function of the position. The electric field and the magnetic field are considered separately in different areas.

Both of the essential and Neumann boundary conditions can be replaced by single or double layer sources. These equivalent single and double layer sources are very useful in integral equation methods. The characteristic of the single layer is that the potential is continuous on the both sides of the layer but the discontinuity of the normal derivative of the potential is σ/ε_0. σ is the density of

Appendix 1.1 The integral equation of 3-D magnetic fields

a single layer source. For the double layer source, however, the normal derivative of the potential is continuous but the potential exhibits a discontinuity of τ/ε_0, where τ is the density of the double layer source.

Green's theorem

$$\int_\Omega (u\nabla^2 v - v\nabla^2 u)\,d\Omega = \oint_s \left(u\frac{\partial v}{\partial n} - v\frac{\partial u}{\partial n}\right) dS$$

is a basic theorem for deriving the various integral equations. One example is given in Sect. 1.4.1.

Green's function and fundamental solution are defined as:

$\mathscr{L} G(\mathbf{r}, \mathbf{r}') = -\delta(\mathbf{r}, \mathbf{r}')$ subject to specific boundary conditions

$\mathscr{L} F(\mathbf{r}, \mathbf{r}') = -\delta(\mathbf{r}, \mathbf{r}')$ in free space.

They are the basic tools in solving integral equations.

References

1. Adolf J. Schwab: *Field Theory Concepts*. Springer-Verlag, 1988.
2. C.J. Capenter: Comparison of Alternative Formulation of 3-Dimensional Magnetic-Field and Eddy Current Problem at Power Frequency. *Proc. IEE*, **124**(11), 1026–1034, 1977.
3. Wang Xianchong: *Electromagnetic Field Theory and Applications* (in Chinese). Science Press, Beijing, 1986.
4. J. Van Bladel: *Electromagnetic Fields*. Hemisphere Publishing Corporation, 1985.
5. I. Stakgold: *Green's Functions and Boundary Value Problems*. John Wiley & Sons, 1979.
6. Herbert P. Neff, Jr: *Basic Electromagnetic Fields*. Harper & Row, Ch. 1, 1981.
7. Zhou Keding: *Special Topic on Engineering Electromagnetic Fields* (in Chinese). Hua Zhong University Press, 1986.
8. J.H. McWhirter, J.J. Oravec, Three-Dimensional Electrostatic Field Solutions in A Rod Gap by A Fredholm Integral Equation. *Proc. of 3rd International Symposium on High Voltage Engineering*, 11. 14, 1979.
9. M.H. Lean: Dual Simple-layer Source Formulation for Two-dimensional Eddy Current and Skin Effect Problems. *J. Appl. Phys.* April **57**(8), 3844–3846, 1985
10. J.A. Stratton: *Electromagnetic Theory*. McGraw-Hill, 1941.

Appendix 1.1 The integral equation of 3-D magnetic fields

In a static magnetic field the vector potential **A** satisfies Poisson's equation as

$$\nabla \times \nabla \times \mathbf{A} = \mu \mathbf{J} \qquad (A.1.1)$$

where μ is the permeativity of a homogeneous and isotropic medium. **A** satisfies the Coulomb gauge. Similar to the scalar fundamental solution G a vectorial fundamental solution **Q** is chosen to represent the vector potential at the point (**r**) produced by a unit current density **J** located at point (**r**′). For example

$$\mathbf{Q}(\mathbf{r}, \mathbf{r}') = \frac{\mathbf{a}}{4\pi R}. \tag{A.1.2}$$

Then

$$\nabla \cdot \mathbf{Q} = \frac{1}{4\pi} \mathbf{a} \cdot \nabla\left(\frac{1}{R}\right) \tag{A.1.3}$$

$$\nabla \times \mathbf{Q} = \nabla\left(\frac{1}{4\pi R}\right) \times \mathbf{a} \tag{A.1.4}$$

where **a** is a unit vector along the positive direction of the current. Using vector identity

$$\nabla \times \nabla \times \mathbf{Q} = \nabla(\nabla \cdot \mathbf{Q}) - \nabla^2 \mathbf{Q}. \tag{A.1.5}$$

Due to $\nabla^2 \mathbf{Q} = 0$, one then obtains

$$\nabla \times \nabla \times \mathbf{Q} = \nabla(\nabla \cdot \mathbf{Q}) = \frac{1}{4\pi} \nabla\left[\mathbf{a} \cdot \nabla\left(\frac{1}{R}\right)\right]. \tag{A.1.6}$$

Multiplying the vector **A** to both sides of Eq. (A.1.6) leads to

$$\mathbf{A} \cdot \nabla \times \nabla \times \mathbf{Q} = \frac{1}{4\pi} \mathbf{A} \cdot \nabla\left[\mathbf{a} \cdot \nabla\left(\frac{1}{R}\right)\right] = \frac{1}{4\pi} \nabla \cdot \left[\mathbf{a} \cdot \nabla\left(\frac{1}{R}\right) \mathbf{A}\right] \tag{A.1.7}$$

on the other hand,

$$\mathbf{Q} \cdot \nabla \times \nabla \times \mathbf{A} = \frac{\mathbf{a}}{4\pi R} \mu \mathbf{J}. \tag{A.1.8}$$

Recalling Green's vector identity

$$\int_\Omega (\mathbf{Q} \cdot \nabla \times \nabla \times \mathbf{P} - \mathbf{P} \cdot \nabla \times \nabla \times \mathbf{Q})\, d\Omega = \oint_S (\mathbf{P} \times \nabla \times \mathbf{Q} - \mathbf{Q} \times \nabla \times \mathbf{P}) \cdot \mathbf{n}\, dS$$

let **P** be **A**, then

$$(\mathbf{P} \times \nabla \times \mathbf{Q}) \cdot \mathbf{n} = \left(\mathbf{A} \times \left[\nabla\left(\frac{1}{4\pi R}\right) \times \mathbf{a}\right]\right) \cdot \mathbf{n} = \mathbf{n} \cdot \left(\mathbf{A} \times \left[\nabla\left(\frac{1}{4\pi R}\right) \times \mathbf{a}\right]\right)$$

$$= \left[\nabla\left(\frac{1}{4\pi R}\right) \times \mathbf{a}\right] \cdot (\mathbf{n} \times \mathbf{A})$$

$$= \mathbf{a} \cdot (\mathbf{n} \times \mathbf{A}) \times \nabla\left(\frac{1}{4\pi R}\right) = \mathbf{a} \cdot \nabla\left(\frac{1}{4\pi R}\right) \times (\mathbf{A} \times \mathbf{n}) \tag{A.1.9}$$

$$(\mathbf{Q} \times \nabla \times \mathbf{P}) \cdot \mathbf{n} = \left(\frac{\mathbf{a}}{4\pi R} \times \nabla \times \mathbf{A}\right) \cdot \mathbf{n} = \mathbf{n} \cdot \left(\frac{\mathbf{a}}{4\pi R} \times \mathbf{B}\right) = \frac{\mathbf{a}}{4\pi R} \cdot \mathbf{B} \times \mathbf{n}$$

$$= \mathbf{a} \cdot \frac{\mathbf{B} \times \mathbf{n}}{4\pi R}. \tag{A.1.10}$$

Appendix 1.1 The integral equation of 3-D magnetic fields

Substituting Eqs. (A.1.7), (A.1.8) into the LHS of Green's vector identity and Eq. (A.1.9), with Eq. (A.1.10) into the RHS of Eq. (1.2.16), the result is:

$$\int_\Omega \left(\frac{\mathbf{a}}{4\pi R}\mu \mathbf{J} - \frac{1}{4\pi}\nabla\cdot\left[\mathbf{a}\cdot\nabla\left(\frac{1}{R}\right)\mathbf{A}\right]\right) d\Omega = \frac{\mathbf{a}}{4\pi}\mu\int_\Omega \frac{\mathbf{J}}{R} d\Omega$$

$$- \frac{\mathbf{a}}{4\pi}\oint_S \left[\nabla\left(\frac{1}{R}\right)\mathbf{A}\cdot\mathbf{n}\right] dS . \tag{A.1.11}$$

Thus one obtains:

$$\int_\Omega \frac{\mu \mathbf{J}}{4\pi R} d\Omega = \oint_S \left[\nabla\left(\frac{1}{4\pi R}\right)\mathbf{A}\cdot\mathbf{n}\right] dS + \oint_S \left[\nabla\left(\frac{1}{4\pi R}\right)\times(\mathbf{A}\times\mathbf{n})\right] dS$$

$$+ \oint_S \frac{\mathbf{n}\times\mathbf{B}}{4\pi R} dS . \tag{A.1.12}$$

This singularity is dealt with as the same way which used in Sect. 1.4.1. The singular point at $r = 0$ is circumscribed by a small sphere of radius r_0. The volume V is now bounded by surface S_0 and S. Due to $\nabla(1/R) = -\mathbf{R}^0/R^2$, the surface integral of Eq. (A.1.12) on S_0 is

$$\frac{1}{4\pi r_0^2}\int_{S_0} \mathbf{R}^0(\mathbf{A}\cdot\mathbf{n}) ds + \frac{1}{4\pi r_0^2}\oint_{S_0} \mathbf{R}^0\times(\mathbf{A}\times\mathbf{n}) dS + \frac{1}{4\pi r_0}\oint_{S_0} \mathbf{n}\times\mathbf{B} dS \tag{A.1.13}$$

and the integrand of the middle term can be transformed to

$$\mathbf{R}^0\times(\mathbf{A}\times\mathbf{n}) = (\mathbf{R}^0\cdot\mathbf{n})\mathbf{A} - (\mathbf{A}\cdot\mathbf{n})\mathbf{R}^0 + \mathbf{A}\times(\mathbf{R}^0\times\mathbf{n}) . \tag{A.1.14}$$

Since $\mathbf{R}^0\cdot\mathbf{n} = 1$, $\mathbf{R}^0\times\mathbf{n} = 0$ the surface integrals over S_0 reduce to

$$\frac{1}{4\pi r_0^2}\oint_{S_0} \mathbf{A} dS + \frac{1}{4\pi r_0}\oint_{S_0} \mathbf{n}\times\mathbf{B} dS . \tag{A.1.15}$$

Assuming A and B are constants over the small sphere, the result of the integral of Eq. (A.1.15) reduces to $\mathbf{A}(\mathbf{r})$. Introducing this result to Eq. (A.1.12), the result is

$$\mathbf{A}(\mathbf{r}) = \frac{\mu}{4\pi}\int_\Omega \frac{\mathbf{J}(\mathbf{r}')}{R} d\Omega - \frac{1}{4\pi}\oint_S \frac{\mathbf{n}\times\mathbf{B}}{R} dS - \frac{1}{4\pi}\oint_S \left[(\mathbf{n}\times\mathbf{A})\times\nabla\left(\frac{1}{R}\right)\right] dS$$

$$- \frac{1}{4\pi}\oint_S (\mathbf{n}\cdot\mathbf{A})\nabla\left(\frac{1}{R}\right) dS . \tag{A.1.16}$$

This general expression of the vector potential $\mathbf{A}(\mathbf{r})$ in integral form includes contributions from all sources. It is a Fredholm integral equation of the second kind while $\mathbf{A}(\mathbf{r})$ is unknown. The three terms of the surface integral represent contributions from the sources outside the surface S: the effect of the surface current density $\mathbf{n} \times \mathbf{B} = \mu \mathbf{K}$: the effect of the dipole moment $\mathbf{n} \times \mathbf{A} = -\mu \mathbf{M}$; and the effect of the equivalent magnetic charge density $\mathbf{n} \cdot \mathbf{A}$. The vector potential $\mathbf{A}$ within volume Ω is continuous and has continuous derivatives of all orders; however, the $\mathbf{A}$ and its derivatives exhibit a certain discontinuity across the surface. It can be proved as follows: Suppose $\mathbf{B}_-$ and $\mathbf{B}_+$ denote the vector $\mathbf{B}$ just inside and outside the surface S which satisfy the boundary condition

$$\mathbf{n} \times (\mathbf{B}_+ - \mathbf{B}_-) = \mu \mathbf{K} \,. \tag{A.1.17}$$

The first surface integral of Eq. (A.1.16) can be regarded as the equivalent surface current, i.e. $-\mathbf{n} \times \mathbf{B}_- = \mathbf{K}$. Comparing Eq. (A.1.16) and Eq. (A.1.15) it is clear that $\mathbf{n} \times \mathbf{B}_+ = 1$. The second term of the surface integral is equivalent to the vector potential produced by a surface polarization density, e.g. $\mathbf{A}_- \times \mathbf{n} = \mu \mathbf{M}$. Note that the tangential component of $\mathbf{A}$ is continuous across surface S only in the case of the magnetization being normal to the surfac, i.e.

$$\mathbf{n} \times (\mathbf{A}_+ - \mathbf{A}_-) = \mu \mathbf{M} - \mu(\mathbf{n} \cdot \mathbf{M}) \cdot \mathbf{n} = \mathbf{A} \times \mathbf{n} - \mu(\mathbf{n} \cdot \mathbf{M})\mathbf{n} \,.$$

Thus

$$\mathbf{n} \times \mathbf{A}_+ = 0 \,. \tag{A.1.18}$$

The last term of the surface integral of Eq. (A.1.16) is related to the field intensity of a surface charge density $(\mathbf{A} \cdot \mathbf{n})$, i.e. $\mathbf{n} \cdot (\mathbf{A}_+ - \mathbf{A}_-) = \mathbf{n} \cdot \mathbf{A}_-$. Thus

$$\mathbf{n} \cdot \mathbf{A}_+ = 0 \,. \tag{A.1.19}$$

So far it is proved that on the positive side of the surface S the normal and the tangential components of $\mathbf{A}$ and the tangential components of B are zero everywhere. Furthermore the normal component of $\mathbf{B}$ must be zero over the positive side of S, because the normal component of the curl $\mathbf{A}$ involves only partial derivatives in those directions tangential to S. If we apply Eq. (A.1.16) to the region externally to surface S, then $\mathbf{A}$ and consequently $\mathbf{B}$ are zero everywhere. If we let $\mathbf{Q} = \nabla(1/R) \times \mathbf{a}$ as the Green's function instead of Eq. (1.4.13), the following equation is obtained:

$$\mathbf{B}(\mathbf{r}) = \frac{\mu}{4\pi} \int_{\Omega} \mathbf{J} \times \nabla \left(\frac{1}{R}\right) d\Omega - \frac{1}{4\pi} \oint (\mathbf{n} \times \mathbf{B}) \times \nabla \left(\frac{1}{R}\right) dS$$

$$- \frac{1}{4\pi} \oint_S (\mathbf{n} \cdot \mathbf{B}) \nabla \left(\frac{1}{R}\right) dS \,. \tag{A.1.20}$$

Chapter 2

General Outline of Numerical Methods

2.1 Introduction

According to Maxwell's equations, all electromagnetic field problems can be expressed in partial differential equations which are subject to specific boundary conditions. By using Green's function, the partial differential equations can be transformed into integral equations or differential-integral equations. The analytical solution of these equations can only be obtained in very simple cases. Therefore numerical methods are significant for the solution of practical problems. In numerical solutions the following aspects have to be considered.

(1) A mathematical model expressed by differential equations, integral equations, or variational expressions is provided to describe physical states.

(2) A discretized model is suggested to approximate the solution domain, so that a set of algebraic equations is obtained.

(3) A computer program is designed to complete the computation.

In designing these steps one should consider:

(1) Does the mathematical model describe the physical state well?
(2) Does the approximate solution satisfy the desired accuracy?
(3) Does the method use the computer sources economically?

In order to obtain a good method for various engineering problems many methods have been developed.

The purpose of various numerical methods that are used to obtain solutions for electromagnetic field problems is to transfer an operator equation (differential or integral equation) into a matrix equation.

In solving field problems the problem can be described by differential or integral equations. Consequently, there are two different kinds of solution methods: using either differential equations or the integral equations. The former is known as the "field" approach or domain method and the second is known as a source distribution technique or the boundary method. Hammond has interpreted these dual approaches in a historical perspective: 'The history of electromagnetic investigation is the history of the interplay of two fundamentally different modes of thought. The first of these, the method of electromagnetic fields which ascribes the action of a continuum, is associated with such thinkers as Gilbert, Faraday and Maxwell. The second, the method of electromagnetic sources, concentrates the attention on the forces between electric and magnetic bodies and is associated with Franklin, Cavendish and Ampere. .... Field problems are conveniently handled by differential equations and sources by integral equations.' [1]. Both of these two methods have advantages and drawbacks. Reference [2] describes the optimal combination of these two methods. No matter of which methods are applied, numerical solution methods consist of the following steps:

The first step is to express the unknown function $u(\mathbf{r})$ contained in the operator equation by the summation of a set of linear independent functions with undetermined parameters of a complete set sequence, e.g.

$$u(\mathbf{r}) = \sum_{i=1}^{N} C_i \psi_i \qquad (2.1.1)$$

where C_i are undetermined parameters and ψ_i are the terms of basis functions. Equation (2.1.1) is called the trial function or approximate solution. If N goes to infinity, the approximate solution will tend towards the real solution.

The second step is to cast the continuous solution domain into a discrete form. The resulting set of discretized subdomains consists of a finite number of elements and nodes. In this fashion the unknown function with infinite degrees of freedom is replaced by an approximate function with finite degrees of freedom.

The third step is to choose a principle of error minimum in order to determine the unknown parameters contained in the trial function. This can be achieved by employing either variational principles or the principle of weighted residuals. After this step is executed, the operator equation is transformed to a matrix equation.

Finally, the approximate solution of a given problem is obtained by solving the linear or non-linear matrix equation derived from the third step.

The finite difference method was the first to be developed [3] from among the well-known numerical methods. *Here the solution domain is subdivided into many nodes in a regular grid. The values of a continuous function within the domain are represented by the values in the finite grid nodes. This method can be interpreted as a method in which the differential operator is replaced by the difference operator.* The finite difference method has been used to solve many engineering problems since the 1950s. But because of the need generally to use regular grids the application of this method is limited.

By the end of the 1950s, the finite element method was introduced, firstly in structural mechanics [4]. The significant difference between the finite difference method and the finite element method is that, *in this method, the domain is discretized by employing a set of small elements with different shapes and sizes.*

With this approach it is easy to solve a problem having complex geometry and different interfacial boundaries to a high degree accuracy. It seems to be one of the most efficient methods for the solution of electromagnetic field problems. As both the finite difference method and the finite element method are based on differential equations and domain discretization, they are called *differential methods or domain methods*.

Almost in the same period volume integral equation methods were developed for solving static magnetic field and eddy current problems [5]. *The volume integral method is based on the principle of superposition.* First, the source area is subdivided into small areas; then the solution in terms of the sum of all such elements is sought. This method is simple to understand and easy to solve in the case of 2-dimensional problems. However, it is limited to just linear problems.

In order to reduce the region of discretization boundary integral equation methods were rapidly developed. The most typical of these based on the boundary integral equation is the boundary element method [6]. *The advantage of the boundary element method is that only the boundary values are treated as being unknowns and only the boundary of the solution domain is discretized. Hence this method reduces the dimensions of the size of the problem by one.* The pre- and the post-data processing is therefore much easier than with the finite element method, especially in the case of 3-D problems.

So far, any 2-D problem can be solved efficiently by one of these methods, and there are many well-designed commercial software packages for analysing and designing purposes. However, for solving 3-D vector fields, especially in the case of problems containing non-linear material or having time-dependent solutions, efficient solution methods are still being developed. Another aspect which is of interest to engineers is to establish efficient software packages that can be used to model complex systems in designing practical electromagnetic devices. Reference [7] discusses such prospects of electromagnetic computing.

As indicated, any such numerical method gives an approximate solution. To ensure this ultimately derives to a real solution, the principle of error minimization should be observed.

The essential purpose of this chapter is to develop a unified framework for the discussion of various numerical methods which are based on the principle of error minimization. The differences and the relationships between various numerical methods are classified in this chapter. The approximate notations are interpreted using concepts of space and operators.

2.2 Operator equations [8, 9]

An operator $\mathscr{L}$ provides a mapping or transformation, according to which a particular element u belonging to a subset of the domain Ω of a space R is uniquely associated with element f belonging to another subset W of space S. This is expressed as

$$\mathscr{L}u = f \tag{2.2.1}$$

where Ω is the domain and W is the range, respectively, of the operator $\mathscr{L}$. That is $\mathscr{L}$ maps Ω onto W; it represents the mapping between two functions. For example, the derivation, integral, gradient, divergence, curl, Laplacian and matrix transformations are operators. Basically an operator is simply a certain type of function, just as the simple function $y = f(x)$ maps the variable x into the variable y. By using the operator notation any differential or integral equation can be expressed in a simple, compact notation. In order to understand the mapping of the operator the concepts of the 'space' are reviewed in the first section.

2.2.1 Hilbert space

A space is a collection of elements considered as a whole. The dimension of space S is the maximum number of linearly independent elements contained within S. Alternatively, the dimension n of S is the number of independent elements required to form a basis for S. Any $n + 1$ element is a dependent set. The space is infinite-dimensional if it contains an infinite number of independent elements.

Metric space

A space R having the following properties is termed a metric space. Assuming **f** and **g** are any two elements belonging to space R, there is a distance d between **f** and **g** which is defined as

$$d = d(\mathbf{f}, \mathbf{g}) \tag{2.2.2}$$

and which satisfies the following properties

$$d(\mathbf{f}, \mathbf{g}) = d(\mathbf{g}, \mathbf{f}) \tag{2.2.3}$$

$$d(\mathbf{f}, \mathbf{g}) \geq 0 \tag{2.2.4}$$

$$d(\mathbf{f}, \mathbf{g}) = 0 \quad \text{if and only if } \mathbf{f} = \mathbf{g} \tag{2.2.5}$$

$$d(\mathbf{f}, \mathbf{g}) \leq d(\mathbf{f}, \mathbf{g}) + d(\mathbf{f}, \mathbf{h}) \quad \mathbf{f}, \mathbf{g}, \mathbf{h} \in R \tag{2.2.6}$$

where **h** is another element that belongs to R.

Linear space

A space is called linear if all operations of the elements of the space satisfy the rules of vector algebra. The usual function space and vector space are linear spaces. The elements $f_1, f_2, \ldots, f_n$ of a linear space S are said to be linearly independent

2.2 Operator equations

if and only if

$$\alpha_1 f_1 + \alpha_2 f_2 + \ldots + \alpha_n f_n = 0 \tag{2.2.7}$$

which implies $\alpha_1 = \alpha_2 = \ldots = \alpha_n = 0$. Otherwise the elements are linearly dependent.

Inner product space (unitary space)

Unitary space is a linear space in which, for any two elements $(\mathbf{f}, \mathbf{g})$, there is an associated real or complex number defined as

$$\langle \mathbf{f}, \mathbf{g} \rangle = \int_\Omega f(\mathbf{r}) g^*(\mathbf{r}') \, d\Omega \tag{2.2.8}$$

where $\langle , \rangle$ denotes the inner product and $\mathbf{g}^*$ is the complex conjugate of $\mathbf{g}$. $\langle \mathbf{f}, \mathbf{g} \rangle$ is the inner product of $\mathbf{f}$ and $\mathbf{g}$ in space S and the following relations are satisfied:

$$\langle c\mathbf{f}, \mathbf{g} \rangle = c \langle \mathbf{f}, \mathbf{g} \rangle \tag{2.2.9}$$

$$\langle \mathbf{f} + \mathbf{g}, \mathbf{h} \rangle = \langle \mathbf{f}, \mathbf{h} \rangle + \langle \mathbf{g}, \mathbf{h} \rangle \tag{2.2.10}$$

$$\langle \mathbf{f}, \mathbf{g} \rangle = \langle \mathbf{g}, \mathbf{f} \rangle^* = \int_\Omega g(\mathbf{r}) f^*(\mathbf{r}') \, d\Omega \tag{2.2.11}$$

$$\langle \mathbf{f}, \mathbf{f} \rangle > 0 \quad \text{if } \mathbf{f} \neq 0 \tag{2.2.12}$$

$$\langle \mathbf{f}, \mathbf{f} \rangle = 0 \quad \text{if } \mathbf{f} = 0 \tag{2.2.13}$$

where c is constant. If $\langle \mathbf{f}, \mathbf{g} \rangle = 0$, the two elements $\mathbf{f}, \mathbf{g}$ are orthogonal. If $\mathbf{f}$ and $\mathbf{g}$ are orthogonal and normalized (the norm of the vector which equals 1 is normalized), they are orthonormal. *An orthogonal set of non-zero elements is independent.*

Normed linear space

In a linear space, the real-value function $\|\mathbf{f}\|$ is defined as the norm of function f as follows

$$\|\mathbf{f}\| = \langle \mathbf{f}, \mathbf{f}^* \rangle^{1/2} \quad \|\mathbf{f}\| > 0$$

$$\|\mathbf{f}\| = 0 \quad \text{if and only if } \mathbf{f} = 0. \tag{2.2.14}$$

If $\|\mathbf{f}\| = 1$, the element $\mathbf{f}$ is normalized. A normed linear space has the following properties

$$\|c\mathbf{f}\| = |c| \, \|\mathbf{f}\| \tag{2.2.15}$$

$$\|\mathbf{f}_1 + \mathbf{f}_2\| \leq \|\mathbf{f}_1\| + \|\mathbf{f}_2\| \tag{2.2.16}$$

$$|\langle \mathbf{u}, \mathbf{f} \rangle| \leq \|\mathbf{u}\| \cdot \|\mathbf{f}\|. \tag{2.2.17}$$

Equation (2.2.17) is the Schwarz inequality. It is an important inequality in linear space.

Complete inner product space [10]

Let (f_n) be a convergent sequence of points in an inner product space. If for each $\varepsilon > 0$ there exists some $N = N(\varepsilon)$ such that for all $n, m \geq N$ the following expressions

$$\| f_n - f_m \| < \varepsilon \tag{2.2.18}$$

and

$$\lim_{n \to \infty} \| f_n - f \| = 0 \tag{2.2.19}$$

are satisfied then the space is called complete inner product space. $\{f_n\}$ is a Cauchy sequence. A complete inner product space is one in which all Cauchy sequences are convergent sequences. A complete inner product space is called a **Hilbert** space. The Schwarz inequality ensures that the inner product space is complete. All elements in a Hilbert space are square integrable.

Subspace

The space A is a subspace of S if each element of A is also an element of space S. A subspace A of a linear space S is called a linear manifold in S.

2.2.2 Definition and properties of operators

An operator represents the relationship between two functions as shown in Eq. (2.2.1). The properties of the operator determine the methods used for solving the operator equations numerically.

Linear operator

If

$$\mathscr{L}(f + g) = \mathscr{L}f + \mathscr{L}g \tag{2.2.20}$$

and

$$\mathscr{L}(cf) = c\mathscr{L}(f) \tag{2.2.21}$$

and the domain Ω and range W are linear spaces, then the operator $\mathscr{L}$ is called a linear operator. In Eqs. (2.2.20) and (2.2.21) f and g are two elements, c is a constant.

Symmetric operator

If

$$\langle \mathscr{L}u, v \rangle = \langle u, \mathscr{L}v \rangle \tag{2.2.22}$$

where u, v are any two functions in the space of $\mathscr{L}$, then $\mathscr{L}$ is a symmetric operator.

2.2 Operator equations

Positive definite operator

If
$$\langle \mathscr{L}\mathbf{u}, \mathbf{u} \rangle > 0 \qquad (2.2.23)$$

for all $\mathbf{u} \neq 0$ in its domain, $\mathscr{L}$ is positive definite. If the sign $>$ is replaced by $\geq$, then $\mathscr{L}$ is positive semi-definite.

Self-adjoint operator

The adjoint operator of $\mathscr{L}$ is the operator $\mathscr{L}^*$ such that
$$\langle \mathscr{L}\mathbf{u}, \mathbf{f} \rangle = \langle \mathbf{f}, \mathscr{L}^*\mathbf{u} \rangle \qquad (2.2.24)$$

where the domain of $\mathscr{L}^*$ is also that of $\mathscr{L}$, then $\mathscr{L}^*$ is the adjoint operator of $\mathscr{L}$. If $\mathscr{L} = \mathscr{L}^*$, $\mathscr{L}$ is termed a self-adjoint operator denoted by $\mathscr{L}^\alpha$. Hence a self-adjoint operator is symmetric but not vice-versa. The bounded operators defined in the whole space are self-adjoint. The operators having even power such as ∇^2 and $\nabla^2 + (\partial^2/\partial t^2)$ are self-adjoint. The operators having odd power cannot be self-adjoint. If the kernel of an integral equation is symmetric, then the integral operator is self-adjoint.

Bounded operator

If
$$\|\mathscr{L}\mathbf{u}\| \leq M\|\mathbf{u}\| \quad M \leq \infty \qquad (2.2.25)$$

then $\mathscr{L}$ is a bounded operator. The smallest M is called the norm of $\mathscr{L}$ and denoted by $\|\mathscr{L}\|$. The operator $\mathscr{L}$ in the Hilbert space is bounded.

Continuous operator

If $\mathbf{u}_n \to \mathbf{u}$, $\mathbf{u}_n$ and $\mathbf{u}$ belong to the same domain and it follows that
$$\mathscr{L}\mathbf{u}_n + \mathscr{L}\mathbf{u} \qquad (2.2.26)$$

then $\mathscr{L}$ is a continuous operator. A bounded linear operator is continuous and vice-versa.

Completely continuous operator

If
$$\lim_{m,n \to \infty} \langle \mathscr{L}\mathbf{u}_n, \mathbf{f}_m \rangle = \langle \mathscr{L}\mathbf{u}, \mathbf{f} \rangle \qquad (2.2.27)$$

or
$$\lim_{n \to \infty} \|\mathscr{L} - \mathscr{L}_n\| = 0$$

then $\mathscr{L}$ is a completely continuous operator (c.c.o.). *Every finite-dimensional* linear operator is a completely continuous operator. A completely continuous operator is bounded and the reverse is true for finite-dimensional spaces, although not for infinite dimensional spaces. If $\mathscr{L}$ is a c.c.o., then the adjoint operator of $\mathscr{L}$ will also be c.c.o.

The identity operator $\mathscr{I}$ which maps

$$\mathscr{I}\mathbf{u}_n = \mathbf{u}_n \qquad (2.2.28)$$

in an infinite dimensional space is therefore not completely continuous.

Inverse operator

If the mapping $\mathscr{L}\mathbf{u} = \mathbf{f}$ is one to one, then the inverse operator exists, i.e.

$$\mathbf{u} = \mathscr{L}^{-1}\mathbf{f} \qquad (2.2.29)$$

The inverse of a linear operator is also linear. Also $\mathscr{L}^{-1}$ is self-adjoint provided that the $\mathscr{L}$ is self-adjoint. *The inverse operator exists if the operator is a bounded positive definite linear operator.*

The Eigen value of an operator

If there is a number λ and an element $\mathbf{u} \neq 0$ and

$$\mathscr{L}\mathbf{u} = \lambda \mathbf{u} \qquad (2.2.30)$$

exists then $\mathbf{u}$ is called an eigen element (or eigenvector) of operator $\mathscr{L}$ while λ is an eigenvalue of $\mathscr{L}$.

Condition number

Define

$$k(\mathscr{L}) = \|\mathscr{L}\| \|\mathscr{L}^{-1}\| \quad k(\mathscr{L}) \geq 1 \qquad (2.2.31)$$

as being the condition number of a linear bounded operator.

Basis

A finite or countably infinite set of vectors $\mathbf{e}_1, \ldots, \mathbf{e}_k \ldots$ is a basis of a space if

(a) the vectors $\mathbf{e}_1, \ldots, \mathbf{e}_k \ldots$ are independent,
(b) each vector $\mathbf{x}$ in the space can be written as a linear combination of a finite or infinite number of basis vectors.

2.2.3 The relationship between the properties of the operators and the solution of the operator equations

Both the approximate approach for the formulation as well as the solution methods for the resulting matrix equation are dependent on the properties of the differential and integral operators.

2.2 Operator equations

(1) If $\mathscr{L}$ is symmetric positive definite, then the operator equation

$$\mathscr{L}\mathbf{u} = \mathbf{f}$$

has only one stable solution [8, 9]. It means that the solution of the above equation is unique.

(2) If $\mathscr{L}$ is completely continuous, the inner product $\langle \mathscr{L}\mathbf{u}, \mathbf{f} \rangle$ exists [11]. Hence, the equivalent functional of the operator equation can be determined.

(3) If $\mathscr{L}$ is a self-adjoint positive definite operator in the Hilbert space then the solution of Eq. (2.2.1) can be approximated by the associated problem which minimizes the quadratic functional $I(\mathbf{u})$ [9, 11]

$$I(\mathbf{u}) = \langle \mathscr{L}\mathbf{u}, \mathbf{u} \rangle - \langle \mathbf{u}, \mathbf{f} \rangle - \langle \mathbf{f}, \mathbf{u} \rangle \tag{2.2.32}$$

where $\mathbf{u}$ is the approximate solution of Eq. (2.2.1). $I(\mathbf{u})$ is a functional. It represents mapping from the function space to the value space.

(4) A separable kernel $k(\mathbf{r}, \mathbf{r}') = \Sigma p(\mathbf{r})q(\mathbf{r}')$ of an integral operator results in a completely continuous operator. Thus the approximate solution of such an integral equation can be found by using the weighted residual principle.

(5) If the kernel of the operator is symmetric, the corresponding operator is self-adjoint. The solution method for the resulting symmetric matrix is thus more convenient than for an asymmetric matrix.

(6) Self-adjoint operators are symmetric and generate a symmetric system matrix which has real eigenvalues.

(7) If $\mathscr{L}$ is a completely continuous operator, then the resulting matrix is positive definite.

(8) If the inverse operator exists, the solution of the original operator equation is unique.

2.2.4 Operator equations of electromagnetic fields

Electromagnetic field problems may be solved by partial differential equations or integral equations. Each approach has its own merits and shortcomings and the selection should be based upon individual requirements of the problem [9]. In this section, only some typical equations are given and the properties of those operators used are analysed.

Static and quasi-static electromagnetic fields

It is well known that if the charge distribution is known the potential satisfies the following equations:

$$-\nabla^2 \varphi = \rho/\varepsilon_0 \tag{2.2.33}$$

or

$$\varphi(\mathbf{r}) = \int_{\Omega'} \frac{\rho(\mathbf{r}')}{4\pi\varepsilon r} d\Omega' = \int_{\Omega'} \rho(\mathbf{r}') G(\mathbf{r}, \mathbf{r}') d\Omega' \tag{2.2.34}$$

where ∇^2 is a differential operator that is both positive definite and self-adjoint. Similarly, $\int_{\Omega'} G(\mathbf{r}, \mathbf{r}') \, d\Omega'$ is a Fredholm integral operator of the first kind, while $G(\mathbf{r}, \mathbf{r}')$ is the corresponding kernel. It is obvious that this kernel is symmetric and separable. Hence the Fredholm integral operator of the first kind is completely continuous and bounded.

In order to show that the Laplacian operator is self-adjoint consider a suitable inner product for the L.H.S. of Eq. (2.2.33)

$$\langle \mathscr{L}\varphi, \psi \rangle = \int_{\Omega'} (-\varepsilon_0 \nabla^2 \varphi) \psi \, d\Omega' \tag{2.2.35}$$

and then use Green's identity:

$$\int_{\Omega} (\psi \nabla^2 \varphi - \varphi \nabla^2 \psi) \, d\Omega = \oint_{\Gamma} \left(\psi \frac{\partial \varphi}{\partial n} - \varphi \frac{\partial \psi}{\partial n} \right) d\Gamma . \tag{2.2.36}$$

In Eq. (2.2.36) let Γ be a sphere of radius r, while φ and ψ are constants up to limit $r \to \infty$. The R.H.S. of the above equation then vanishes within this limit. Equation (2.2.36) therefore reduces to

$$\int_{\Omega} \psi \nabla^2 \varphi \, d\Omega = \int_{\Omega} \varphi \nabla^2 \psi \, d\Omega . \tag{2.2.37}$$

According to the definition of the self-adjoint operator

$$\langle \mathscr{L}\varphi, \psi \rangle = \langle \psi, \mathscr{L}^a \varphi \rangle$$

it is evident that ∇^2 is self-adjoint, i.e.

$$\mathscr{L} = \mathscr{L}^a = -\varepsilon_0 \nabla^2 . \tag{2.2.38}$$

Hence the equivalent functional of the Laplace operator exists and a symmetric matrix equation is obtained.

The self-adjoint property of the integral operator of Eq. (2.2.34) can also be proved by the definition of Eq. (2.2.24) and the symmetry of Green's function of Poisson's equation as below:

$$\begin{aligned}
\langle \mathscr{L}u, f \rangle - \langle u, \mathscr{L}^* f \rangle &= \int_{\Omega} [f^*(\mathscr{L}u) - u(\mathscr{L}f)^*] \, d\Omega \\
&= \int_{\Omega} \int_{\Omega} [f^*(\mathbf{r}) G(\mathbf{r}, \mathbf{r}') u(\mathbf{r}') \\
&\quad - u(\mathbf{r}) G^*(\mathbf{r}', \mathbf{r}) f^*(\mathbf{r}')] \, d\Omega \, d\Omega \\
&= \int_{\Omega} \int_{\Omega} [f^*(\mathbf{r}) G(\mathbf{r}, \mathbf{r}') u(\mathbf{r}') \\
&\quad - f^*(\mathbf{r}') G(\mathbf{r}', \mathbf{r}) u(\mathbf{r})] \, d\Omega \, d\Omega \\
&= 0 .
\end{aligned}$$

This result shows that the integral operator $\int_{\Omega} G(\mathbf{r}, \mathbf{r}') \, d\Omega'$ is self-adjoint.

2.2 Operator equations

In case of the interface between different dielectrics with the permittivity of ε_0 and relative permittivity ε_r, respectively, the equation for the charge density in a single layer is given by [12]:

$$\mathcal{L}\sigma(s) = \frac{\varepsilon_r + 1}{2\varepsilon_0}\sigma(s) + \frac{\varepsilon_r - 1}{\varepsilon_0}\oint_s \sigma(s')\frac{\partial G}{\partial n}(s, s')\,ds' = 0. \tag{2.2.39}$$

This is a Fredholm integral equation of the second kind. Due to the definition of the adjoint operator, the adjoint operator of $\mathcal{L}$ given in Eq. (2.2.39) is obtained by replacing the kernel $\left(\dfrac{\partial G}{\partial n}(s, s')\right)$ by $\left(\dfrac{\partial G}{\partial n'}(s', s)\right)$. In general, the adjoint operator of a complex integral operator is one with the kernel replaced by its complex-conjugate transpose. For a general curve it is given as

$$\frac{\partial G}{\partial n}(s, s') \neq \frac{\partial G}{\partial n'}(s', s). \tag{2.2.40}$$

Hence the integral operator in Eq. (2.2.39) is not self-adjoint. The treatment of this kind of operator is given in reference [12].

Diffusion equations

For time harmonic electromagnetic fields the problems are divided into two kinds: determination and the eigenvalue problems. The determination problem is expressed by in-homogeneous Helmholtz equation:

$$(\nabla^2 + \beta^2)\mathbf{A} = -\mu \mathbf{J}_s \tag{2.2.41}$$

$$\beta^2 = \omega^2\mu\varepsilon \qquad \varepsilon = \varepsilon' - j\varepsilon'' \tag{2.2.42}$$

where the operator is $\mathcal{L} = (\nabla^2 + \beta^2)$. The difference between Eq. (2.2.41) and the Laplacian equation is that the term $\beta^2 \mathbf{A}$ is added and $\mathbf{J}$ and $\mathbf{A}$ are complex functions. In Eq. (2.2.42) ε is an equivalent permittivity. If β^2 is a real constant, it can be proved in the same way that the operator $(\nabla^2 + \beta^2)$ is a linear continuous and symmetric operator. If $\beta^2 > 0$, the operator is bounded-below. Hence the Helmholtz operator $\mathcal{L} = (\nabla^2 + \beta^2)$ is self-adjoint.

If the dielectric medium is lossy, the permittivity ε is a complex quantity then the operator $\mathcal{L} = (\nabla^2 + \beta^2)$ is non-self adjoint. The proof is given in [13]. The conclusion given in [13] is that for the non-self-adjoint complex operator the system equation adopts the same form as in the real self-adjoint case [2, 13]. Mikhlin [10] states that an operator needs to be neither self-adjoint nor positive-bounded-below to ensure convergence. If the operator possesses a component with such properties and given certain uniqueness and completeness conditions, then a convergence holds. From the experience of McDonald and Weler [2] solutions for such problems seem to converge as well as those involving self-adjoint, positive-definite operators.

Fast transient fields

The differential forms of wave equations are

$$\nabla^2 \varphi - \varepsilon\mu \frac{\partial^2 \varphi}{\partial t^2} = -\rho/\varepsilon \tag{2.2.43}$$

$$\nabla^2 \mathbf{A} - \mu\varepsilon \frac{\partial^2 \mathbf{A}}{\partial t^2} = -\mu \mathbf{J}. \tag{2.2.44}$$

The operator of these equations is

$$\mathscr{L} = \nabla^2 - \mu\varepsilon \frac{\partial^2}{\partial t^2}. \tag{2.2.45}$$

The integral expression for Eq. (2.2.43) is

$$\varphi = \frac{1}{\varepsilon} \int_{\Omega'} \rho(\mathbf{r}') G(\mathbf{r}, \mathbf{r}') \, d\Omega' \tag{2.2.46}$$

with

$$G(\mathbf{r}, \mathbf{r}') = \frac{\exp(-jk|\mathbf{r} - \mathbf{r}'|)}{4\pi|\mathbf{r} - \mathbf{r}'|}. \tag{2.2.47}$$

It can be proved that this integral operator is non-self-adjoint[†]. The equivalent functional of non-self-adjoint operators are discussed in references [2, 13, 14].

2.3 Principles of error minimization

An approximate numerical solution of a boundary value problem is one that minimizes the error of approximation. Any approximation contains two different aspects. First, the infinite dimensional space of the real solution is approximated by a discretized domain of finite dimensions. Second, the continuous function of the real solution is replaced by a simple approximate function such as a polynomial. Various approximate formulations (e.g. the finite element method, boundary element method, method of moments and so on) are developed depending on the different choice of error minimization. The approximate function, in terms of a trial function or a basis function, could be a pulse function, δ function or polynomials with different orders as shown in Fig. 2.3.1.

In certain cases the basis functions must be differentiable, integrable and must satisfy several continuous conditions as will be discussed in Chap. 6.

---

* $\langle \mathscr{L}u, v \rangle = \dfrac{1}{4\pi} \displaystyle\int_\Omega \int_\Omega v^*(\mathbf{r}) u(\mathbf{r}') \dfrac{\exp(-jk|\mathbf{r} - \mathbf{r}'|)}{|\mathbf{r} - \mathbf{r}'|} \, d\Omega' \, d\Omega$

$\langle u, \mathscr{L}v \rangle = \dfrac{1}{4\pi} \displaystyle\int_\Omega \int_\Omega v^*(\mathbf{r}) u(\mathbf{r}') \dfrac{\exp(jk|\mathbf{r} - \mathbf{r}'|)}{|\mathbf{r} - \mathbf{r}'|} \, d\Omega' \, d\Omega$

2.3 Principles of error minimization

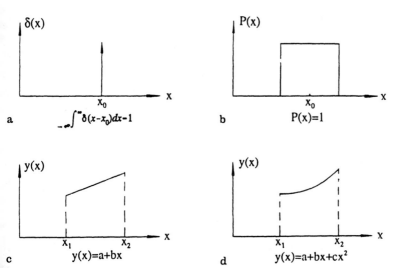

Fig. 2.3.1a–d. Examples of basis functions

2.3.1 Principle of weighted residuals [15]

The principle of weighted residuals is to minimize the error of approximation in an weighted average sense. Consider the following boundary value problems:

$$\mathscr{L}u = f \quad \text{in domain } \Omega \tag{2.3.1}$$

$$u|_{\Gamma_1} = u_0 \quad \text{on boundary } \Gamma_1 \tag{2.3.2}$$

$$g|_{\Gamma_2} = g_0 \quad \text{on boundary } \Gamma_2 \tag{2.3.3}$$

where u and f are elements of the space while g is the normal derivative of the function u, i.e. $g = \partial u/\partial n$. In Eq. (2.3.1) the operator $\mathscr{L}$ represents a differential operation. It may be a positive definite and self-adjoint operator. Alternatively, it may be non-self-adjoint in a Hilbert space.

Using approximate methods, a set of linear independent functions

$$\tilde{u} = \sum_{i=1}^{n} \alpha_i \psi_i = \sum_{i=1}^{n} \alpha_i N_i \tag{2.3.4}$$

is constructed in terms of the exact solution, i.e.

$$u = \tilde{u} . \tag{2.3.5}$$

Then, the residual exists both in the domain and on the boundaries, e.g.

$$\begin{cases} R_\Omega = \mathscr{L}\tilde{u} - f & \text{in } \Omega \\ R_1 = \tilde{u}|_{\Gamma_1} - u_0 & \text{on } \Gamma_1 \\ R_2 = \tilde{g}|_{\Gamma_2} - g_0 & \text{on } \Gamma_2 \end{cases} \tag{2.3.6}$$

where u_0 and g_0 are the given boundary conditions of the first and of the second kind. Γ_1, Γ_2 are the corresponding boundaries and $\Gamma_1 + \Gamma_2 = \Gamma$.

In Eq. (2.3.4) the unknown constants α_i are determined by the principle of weighted residuals, i.e.

$$\int_\Omega R_\Omega W \, d\Omega + \int_{\Gamma_1} R_1 W_1 \, d\Gamma_1 + \int_{\Gamma_2} R_2 W_2 \, d\Gamma_2 = 0 \qquad (2.3.7)$$

This means that the residuals are forced to zero in an average sense, where W, W_1, W_2 are weighting functions of both the domain and boundaries, respectively.

If in Eq. (2.3.7) the approximate solution is chosen to satisfy the boundary conditions, then Eq. (2.3.7) reduces to

$$\int_\Omega R_\Omega W \, d\Omega = 0 \,. \qquad (2.3.8)$$

On the other hand, if the approximate solution is chosen to satisfy the function in the domain, then Eq. (2.3.7) reduces to

$$\int_{\Gamma_1} R_1 W_1 \, d\Gamma_1 + \int_{\Gamma_2} R_2 W_2 \, d\Gamma_2 = 0 \,. \qquad (2.3.9)$$

These two cases correspond to the boundary method and to the domain method, respectively.

The substitution of Eq. (2.3.4) into Eq. (2.3.8) leads to

$$\sum_{i=1}^N \alpha_i \langle \mathscr{L} N_i, w_j \rangle = \langle f, w_j \rangle \,. \qquad (2.3.10)$$

It is possible to obtain, through manipulation, the following algebraic equation:

$$\mathbf{K}\{\alpha\} = \mathbf{B} \,. \qquad (2.3.11)$$

The unknown constants contained in the trial function are obtained by solving the above matrix equation. Thus the approximate solution is found.

Many different methods are derived corresponding to different choices of the criterion of weighted residuals. For instance, the sub-domain method, the collocation method, the least square method, the Galerkin method and the method of moments are all based on the principle of weighted residuals. These several criteria were unified by Crandall [16] as the method of weighted residuals. Collatz [17] called them error distribution principles.

2.3.2 Orthogonal projection principle [18–20]

One view of numerical methods for solving a linear operator equation is that they represent a linear projection of the exact solution onto a certain finite dimensional linear space. Some of the projection methods are orthogonal while others can be termed non-orthogonal. An orthogonal projection is one that minimizes a certain error norm. The reason is explained in the following subsections.

2.3 Principles of error minimization

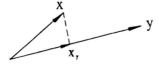

Fig. 2.3.2. Projection of a vector

2.3.2.1 Projection operator

Taking a 2-D case, the projection of **x** onto the line generated by **y** is the vector $\mathbf{x}_p = \langle \mathbf{x}, \mathbf{e} \rangle \mathbf{e}$, where **e** is the unit vector lying on the line **y**. The decomposition $\mathbf{x} = \mathbf{x}_p + \mathbf{z}$ is unique where **z** is orthogonal to **y**, as shown in Fig. 2.3.2.

In an n-D space let **f** be an arbitrary element of a Hilbert space A and let B be a subspace of A. Then **f** can be decomposed uniquely as $\mathbf{f} = \mathbf{g} + \mathbf{k}$, where **g** is in the subspace B which is 'closest' to **f** while **k** is perpendicular to B. The element **g** is called the 'projection' of **f** on B. From a geometrical view point **g** is the point where the 'plumb line' from **f** to B intersects B and δ ($\delta = \|\mathbf{f} - \mathbf{g}\|$) is the length of that line.

Thus it is possible to define the projection operator $\mathscr{P}$ in A by

$$\mathbf{g} = \mathscr{P}\mathbf{f} . \qquad (2.3.12)$$

A linear operator $\mathscr{P}$ which maps the whole of a Hilbert space A onto a particular subspace B is called a projection operator only if it maps the elements of B onto themselves, i.e. $\mathscr{P}(u) = u$ for all $u \in B$.

2.3.2.2 Orthogonal projection

A projection $\mathscr{P}$ is said to be an orthogonal projection if, for all elements u in space A and all v in subspace B, exists then

$$\langle u - \mathscr{P}u, v \rangle = 0 . \qquad (2.3.13)$$

This indicates that the residual $u - \mathscr{P}u$ is orthogonal to all v in the subspace B. The orthogonal projection $\mathscr{P}$ of a Hilbert space onto a subspace is unique. The length of the residual $\|u - \mathscr{P}u\|$ is the minimum distance from u to the subspace B. The proof for this conclusion is referred to in [19].

Concerning the function u in space A, which has a finite norm and constructs its orthogonal series in terms of the functions $\psi_1, \psi_2, \ldots$, it can be shown that the summation

$$\sum_{n=1}^{\infty} a_n \psi_n \qquad (2.3.14)$$

is convergent in the mean. Let

$$u_1 = \sum_{n=1}^{N} a_n \psi_n . \qquad (2.3.15)$$

Therefore u_1 is the orthogonal projection of u onto subspace B only if

$$a_n = \langle u, \psi_n \rangle \qquad (2.3.16)$$

where $\{\psi_1, \ldots, \psi_n\} = S$ is a set of n elements of A. It is assumed that S is a linearly independent set.[1] The set S is said to form a basis of A.

Let the difference between u and u_1 be

$$u_2 = u - u_1 . \qquad (2.3.17)$$

Therefore u_2 is orthogonal to any function from the subspace B since

$$\langle u_2, \psi_k \rangle = \langle u, \psi_k \rangle - \langle u_1, \psi_k \rangle = a_k - \langle \sum a_n \psi_n, \psi_k \rangle$$
$$= a_k - a_k = 0 . \qquad (2.3.18)$$

This means that if the function u is approximated by its orthogonal projection in the subspace, then the error is orthogonal to the subset ψ_k.

Let

$$u_2 = e, \quad w = \psi_k \qquad (2.3.19)$$

then

$$\langle e, w \rangle = 0 . \qquad (2.3.20)$$

This equation illustrates that, *if the basis function ψ_k is chosen as the weighting function, the method is satisfying the condition of orthogonal projection and the error of the approximation is minimum.*

In terms of a geometrical explanation assume that the elements α and β are in space A and B, respectively. Then in the case of $\gamma = \alpha + \beta$, γ is in space $A + B$. Next define a projection operator $P(\gamma) = \alpha$. Since the domain of $\mathscr{P}$ is a linear space of $A + B$, the range of $\mathscr{P}$ is the linear space of A. Therefore lines or planes passing through the origin of the coordinates are subspaces of a three-dimensional space. Suppose A and B are one-dimensional subspaces as shown in Fig. 2.3.3(a), then the space $A + B$ is on the x-y plane. Thus γ is on the x-y

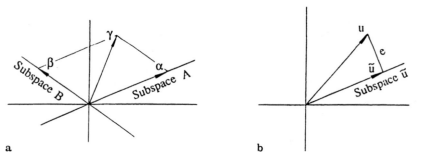

Fig. 2.3.3a, b. Geometrical explanation of the projection method

---

[1] A linearly independent set satisfies $\alpha_1 \psi_1 + \alpha_2 \psi_2 + \ldots \alpha_n \psi_n = 0$

2.3 Principles of error minimization

plane. Recall that the approximate solution $\tilde{u}$ is the orthogonal projection of u onto the subspace $\tilde{u}$. Hence Fig. 2.3.3(b) demonstrates that the residual e is orthogonal to all w in subspace $\tilde{u}$.

2.3.2.3 Orthogonal projection methods

If the approximate solution u and the error e are orthogonal to each other, as shown in Fig. 2.3.3(b), then the error e is minimum. In other words, the error norm for the orthogonal projection is smaller than the error norm for the non-orthogonal projection. This is clear by comparing Fig. 2.3.3(a) and (b). The mathematical proof is given in reference [20]. Most of the weighted residual approach uses an orthogonal projection method which ensures that the error tends to zero.

2.3.2.4 Non-orthogonal projection methods

In the collocation method the Dirac-delta function $\delta(\mathbf{r}, \mathbf{r}')$ is chosen as the weighting function. Since the δ function is not square integrable, it cannot be an element of a Hilbert space. Thus the inner product does not exist. Instead of an inner product a bilinear functional is defined as

$$I(e, w) = \int_\Omega e w \, d\Omega . \tag{2.3.21}$$

Let P_i represent the locations of the N collocation points, therefore substitution of P_i into the operator equation leads to

$$\mathscr{L}u(P_i) - f(P_i) = 0 \quad 1 \le i \le N . \tag{2.3.22}$$

Hence the matrix equation of the collocation method is derived by the bilinear functional[1]

$$\int_\Omega [\mathscr{L}u(P_i) - f(P_i)] \delta(r - r_i) d\Omega = 0 . \tag{2.3.23}$$

Thus the residual is zero at N specified points. As N increases the residual is zero at more and more points and presumably approaches zero everywhere.

2.3.3 Variational principle [15, 21]

Many problems in engineering may be characterized by the variational principle. For example, the electric energy is minimum if the system is stable. This minimum-energy principle is mathematically equivalent to Laplace's equation

---

[1] If the bilinear functional is symmetric and positive definite, then this functional becomes an inner product.

in the sense that a potential distribution which satisfies Laplace's equation also minimizes the energy, and vice-versa. Hence, the field distribution can be obtained using the principle of minimum energy. In general, the variational method is to find an extreme function of a functional corresponding to a certain problem. The functional of a Poissonian problem subject to the Dirichlet boundary condition is expressed by

$$\begin{cases} I(\varphi) = \int_\Omega \frac{1}{2}(\varepsilon |\nabla\varphi|^2 - 2\rho\varphi) d\Omega \\ \varphi|_{\Gamma_1} = \varphi_0 \ . \end{cases} \quad (2.3.24)$$

If the function $\varphi(\mathbf{r})$ minimizes the functional $I(\varphi)$, then $\varphi(\mathbf{r})$ is the solution of the given boundary value problem.

The variational method is in many respects similar to the method of weighted residuals. The solution of Eq. (2.3.24) is expanded in terms of a trial function with unkown constants which are determined by forcing the functional $I(\varphi)$ to be stationary or minimized or maximized with respect to these undetermined constants.

If an operator is symmetric and positive definite, then the function u minimizes the functional [9], i.e.

$$\frac{\partial}{\partial \alpha_i} I(u) = \frac{\partial}{\partial \alpha_i} [\langle \mathscr{L}u, u \rangle - \langle u, f \rangle - \langle f, u \rangle] = 0 \quad (2.3.25)$$

where α_i are the constants included in the approximate solution (Eq. (2.3.4)). By solving Eq. (2.3.25) to determine the parameter α_i, the required approximate solution u is obtained. The operator in Eq. (2.3.25) may be either differential or integral [14, 22].

The variational approach uses proof of convergence for the solution. By contrast, convergence of the solution is not generally guaranteed when using the principle of weighted residuals. However, some problems are not characterized by the variational principle. In electrical engineering both the weighted residuals and the variational principle are used to derive the approximate solutions of a problem.

2.4 Categories of various numerical methods

Most of the numerical methods are based on the principle of weighted residuals or the variational principle. Both principles force the error between the real and the approximate solution to approach zero. The variational method is based on the equivalent functional of the governing equation. The weighted residual approaches are based on the operator equation directly. In fact, these two methods are unified.

As discussed in Sect. 2.3.1, if the unknown function is replaced by the approximate solution, the residuals may be produced both in the domain or on

2.4 Categories of various numerical methods

the boundaries. For convenience, trial functions are chosen so that one of these sets of residuals vanishes. For instance, *using domain methods, the selection requires a trial function that satisfies the boundary conditions exactly so that the boundary residuals vanish. On the other hand, using boundary methods the boundary conditions are not satisfied by the approximate solution. Rather the differential equation itself is satisfied exactly.* The third approach is a mixed one, in which neither the differential equation nor the boundary conditions are satisfied. Based on different choices regarding the principle of minimum error and discretization, many different methods have been developed to date. In the later part of this book all these methods are discussed using either differential equation methods (domain methods) or integral equation methods (boundary methods). In domain methods the discretization will be carried out within all of the domain. By contrast, in the boundary methods only the boudary of the domain needs to be discretized. Therefore in such cases the dimensionality of the problem is reduced by one.

If the category is dependent on the approximate principle, there is another kind of classification as listed in the following subsections. Applying the advantages of both domain and boundary methods these mixed methods are very useful for certain practical problems, as shown in references [23-25].

2.4.1 Methods of weighted residuals

The weighted residual approach covers all numerical methods. *Based on the choice of different weighting functions, basis functions and the approach to discretization different methods are formulated.*

2.4.1.1 Method of moments

The method of moments is a general form of weighted residuals. In the theory of mathematics the method of moments is an interior weighted residual method in which the power series is chosen as the weighting function. In electromagnetic field theory which was originally used by Harrington [26], no matter which kind of weighting functions are chosen, the weighted residual methods in the interior regions are called moment methods. In fact it makes no difference where interior region or boundary region are considered. The process of the method of moments is as follows. After substituting the approximate solution

$$u = \sum_{i=1}^{n} \alpha_i \psi_i \qquad (2.4.1)$$

into an operator equation and assuming that the residuals are zero, then a set of algebraic equations are generated as follows:

$$\sum_{i=1}^{n} \alpha_i \langle w_m, \mathscr{L}\psi_i \rangle = \langle w_m, f \rangle \quad m = 1, 2, \ldots, n. \qquad (2.4.2)$$

This can be written in matrix form:

$$\mathbf{K}\{\alpha_i\} = \{b_n\} \quad (2.4.3)$$

where

$$k_{ij} = \begin{bmatrix} \langle w_1, \mathscr{L}\psi_1 \rangle & \langle w_1, \mathscr{L}\psi_2 \rangle & \cdots & \langle w_1, \mathscr{L}\psi_n \rangle \\ \vdots & & & \\ \langle w_n, \mathscr{L}\psi_1 \rangle & \langle w_n, \mathscr{L}\psi_2 \rangle & \cdots & \langle w_n, \mathscr{L}\psi_n \rangle \end{bmatrix} \quad (2.4.4)$$

$$b_n = \begin{Bmatrix} \langle w_1, f \rangle \\ \langle w_n, f \rangle \end{Bmatrix} \quad (2.4.5)$$

where ψ_i are the basis functions. The weighting function w_j may be a pulse function, a δ function or any other continuous function with a higher order, as described in Chap. 10. This method is widely used for solving fast transient field problems because of the flexible choice of the weighting function. In the method of moments discretization will be executed in the domain or on the boundary. If only the boundary is discretized, the method of moments is the same as the boundary element method. In reference [27] Harrington explains the relationship between the method of moments, the boundary element method, Galerkin's methods and so on.

2.4.1.2 Galerkin's finite element method

If the weighting functions are chosen to be the same as the basis function, i.e.

$$w_j = N_j \quad (2.4.6)$$

then the method is known as Galerkin finite element method. Here N_j is the interpolation function of the approximate solution. This approach was first published by Galerkin Bubnov in 1915. With this method the solution of a partial differential equation can be obtained without using the Ritz method (see Sect. 5.4.1).

According to Eq. (2.4.2), after discretizing the domain, the coefficients of the element matrix for the LHS of Eq. (2.4.3) are

$$k_{ij}^e = \int_\Delta \nabla N_i \cdot \nabla N_j \, dS \quad (2.4.7)$$

where the subscript Δ of the integral represents the subarea of the triangular element and the superscript e represents element.

Using Galerkin finite element method the trial functions are chosen as a set of complete functions. Thus the residuals are forced to be zero by letting the residual be orthogonal to each member of a complete set of functions. The convergence of the Galerkin weighted approximation was proved in reference [18].

Due to the nature of the inner product the Galerkin form is a weak formulation. It reduces the requirements for continuity of the approximate function by one. It was shown in 1940 that the Galerkin method can also be used to solve the Fredholm integral equation. Examples are given in references [28, 29].

2.4.1.3 Collocation methods

The method of collocation is another special case of a moment method. It is also called the point matching method. The coefficients α_i in the approximate solution of Eq. (2.3.4) are determined by satisfying the governing equation at certain specific points. The collocation method corresponds to the method of moments if the Dirac-delta function is chosen as the weighting function, i.e.

$$\int_\Omega [\mathscr{L}u(P) - f(P)]\delta(\mathbf{r} - \mathbf{r}_i)d\Omega = 0 \; .$$

The residual equals zero only in some specific points and not in the average which means in the whole domain. The accuracy depends on the number and the distribution of the matching points. The matching points may be chosen on the boundary or within the domain. Therefore the method may be categorized as being either a domain method or a boundary method. The charge simulation method discussed in Chap. 7 is one of the collocation methods. The advantages of this method are:

(1) There are no inner products to be calculated.
(2) Usually the resultant algebraic equations have fewer terms than the corresponding equations for Galerkin approximation, especially if the matching points are chosen to be on the boundary. The accuracy of the solution is sensitive to the position of the collocation points.

2.4.1.4 Boundary element methods

In the boundary element method the fundamental solution is chosen as the weighting function. If in Eq. (2.3.7) it is

$$W_1 = \frac{\partial W}{\partial n}, \qquad W_2 = W \tag{2.4.8}$$

then Eq. (2.3.7) is simplified to

$$\int_\Omega R(u)W d\Omega = \int_{\Gamma_1} R_1(u)\frac{\partial W}{\partial n}d\Gamma + \int_{\Gamma_2} R_2(u)W d\Gamma \; . \tag{2.4.9}$$

After the boundary of the problem domain is discretized by elements, the matrix form of the boundary element equation is

$$\mathbf{H}\{u\} = \mathbf{G}\left\{\frac{\partial u}{\partial n}\right\} \qquad (2.4.10)$$

where u and $\partial u/\partial n$ are known and unknown variables of the boundary nodes. $\mathbf{H}$, $\mathbf{G}$ are two coefficient matrices of the order $N \times N$ (N is the total number of boundary nodes) estimated by integrations. The detailed procedure is given in Chap. 9.

2.4.2 Variational approach

As indicated in Sect. 2.3.4, an equivalent functional exists for a self-adjoint operator. *The function which minimizes the equivalent functional is the solution of the corresponding operator equation.* Thus the first step of a variational method is to construct an equivalent functional $I(u)$ by Eq. (2.2.32). The next step is to assume an interpolation function

$$u = \sum_{i=1}^{n} \alpha_i N_i$$

in terms of the unknown function contained within the functional

$$I(u) = \int_\Omega f(x, y, z, u, u'_x, u'_y, u''_{xx}, u''_{yy}, \ldots)\,d\Omega \ . \qquad (2.4.11)$$

The undetermined parameters α_i are obtained by minimizing the functional $I(u)$, i.e.

$$\frac{\partial I(u)}{\partial \alpha_i} = 0 \ . \qquad (2.4.12)$$

For example, let the operator be Laplacian operator and let it be subject to the inhomogeneous boundary conditions of the second kind, i.e.

$$-\nabla^2 u = \rho/\varepsilon \qquad (2.4.13)$$

$$\frac{\partial u}{\partial n} = g(s) \ . \qquad (2.4.14)$$

According to Eq. (2.2.32) the equivalent functional is derived by

$$\begin{aligned}I(u) &= \langle -\nabla^2 u, u \rangle - 2\langle u, f \rangle \\ &= -\int_\Omega u(\nabla \cdot \nabla u)\,d\Omega - 2\int_\Omega uf\,d\Omega \ .\end{aligned} \qquad (2.4.15)$$

2.4 Categories of various numerical methods

Applying the vector identity $\nabla \cdot (uv) = \nabla u \cdot v + u\nabla \cdot v$ and letting $v = \nabla u$, then Eq. (2.4.15) becomes

$$I(u) = -\int_\Omega \nabla \cdot (u\nabla u) d\Omega + \int_\Omega \nabla u \cdot \nabla u d\Omega - 2\int_\Omega uf d\Omega$$
$$= \int_\Omega |\nabla u|^2 d\Omega - 2/\varepsilon \int_\Omega u\rho d\Omega - \oint_\Gamma ug(s) d\Gamma . \tag{2.4.16}$$

Equation (2.4.16) is the equivalent functional for the boundary value problem described by Eqs. (2.4.13) and (2.4.14). After discretization and using the extremum principle (Eq. (2.4.12)) the following matrix equation is obtained:

$$[\mathbf{K}]\{\alpha\} = \{b\} . \tag{2.4.17}$$

The coefficients of the matrix and the column vector of Eq. (2.4.17) are

$$k_{ij}^e = \int_\Delta \nabla N_i \cdot \nabla N_j dS \tag{2.4.18}$$

$$b_i = \frac{2}{\varepsilon}\rho_i \int_\Delta N_i dS . \tag{2.4.19}$$

Equation (2.4.18) is exactly the same as Eq. (2.4.7). It is apparent that the boundary conditions of the second kind are automatically satisfied in the process of variation.

If Eqs. (2.4.13) and (2.4.14) are expressed in integral form, i.e.

$$\varphi(\mathbf{r}) = \int_\Omega \frac{1}{\varepsilon} G(\mathbf{r}, \mathbf{r}')\rho(\mathbf{r}') dV' + \int_\Gamma \frac{1}{\varepsilon} G(\mathbf{r}, \mathbf{s}')\sigma(s) d\Gamma' \tag{2.4.20}$$

the second term of the RHS of Eq. (2.4.20) is given as the boundary condition, i.e.

$$\int_s \frac{1}{\varepsilon} G(s, s')\sigma(s) dS' = g(s) . \tag{2.4.21}$$

Because the kernel $G(\mathbf{r}, \mathbf{r}') = (1/4\pi|\mathbf{r} - \mathbf{r}'|)$ is symmetric, by using the same procedure as was used to derive Eq. (2.4.16) the equivalent functional is

$$I(\varphi) = \int_s \sigma(s) \int_{s'} \frac{1}{\varepsilon} G(s, s')\sigma(s') dS' dS - 2\int_s \sigma(s)g(s) dS . \tag{2.4.22}$$

Supposing

$$\sigma(s) = \sum_{i=1}^n \alpha_i \sigma_i \tag{2.4.23}$$

the coefficients of the element matrix and the RHS of the matrix equation are obtained thus:

$$k_{ij}^e = \int_{S_e} \alpha_i(s) \int_{S_e} \frac{1}{\varepsilon} G(s, s') \alpha_j(s') dS' \, dS \qquad (2.4.24)$$

$$b_i = \int_{S_e} \alpha_i g(s) dS \, . \qquad (2.4.25)$$

In the integral equation method the singularities of the integral must be handled specifically [28].

If the operator is non-self-adjoint there are several ways to find a generalized functional. McDonald [2] defined a modified self-adjoint operator $\mathscr{L}'$ to replace a non-self-adjoint operator of an interface problem where

$$\mathscr{L}'u = \langle \mathscr{L}u, G \rangle \qquad (2.4.26)$$

and G is the Green function of the problem concerned.

In reference [13] an adjoint formulation derived from the quadratic functional for a self-adjoint operator was extended to solve the integral equation with a non symmetric kernel and for non-self-adjoint partial differential equations.

2.5 Summary

1. *The numerical solution of an operator equation is to approximate the continuous information contained in the exact solution using discrete values.* Thus it is appropriate to refer to numerical methods as being discretization methods.

2. The approximate solution of an operator equation

$$\mathscr{L}u = f \quad u \in \Omega \subset A$$

is constructed by using a set of linearly independent elements ψ_k in the subspace of A, i.e.

$$\tilde{u} = \sum_{k=1}^{n} \alpha_k \psi_k \, .$$

The set $S = \{\psi_1, \ldots \psi_k, \ldots \psi_n\}$ is said to form a basis of space A. *In both the weighted residual approach and the variational principle the approximate solution is an orthogonal projection of the real solution from the original space A onto subspace B. Hence they satisfy the minimum error principle.*

3. Based on the type of governing equation both the differential equation methods and integral equation methods can be developed.

4. The weighted residual approach or the variational principle provide two means for satisfying the minimum error principle. *By using different choices for the weighting function a series of numerical methods have been developed.*

5. Another approach in developing numerical solution methods is to replace the continuous function by discrete values at a large number of grid nodes. The grid nodes may be on the boundary of the solution domain or on the whole domain. This gives rise both to boundary methods and domain methods, respectively. In the domain methods the residuals are zeroes on the boundary. In boundary methods the residuals are zeroes in the domain.

6. The properties of the discretized algebraic equation are dependent on the properties of the operator. *The symmetric positive definite operator leads to a symmetric matrix equation and good results can easily be obtained.*

7. From the theoretical viewpoint the functional exists only for the self-adjoint operator. However, when dealing with the problems of electromagnetic fields, the equivalent functional can be developed for non-self-adjoint operators by several means.

References

1. P. Hammond, Sources and Fields—Some Thoughts on Teaching the Principles of Electromagnetism, *Int. J. Elect. Eng.*, 7, 65–76, 1969.
2. B.H. Mcdonald, A. Wexler, Mutually Constrained Partial Differential and Intergral Equation Field Formulations. In *Finite Elements in Electrical and Magnetic Field Problems*. (Ed. M.V.K. Chari and P.P. Silvester), John Wiley & Sons, 1980, Ch. 9
3. L.F. Richardson, The Approximate Arithmetical Solution by Finite Differences of Physical Problems, *Trans. Roy. Soc. (London) A*, 210, 305–357, 1910
4. M.J. Turner, R.W. Clough et al. Stiffness and Deflection Analysis of Complex Structures, *J. Aero. Sci.* 23, 805–823, 1956
5. G.T. Symm, Integral Equation Methods in Potential Theory II, *Proc. Roy. Soc. (London) A* 275, 33–46, 1963
6. C.A. Brebbia, *Boundary Element Method for Engineers*, Pentech Press, 1978
7. C.W. Trowbridge, Electromagnetic Computing: The Way Ahead?, *IEEE Trans. Mag.* 24(1), 13–18, 1988
8. L. Collatz, *Functional Analysis and Numerical Mathematics* (English edn) Academic Press, 1966
9. I. Stakgold, *Boundary Value Problem in Mathematic Physics*, Macmillan, 1967
10. S.G. Mikhlin, *Variational Methods in Mathematical Physics*, Pergamon Press, 1964
11. S.G. Mikhlin, *Mathematic Physics, An Advanced Course*, North-Holland, 1970
12. B.H. McDonald, M. Friedman, A. Wexler, Variational Solution of Integral Equations, *IEEE Trans. MTT* 22(3), 237–248, 1974
13. G. Jeng, A. Wexler, Self-adjoint Variational Formulation of Problems Having Non-self-adjoint Operator, *IEEE Trans. MTT*, 26(2), 91–94, 1978
14. Zhang Wenzun, *Functional Methods in Electromagnetic Engineering* (in Chinese), Science Press, Shanghai, 1984
15. Bruce A. Finlayson, *The Method of Weighted Residuals and Variational Principles*, Academic Press, 1972
16. S.H. Crandall, *Engineering Analysis*, McGraw-Hill, New York, 1956
17. L. Collatz, *The Numerical Treatment of Differential Equations*, Springer-Verlag, 1960
18. R. Wait, A.R. Mitchell, *Finite Element Analysis and Applications*, John Wiley & Sons, 1985
19. C.W. Steele, *Numerical Computation of Electric and Magnetic Fields*, VNR Company, 1987
20. Wang Xinzhong, *Electromagnetic Field Theory and Applications* (in Chinese), Science Press, Bejing, 1986
21. L. Cairo, T. Kahan, *Variational Techniques in Electromagnetism* (English edn trans. by G.D. Sims), Blackie and Son Limited, 1965
22. P.P. Silvester, R.L. Ferrari, *Finite Elements For Electrical Engineers*, Cambridge University Press, 1983

23. Yukio Kagawa, Tadakuni Murai, Shinji Kitagami, On the Compatibility of Finite Element-Boundary Element Coupling in Field Problems, *COMPEL*, 1(4), 197–217, 1982
24. S.J. Salon, J.M. Schnedia, A. Hybrid Finite Element-Boundary Integral Formulation of the Eddy Current Problem, *IEEE Trans. on Mag.*, 18(2), 461–466, 1982
25. S.J. Salon, J. D'Angelo, Application of the Hybrid Finite Element Boundary Element Method in Electromagnetics, *IEEE Trans. on Mag.* 24(1), 80–85, 1988
26. R.F. Harrington, *Field Computation by Moment Methods*, Macmillan, 1968
27. R.F. Harrington, Boundary Element and the Method of Moment. In *Boundary Elements, Proc. of International Conference of Boundary Element Method* (Ed. C.A. Brebbia), 31–40, 1980
28. G. Jeng, A. Wexler, Isoparametric, Finite Element, Variational Solution of Integral Equation for Three-Dimensional Fields, *Int. J. for Numer. M. in Eng.*, 11(9), 1455–1471, 1977
29. W. Lipinski, P. Krason, Integral Equations Methods of Analysis Describing the Skin Effect in Conductors, *IEEE Trans. on Mag.* 18(2), 473–475, 1982

Part Two
Domain Methods

Domain methods are based on differential equations and on discretization of the whole domain by regular grids or elements. The finite difference method (FDM) and the finite element method (FEM) are the most familiar domain methods. In both FDM and FEM variational principles or the principle of weighted residuals are used to derive algebraic equations for the partial differential equations corresponding to a specific problem.

There are four chapters in this part. Chapter 3 is concerned with the finite difference method. The finite element method is discussed in three chapters. Chapter 4 describes the general procedures used in the finite element method. The discretization equations at nodes are derived by using weighted residuals. Chapter 5 concentrates on using the variational principle to derive the equivalent functional and the finite element equations of various boundary value problems. The properties of the finite element method are revealed clearly by using the minimization principle of the functional. Further applications such as the open boundary problem are also described in Chap. 5. Important techniques of element discretization are classified in Chap. 6.

Chapter 3

Finite Difference Method

3.1 Introduction

The finite difference method (FDM) is an approximate method for solving partial differential equations. It has been used to solve a wide range of problems. These include linear and non-linear, time independent and dependent problems. This method can be applied to problems with different boundary shapes, different kinds of boundary conditions, and for a region containing a number of different materials. Even though the method was known by such workers as Gauss and Boltzmann, it was not widely used to solve engineering problems until the 1940s. The mathematical basis of the method was already known to Richardson in 1910 [1] and many mathematical books such as references [2 and 3] were published which discussed the finite difference method. Specific reference concerning the treatment of electric and magnetic field problems is made in [4]. The application of FDM is not difficult as it involves only simple arithmetic in the derivation of the discretization equations and in writing the corresponding programs. During 1950–1970 FDM was the most important numerical method used to solve practical problems ([5–7]). With the development of high speed computers having large scale storage capability many numerical solution techniques appeared for solving partial differential equations. However, due to the ease of application of the finite difference method it is still a valuable means of solving these problems ([8–11]).

Similar to other numerical methods, the aim of finite difference is to replace a continuous field problem with infinite degrees of freedom by a discretized field with finite regular nodes. The partial derivatives of the unknown function are approximated by the difference quotients at a set of finite discretization points. The original partial differential equation is then transformed in to a set of algebraic equations. The solution of these simultaneous equations is the approximate solution of the original boundary value problem.

In this chapter discretization equations of Poissonian problems both in translational symmetrical and axi-symmetrical coordinates are discussed. The solution of a diffusion equation with linear and non-linear parameters is presented.

Several specific discretization formulations for various types of boundary conditions and interfacial conditions are developed. Iterative methods to solve matrix equations derived by FDM are introduced. Examples are given for solving electrostatic and diffusion problems. Finally, the relationships between finite difference equations and the variational principle, together with the approaches weighted residuals, are discussed in the last section.

3.2 Difference formulation of Poisson's equation

Finite difference methods are used in both 2-D and 3-D cases. Difference equations for 3-D cases are the extensions of 2-D problems. Hence the difference equations for 2-D cases are developed first in this section.

3.2.1 Discretization mode for two-dimensional problems

The basic idea of FDM is to replace the derivatives of an unknown function by the difference quotients of unknown functions. The form of finite difference equations depends on the form of the domain discretization. Assume that a two-dimensional area Ω is bounded by the contour Γ, the potential function u within the domain Ω satisfies Poisson's equation and is subject to the Dirichlet boundary condition as shown below:

$$\nabla^2 u = F(x, y) \quad \text{in domain } \Omega$$

$$u|_\Gamma = g(\Gamma) \quad \text{on boundary } \Gamma . \tag{3.2.1}$$

In principle Ω can be divided into an arbitrary grid as shown in Fig. 3.2.1(a)–(e). In order to simplify matters the square grid shown in Fig. 3.2.1(a) is adopted. Depending on the application uniform or non-uniform rectangular grids shown in Fig. 3.2.1(b) and (c) are used. For some specific demands the triangular grids (Fig. 3.2.1(d)) are considered. For problems with circular boundaries the polar grids shown in Fig. 3.2.1(e) are used.

In Fig. 3.2.1 *the distances between grid lines are called steps or mesh lengths while each intersection of grid lines is called a node*. After the domain is subdivided into grids the continuous function is replaced by a great number of discretized values at these nodes.

3.2.2 Difference equations in 2-D Cartesian coordinates

After the grid has been specified, the derivatives of the unknown function are approximated by taking the difference quotients of the function related to several adjacent nodes. Based on the definition of the difference and the difference

3.2 Difference formulation of Poisson's equation

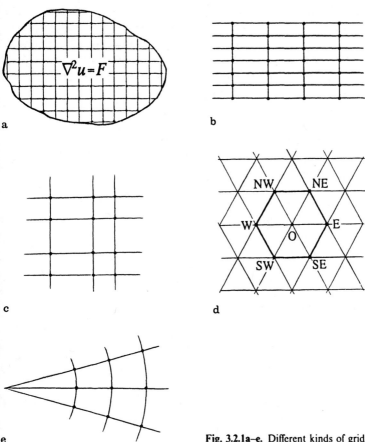

Fig. 3.2.1a–e. Different kinds of grids

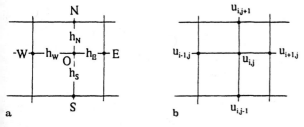

Fig. 3.2.2a, b. An asymmetric star O–ENWS

quotient several methods can be used to derive the discretization formulations. In this section Taylor's series are used to derive the difference equations.

In order to obtain a general discretization formula and to simplify this discussion for boundary discretization a non-uniform mesh shown in Fig. 3.2.2(a) is considered first.

Assume that the distances between two branches of parallel lines are different as shown in Fig. 3.2.2(a). Any node $O(x_0, y_0)$ together with its four adjacent nodes $E(x_E, y_E)$, $N(x_N, y_N)$, $W(x_W, y_W)$, $S(x_S, y_S)$ constructs an asymmetric five-pointed star. Let $u_0(x_0, y_0)$ be the potential value of node O and u_E, u_N, u_W, u_S the potentials at the nodes E, N, W, S, the potential u_0 of node O can then be expressed in terms of the potential values of the nodes E, N, W, S. The result is derived as follows.

If the function $u(x)$ and its derivatives $\partial u/\partial x$, $\partial^2 u/\partial x^2, \ldots$ are continuous in x, then $u(x)$ can be expanded by the Taylor's series:

$$u(x + \Delta x) = u(x) + \left(\frac{\partial u}{\partial x}\right)\Delta x + \frac{1}{2!}\left(\frac{\partial^2 u}{\partial x^2}\right)(\Delta x)^2 + \frac{1}{3!}\left(\frac{\partial^3 u}{\partial x^3}\right)(\Delta x)^3$$
$$+ \frac{1}{4!}\left(\frac{\partial^4 u}{\partial x^4}\right)(\Delta x)^4 + \ldots \tag{3.2.2}$$

or

$$u(x - \Delta x) = u(x) - \left(\frac{\partial u}{\partial x}\right)\Delta x + \frac{1}{2!}\left(\frac{\partial^2 u}{\partial x^2}\right)(\Delta x)^2 - \frac{1}{3!}\left(\frac{\partial^3 u}{\partial x^3}\right)(\Delta x)^3$$
$$+ \frac{1}{4!}\left(\frac{\partial^4 u}{\partial x^4}\right)(\Delta x)^4 - \ldots \tag{3.2.3}$$

where Δx is a small increment of the variable x.

Let

$$\Delta x_E = x_E - x_0 = h_E, \qquad \Delta x_W = x_0 - x_W = h_W. \tag{3.2.4}$$

Substitution of the above two expressions into Eqs. (3.2.2) and (3.2.3), respectively, yields:

$$u_E = u_0 + \left(\frac{\partial u}{\partial x}\right)_0 h_E + \frac{1}{2!}\left(\frac{\partial^2 u}{\partial x^2}\right)_0 (h_E)^2 + \frac{1}{3!}\left(\frac{\partial^3 u}{\partial x^3}\right)_0 (h_E)^3$$
$$+ \frac{1}{4!}\left(\frac{\partial^4 u}{\partial x^4}\right)_0 (h_E)^4 \tag{3.2.5}$$

$$u_W = u_0 - \left(\frac{\partial u}{\partial x}\right)_0 h_W + \frac{1}{2!}\left(\frac{\partial^2 u}{\partial x^2}\right)_0 (h_W)^2 - \frac{1}{3!}\left(\frac{\partial^3 u}{\partial x^3}\right)_0 (h_W)^3$$
$$+ \frac{1}{4!}\left(\frac{\partial^4 u}{\partial x^4}\right)_0 (h_W)^4 \tag{3.2.6}$$

where h_E is a forward step while h_W is a backward step. Similarly, if u is a function of the variable y, the following two expressions are obtained:

$$u_N = u_0 + h_N\left(\frac{\partial u}{\partial x}\right)_0 + \frac{1}{2}h_N^2\left(\frac{\partial^2 u}{\partial y^2}\right)_0 + \frac{1}{3!}\left(\frac{\partial^3 u}{\partial y^3}\right)_0 (h_N)^3$$
$$+ \frac{1}{4!}\left(\frac{\partial^4 u}{\partial x^4}\right)_0 (h_N)^4 \tag{3.2.7}$$

3.2 Difference formulation of Poisson's equation

$$u_S = u_0 - h_S\left(\frac{\partial u}{\partial x}\right)_0 + \frac{1}{2}h_S^2\left(\frac{\partial^2 u}{\partial y^2}\right)_0 - \frac{1}{3!}\left(\frac{\partial^3 u}{\partial y^3}\right)_0 (h_S)^3$$

$$+ \frac{1}{4!}\left(\frac{\partial^4 u}{\partial x^4}\right)_0 (h_N)^4 \tag{3.2.8}$$

where h_N, h_S are forward and backward steps in the y direction. Forming the sum of α times Eq. (3.2.5) and $-\frac{h_E^2}{h_W^2}\alpha$ times Eq. (3.2.6), yields

$$\left(u_E - \frac{h_E^2}{h_W^2}u_W\right) = \left(1 - \frac{h_E^2}{h_W^2}\right)u_0 + \left(h_E + \frac{h_E^2}{h_W}\right)\left(\frac{\partial u}{\partial x}\right)_0$$

$$+ \frac{1}{3!}h_E(h_E^2 + h_W h_E)\left(\frac{\partial^3 u}{\partial x^3}\right)_0 + \frac{1}{4!}h_E^2(h_E^2 - h_W^2)\left(\frac{\partial^4 u}{\partial x^4}\right)_0. \tag{3.2.9}$$

Neglecting the terms containing the partial derivatives of the order higher than three, which is valid if h_E, h_W, h_N, h_S are small, results in

$$\left(\frac{\partial u}{\partial x}\right)_0 \cong \frac{h_W}{h_E(h_E + h_W)}u_E + \frac{h_E - h_W}{h_E h_W}u_0 - \frac{h_E}{h_W(h_E + h_W)}u_W \tag{3.2.10}$$

and

$$\left(\frac{\partial u}{\partial y}\right)_0 \cong \frac{h_S}{h_N(h_N + h_S)}u_N + \frac{h_N - h_S}{h_N h_S}u_0 - \frac{h_N}{h_S(h_N + h_S)}u_S. \tag{3.2.11}$$

These expressions show that the partial derivatives of the first order of the functions $u(x)$ and $u(y)$ at any node O are approximated by algebraic equations. If

$$h_E = h_W = h_x \qquad h_N = h_S = h_y \tag{3.2.12}$$

then Eqs. (3.2.10) and (3.2.11) reduce to

$$\left(\frac{\partial u}{\partial x}\right)_0 \cong \frac{u_E - u_W}{2h_x} \tag{3.2.13}$$

$$\left(\frac{\partial u}{\partial y}\right)_0 \cong \frac{u_N - u_S}{2h_y}. \tag{3.2.14}$$

These equations indicate that the first order derivatives of the function are approximated by expressions of the central difference quotients of the first order. It means that the first order partial derivatives of the function u at any point '0' is dependent on the function value of its neighbouring nodes and the step length. The accuracy of the central difference quotient is higher than the forward and the backward difference quotient $\left(\frac{\partial u}{\partial x}\right)_0 \cong \frac{u_E - u_0}{h}$ or $\left(\frac{\partial u}{\partial x}\right)_0 \cong \frac{u_W - u_0}{h}$ (these are obtained directly from Eqs. (3.2.5) or (3.2.6)). The reason will be interpreted in Sect. 3.5.2. In Eqs. (3.2.13) and (3.2.14) the accuracy is higher if the length of h is smaller.

On the other hand the summation of α times Eq. (3.2.5) and $(h_E/h_W)\alpha$ times Eq. (3.2.6) is

$$\left(u_E + \frac{h_E}{h_W}u_W\right) = \left(1 + \frac{h_E}{h_W}\right)u_0 + \frac{1}{2}h_E(h_E + h_W)\left(\frac{\partial^2 u}{\partial x^2}\right)_0$$
$$+ \frac{1}{3!}h_E(h_E^2 - h_W^2)\left(\frac{\partial^3 u}{\partial x^3}\right)_0 + \frac{1}{4!}h_E(h_E^3 + h_W^3)\left(\frac{\partial^4 u}{\partial x^4}\right)_0. \quad (3.2.15)$$

Ignoring the terms containing h_E and h_W to the power higher than three the following expressions are obtained:

$$\left(\frac{\partial^2 u}{\partial x^2}\right)_0 \cong \frac{2u_E}{h_E(h_E + h_W)} + \frac{2u_W}{h_W(h_E + h_W)} - \frac{2u_0}{h_E h_W} \quad (3.2.16)$$

$$\left(\frac{\partial^2 u}{\partial y^2}\right)_0 \cong \frac{2u_N}{h_N(h_N + h_S)} + \frac{2u_S}{h_S(h_S + h_N)} - \frac{2u_0}{h_N h_S} \quad (3.2.17)$$

If the forward and the backward steps in the x, y directions are identical, respectively, e.g. $h_E = h_W = h_x$, $h_N = h_S = h_y$, then the above two equations reduce to

$$\frac{\partial^2 u}{\partial x^2} \cong \frac{u_E - 2u_0 + u_W}{h_x^2} \quad (3.2.18)$$

$$\frac{\partial^2 u}{\partial y^2} \cong \frac{u_N - 2u_0 + u_S}{h_y^2}. \quad (3.2.19)$$

Equations (3.2.18) and (3.2.19) express that *the second order partial derivatives of the function are simplified by the difference quotients.*

By introducing Eqs. (3.2.16) and (3.2.17) into Eq. (3.2.1) the discretization form of Poisson's equation is

$$\alpha_E u_E + \alpha_N u_N + \alpha_W u_W + \alpha_S u_S + \alpha_0 u_0 - \tfrac{1}{2}F_0 = 0 \quad (3.2.20)$$

with

$$\begin{cases} \alpha_E = \dfrac{1}{h_E(h_E + h_W)} & \alpha_W = \dfrac{1}{h_W(h_E + h_W)} \\[6pt] \alpha_N = \dfrac{1}{h_N(h_N + h_S)} & \alpha_S = \dfrac{1}{h_S(h_N + h_S)} \\[6pt] \alpha_0 = -\left(\dfrac{1}{h_E h_W} + \dfrac{1}{h_N h_S}\right). \end{cases} \quad (3.2.21)$$

3.2 Difference formulation of Poisson's equation

F_0 is the source density of point O. Equation (3.2.20) is the general discretization form of Poisson's equation corresponding to the discretization form of an asymmetric five-pointed star. This form is rarely required, but any simple expressions can be derived immediately from Eq. (3.2.20). In the case of $h_E = h_W = h_x$ and $h_N = h_S = h_y$ the discretization equation at node O is

$$\frac{u_E - 2u_0 + u_W}{h_x^2} + \frac{u_N - 2u_0 + u_S}{h_y^2} = F_0 . \tag{3.2.22}$$

If

$$h_x = h_y = h \tag{3.2.23}$$

then

$$u_0 = \tfrac{1}{4}(u_E + u_N + u_W + u_S - h^2 F_0) . \tag{3.2.24}$$

This is the commonly used difference equation of the Poissonian problem at node O discretized by a symmetric five-pointed star. For generalization purposes Eq. (3.2.24) can be written in a generic form:

$$u_{i,j} = \tfrac{1}{4}(u_{i+1,j} + u_{i,j+1} + u_{i-1,j} + u_{i,j-1} - h^2 F_0) . \tag{3.2.25}$$

The subscripts of the sequence i, j are shown in Fig. 3.2.2(b).

For the Laplace's equation Eq. (3.2.24) reduces to

$$u_0 = \tfrac{1}{4}(u_E + u_N + u_W + u_S)$$
$$= \tfrac{1}{4}(u_{i+1,j} + u_{i,j+1} + u_{i-1,j} + u_{i,j-1}) . \tag{3.2.26}$$

This equation shows that the potential value of any point is the average of the potentials of its four neighbouring points. Equation (3.2.26) is the symmetric five-point difference equation of Laplacian problems.

If triangular grids are adopted (Fig. 3.2.1(d)), the discretization formula at point O is

$$u_0 = \frac{1}{6}(u_E + u_{NE} + u_{NW} + u_W + u_{SW} + u_{SE} - h^2 F_0) . \tag{3.2.27}$$

If 3-D cases are considered, the above classification may be easily extended to a seven-node star, then

$$u_0 = \tfrac{1}{6}(u_E + u_N + u_W + u_S + u_F + u_B) \tag{3.2.28}$$

where u_F and u_B are the potentials of the front and the back points of point O.

All of the above discretization equations are derived using Cartesian coordinates. The next subsection discusses the problem in the case of a discretization in polar-coordinates.

3.2.3 Discretization equation in polar coordinates

If the problem involves circular boundaries, the polar coordinates (shown in Fig. 3.2.3) are applied for convenience.

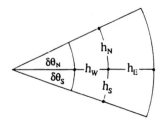

Fig. 3.2.3. Five-pointed star in polar coordinates

In polar coordinates Laplace's equation is expanded to

$$\frac{\partial^2 u}{\partial r^2} + \frac{1}{r}\frac{\partial u}{\partial r} + \frac{1}{r^2}\frac{\partial^2 u}{\partial \theta^2} = 0 \ . \tag{3.2.29}$$

The approximate expressions of $\partial u/\partial r$, $\partial^2 u/\partial r^2$, $\partial^2 u/\partial \theta^2$ can be obtained by replacing variables r and θ in the approximate expressions of $\partial u/\partial x$, $\partial^2 u/\partial x^2$ and $\partial^2 u/\partial y^2$. Introducing these approximations into Eq. (3.2.29) the following equation is obtained:

$$\alpha_E(u_E - u_0) + \alpha_W(u_W - u_0) + \alpha_N(u_N - u_0)\left(1 + \frac{h_S}{2r}\right)$$

$$+ \alpha_S(u_S - u_0)\left(1 - \frac{h_N}{2r}\right) = 0 \tag{3.2.30}$$

where $\alpha_E, \ldots, \alpha_S$ have the same values as given in Eq. (3.2.21). In general the position of a five-pointed star is expressed by double subscripts i, j. Let $r = ih$ ($i = 1, 2, \ldots$) and $\theta = j\delta\theta$ ($j = 0, 1, 2, \ldots$) and by using a similar process as in Eq. (3.2.20), the following equation is obtained:

$$\frac{(u_{i+1,j} - 2u_{i,j} + u_{i-1,j})}{h^2} + \frac{1}{ih}\frac{(u_{i+1,j} + u_{i-1,j})}{2h}$$

$$+ \frac{1}{(ih)^2}\frac{(u_{i,j+1} - 2u_{i,j} + u_{i,j-1})}{(\delta\theta)^2} = 0 \tag{3.2.31}$$

This equation can be rewritten to

$$\left(1 - \frac{1}{2i}\right)u_{i-1,j} + \left(1 + \frac{1}{2i}\right)u_{i+1,j} - 2\left[1 + \frac{1}{(i\delta\theta)^2}\right]u_{i,j}$$

$$+ \frac{1}{(i\delta\theta)^2}u_{i,j-1} + \frac{1}{(i\delta\theta)^2}u_{i,j+1} = 0 \tag{3.2.32}$$

If the region is far away from the origin of the coordinates and it is assumed that the angle between any two radical lines is an equalized small value and the

3.2 Difference formulation of Poisson's equation

radius r varies as a geometric series which satisfies the relationship

$$\frac{r + h_E}{r} = \frac{r}{r - h_W} = \beta \quad (3.2.33)$$

then Eq. (3.2.30) is reduced to Eq. (3.2.26).

The reason is explained as follows [12].

If in Fig. 3.2.3 $\theta_N = \theta_S = \theta$ (θ is small), then $h_N = h_S = h = r\theta$. Assume that the coefficient of $(u_S - u_0)$ and $(u_N - u_0)$ are equal to the coefficients of $(u_W - u_0)$ and $(u_E - u_0)$, respectively, i.e.

$$\begin{cases} \dfrac{1}{2(r\theta)^2} = \dfrac{1 + \dfrac{h_W}{2r}}{h_W(h_W + h_E)} \\ \dfrac{1}{2(r\theta)^2} = \dfrac{1 - \dfrac{h_E}{2r}}{h_E(h_E + h_W)} \end{cases} \quad (3.2.34)$$

the solutions of the above equations are then

$$\begin{cases} h_E = \dfrac{r\theta}{2}[\theta + (4 - \theta^2)^{1/2}] \\ h_W = \dfrac{r\theta}{2}[-\theta + (4 - \theta^2)^{1/2}] \end{cases} \quad (3.2.35)$$

Now the terms $\dfrac{r + h_E}{r}$ and $\dfrac{r}{r - h_W}$ can be expressed as

$$\begin{cases} \dfrac{r + h_E}{r} = 1 + \dfrac{\theta}{2}[\theta + (4 - \theta^2)^{1/2}] \\ \dfrac{r}{r - h_W} = \dfrac{1}{1 - \dfrac{\theta}{2}[-\theta + (4 - \theta^2)^{1/2}]} \end{cases} \quad (3.2.36)$$

Comparing these two expressions, if θ is small enough, the difference between $\dfrac{r + h_E}{r}$ and $\dfrac{r}{r - h_W}$ becomes very small and it follows that

$$\frac{r + h_E}{r} \cong \frac{r}{r - h_W} = \beta = 1 + \theta\left(1 + \frac{\theta}{2}\right). \quad (3.2.37)$$

The error of Eq. (3.2.37) depends on the value of θ. If $\theta = 15°$, then the error is less than 0.41%; if $\theta = 7.5°$, then the error is less than 0.039%. *Thus if an appropriate value of θ is chosen and the radius varies as a geometric series subject to ratio β, then the difference formulation in the area far away from the origin of the coordinates is the same as the one obtained in rectangular coordinates.*

3.2.4 Discretization formula of axisymmetric fields

In cylindrical coordinates it can be assumed that the z-axis of the coordinates coincides with the axis of symmetry. The potential distribution is independent of the coordinate θ, i.e. $\partial u/\partial \theta = 0$. Thus the expression of Laplace's equation becomes

$$\frac{\partial^2 u}{\partial r^2} + \frac{1}{r}\frac{\partial u}{\partial r} + \frac{\partial^2 u}{\partial z^2} = 0 . \tag{3.2.38}$$

The mesh discretization in the r–z plane shown in Fig. 3.2.4 is similar to that of the x–y plane.

By using the same method in deriving the difference equation as in Sect. 3.2.2 and supposing $h_r = h_z = h$ the difference equation in the r–z plane is

$$\left(1 + \frac{h}{2r_0}\right)u_E + u_N + \left(1 - \frac{h}{2r_0}\right)u_W + u_S - 4u_0 = h^2 F_0 . \tag{3.2.39}$$

Let $r = ih (i = 1, 2, \ldots)$. The general formula of the node i, j is then

$$u_{i,j} = \frac{1}{4}\left[\left(1 + \frac{1}{2i}\right)u_{i+1,j} + u_{i,j+1} + \left(1 - \frac{1}{2i}\right)u_{i-1,j} + u_{i,j-1} - h^2 F_0\right] . \tag{3.2.40}$$

If i is very large, Eq. (3.2.40) is identical with Eq. (3.2.25). It shows that, for *axisymmetrical problems if the area is far away from the axis, the field distribution is almost identical to the field distribution in Cartesian coordinates.*

Concerning the points located on the axis of symmetry, i.e. $r = 0$, the term $\frac{1}{r}\frac{\partial u}{\partial r}$ then becomes indefinite. By using the extremum principle

$$\lim_{r \to 0}\left(\frac{1}{r}\frac{\partial u}{\partial r}\right) = \lim_{r \to 0}\frac{\left(\frac{\partial u}{\partial r}\right)}{(r)} = \left(\frac{\partial^2 u}{\partial r^2}\right)_{r=0} .$$

Equation (3.2.38) reduces to

$$2\frac{\partial^2 u}{\partial r^2} + \frac{\partial^2 u}{\partial z^2} = F . \tag{3.2.41}$$

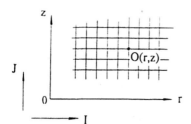

Fig. 3.2.4. Grid nodes in r–z plane

3.2 Difference formulation of Poisson's equation

Since the field is symmetric to the axis, the difference formula of the points located on the axis is

$$u_0 = \tfrac{1}{6}(4u_E + u_N + u_S - h^2 F_0). \qquad (3.2.42)$$

All the above formulations are derived for solving problems in linear and uniform materials.

3.2.5 Discretization formula of the non-linear magnetic field

If the reluctivity $v(v = 1/\mu)$ of a given material is a function of the magnetic flux density, Poisson's equation for the magnetic vector potential in a 2-D case is

$$\frac{\partial}{\partial x}\left(v\frac{\partial A}{\partial x}\right) + \frac{\partial}{\partial y}\left(v\frac{\partial A}{\partial y}\right) = -J. \qquad (3.2.43)$$

Assuming that the current density and the reluctivity of the medium are constant in the area of each mesh, the discretization equation can be derived by the same method as discussed in Sect. 3.2.2. The result [13] is

$$A_0 = \left(J_0 + \sum_{i=1}^{4} \alpha_i A_i\right) \Big/ \sum_{i=1}^{4} \alpha_i \qquad (3.2.44)$$

with

$$\begin{cases} \alpha_E = \dfrac{1}{2h_E}(v_E h_S + v_N h_N) & \alpha_N = \dfrac{1}{2h_N}(v_N h_E + v_w h_w) \\[6pt] \alpha_w = \dfrac{1}{2h_E}(v_w h_N + v_S h_S) & \alpha_S = \dfrac{1}{2h_E}(v_S h_w + v_E h_E) \end{cases} \qquad (3.2.45)$$

$$J_0 = \tfrac{1}{4}(J_E h_S h_E + J_N h_E h_N + J_w h_N h_w + J_S h_w h_S). \qquad (3.2.46)$$

In the solution of a non-linear problem an under-relaxation factor β of v may be introduced in some cases to accelerate the iteration as in

$$v^{(n+1)} = (v^{(n+1)} - v^{(n)})\beta + v^{(n)}. \qquad (3.2.47)$$

The experience of solving the problem as shown in reference [14] indicates that an appropriate relaxation factor of v yields a better rate of convergence than a constant relaxation factor such as $\beta = 0.1$. The relaxation factor depends on the value of the flux density.

3.2.6 Difference equations for time-dependent problems

Applying the difference quotient of time, a 1-D diffusion equation is considered as an example, i.e.

$$\frac{\partial^2 u}{\partial x^2} = a\frac{\partial u}{\partial t}. \qquad (3.2.48)$$

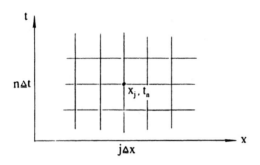

Fig. 3.2.5. Discretization of a 1-D time dependent problem

The variables of x and t at any point are shown in Fig. 3.2.5. The time iteration and the position iteration are indicated by the superscript and the subscript, respectively. Thus the expressions of the Taylor's series for t and x are

$$u_j^{n+1} = u_j^n + \Delta t \left(\frac{\partial u}{\partial t}\right)_j^n + \frac{(\Delta t)^2}{2!}\left(\frac{\partial^2 u}{\partial t^2}\right)_j^n \tag{3.2.49}$$

$$u_{j+1}^n = u_j^n + \Delta x \left(\frac{\partial u}{\partial x}\right)_j^n + \frac{(\Delta x)^2}{2!}\left(\frac{\partial^2 u}{\partial x^2}\right)_j^n. \tag{3.2.50}$$

By using the forward time difference quotient of the first order and the centred-space difference quotient of the second order, Eq. (3.2.48) is transferred to

$$\frac{u_j^{n+1} - u_j^n}{\Delta t} = \frac{1}{a}\frac{u_{j+1}^n - 2u_j^n + u_{j-1}^n}{(\Delta x)^2}. \tag{3.2.51}$$

This equation can be rearranged to

$$u_j^{n+1} = C u_{j+1}^n + (1 - 2C)u_j^n + C u_{j-1}^n \tag{3.2.52}$$

with

$$C = \frac{\Delta t}{a(\Delta x)^2}. \tag{3.2.53}$$

Equation (3.2.52) is known as an explicit formulation; the new value u_j^{n+1} can be evaluated by the three nodal values at the nth iteration. The steps Δt and Δx must satisfy Eq. (3.2.53). Notice that the constant $2C$ must be less than 1, otherwise the iteration will be broken. In addition, if $C < 1/2$ is satisfied, the solution can be convergent but the error could be oscillated. If $C < 1/4$, the solution will not oscillate but the penalty is that the Δt must be very small [15]. Comparison with Eq. (3.2.51) shows that a stable solution is obtained if an implicit formulation [15] is used.

If the spatial derivative of a function is approximated at time $n + 1$, i.e.

$$\frac{\partial^2 u}{\partial x^2} = \frac{u_{j+1}^{n+1} - 2u_j^{n+1} + u_{j-1}^{n+1}}{(\Delta x)^2} \tag{3.2.54}$$

3.2 Difference formulation of Poisson's equation

then Eq. (3.2.51) is transformed to

$$\frac{u_j^{n+1} - u_j^n}{\Delta t} = \frac{1}{a} \frac{u_{j+1}^{n+1} - 2u_j^{n+1} + u_{j-1}^{n+1}}{(\Delta x)^2}. \tag{3.2.55}$$

This is an implicit iterative formulation. The fundamental difference between Eqs. (3.2.51) and (3.2.55) is that in Eq. (3.2.51) there is only a single unknown u_j^{n+1}. It can be evaluated by a simple iteration. In Eq. (3.2.55), at the $(n + 1)$th iteration of time, the nodal values of three neighbouring nodes are unknowns. They cannot be solved directly from Eq. (3.2.55). Hence Eq. (3.2.55) is called the implicit form of a difference equation. The implicit formulation contains a set of algebraic equations. Introducing Eq. (3.2.55) into Eq. (3.2.48) yields

$$Cu_{j-1}^{n+1} + (1 + 2C)u_j^{n+1} - Cu_{j+1}^{n+1} = u_j^n. \tag{3.2.56}$$

This equation is applied to all interior nodes except the points lying on the boundary which must be modified to reflect the boundary conditions. The solution of Eq. (3.2.56) may be derived using different methods.

The well known Crank-Nicolson equation provides an alternative implicit scheme which has the accuracy of the second order in both space and time. The difference approximation is developed at the midpoint of the time increment, i.e. the temporal derivative of the first order is approximate at $t^{n+\frac{1}{2}}$ by

$$\frac{\partial u}{\partial t} = \frac{u_j^{n+1} - u_j^n}{\Delta t}. \tag{3.2.57}$$

The second order derivative of space is determined on the midpoint by averaging the difference approximation at the beginning (t^n) and the end (t^{n+1}) of the time increment, i.e.

$$\frac{\partial^2 u}{\partial x^2} = \frac{1}{2}\left[\frac{u_{j+1}^n - 2u_j^n + u_{j-1}^n}{(\Delta x)^2} + \frac{u_{j+1}^{n+1} - 2u_j^{n+1} + u_{j-1}^{n+1}}{(\Delta x)^2}\right]. \tag{3.2.58}$$

Introducing Eq. (3.2.57) and (3.2.58) into Eq. (3.2.48) the Crank-Nicolson equation can be expressed as [16]:

$$\beta \frac{\partial^2 u}{\partial x^2}\bigg|^{n+1} + (1 - \beta)\frac{\partial^2 u}{\partial x^2}\bigg|^n = a(u^{n+1} - u^n). \tag{3.2.59}$$

When $\beta = \frac{1}{2}$, Eq. (3.2.59) becomes

$$2\left[a\frac{(\Delta x)^2}{\Delta t} + 1\right]u_j^{n+1} - u_{j+1}^{n+1} - u_{j-1}^{n+1} = B^n \tag{3.2.60}$$

where

$$B^n = u_{j+1}^n - 2u_j^n + u_{j-1}^n + 2a\frac{(\Delta x)^2}{\Delta t}u_j^n. \tag{3.2.61}$$

Therefore, if the implicit equation is adopted, algebraic equations have to be solved in the process of each iteration, but a more accurate result is obtained.

In Eq. (3.2.59), if $\beta = 0$, Eq. (3.2.59) is the same as Eq. (3.2 51). If $\beta = 1$, Eq. (3.2.54) is called an O'Brien equation.

3.3 Solution methods for difference equations

In order to select the solution method to solve the discretization equations the characteristics of the algebraic equations have to be examined first.

3.3.1 Properties of simultaneous equations

A Dirichlet problem as shown in Fig. 3.3.1 is used as an example to classify the properties of the matrix equation derived by using the finite difference method.

In Fig. 3.3.1 the solution region Ω is subdivided into a square grid and boundary nodes are assumed to be identical with the nodes of the grid. For simplicity, the source term is considered as a constant. The algebraic equations of the nine interior nodes are

$$\begin{cases} 4u_1 & -u_2 & 0 & -u_4 & 0 & 0 & 0 & 0 & 0 = -h^2 F_0 \\ -u_1 & 4u_2 & -u_3 & 0 & -u_5 & 0 & 0 & 0 & 0 = -h^2 F_0 \\ & -u_2 & 4u_3 & 0 & 0 & -u_6 & 0 & 0 & 0 = 100 - h^2 F_0 \\ -u_1 & 0 & 0 & 4u_4 & -u_5 & 0 & -u_7 & 0 & 0 = -h^2 F_0 \\ 0 & -u_2 & 0 & -u_4 & 4u_5 & -u_6 & 0 & -u_8 & 0 = -h^2 F_0 \\ 0 & 0 & -u_3 & 0 & -u_5 & 4u_6 & 0 & 0 & -u_9 = 100 - h^2 F_0 \\ 0 & 0 & 0 & -u_4 & 0 & 0 & 4u_7 & -u_8 & 0 = -h^2 F_0 \\ 0 & 0 & 0 & 0 & -u_5 & 0 & -u_7 & 4u_8 & -u_9 = -h^2 F_0 \\ 0 & 0 & 0 & 0 & 0 & -u_6 & 0 & -u_8 & 4u_9 = 100 - h^2 F_0 \end{cases}$$

(3.3.1)

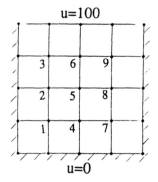

Fig. 3.3.1. A Dirichlet problem subdivided by square meshes

3.3 Solution methods for difference equations

These equations can be written in a matrix form:

$$\mathbf{Ku} = \mathbf{B} \qquad (3.3.2)$$

where

$$\mathbf{K} = \begin{bmatrix} 4 & -1 & & -1 & & & & & \\ -1 & 4 & -1 & & -1 & & & & \\ & -1 & 4 & & & -1 & & & \\ -1 & & & 4 & -1 & & -1 & & \\ & -1 & & -1 & 4 & -1 & & -1 & \\ & & -1 & & -1 & 4 & & & -1 \\ & & & -1 & & & 4 & -1 & \\ & & & & -1 & & -1 & 4 & -1 \\ & & & & & -1 & & -1 & 4 \end{bmatrix} \qquad (3.3.3)$$

and $\mathbf{u}$ is a column vector consisting of nine unknown potentials at nine interior nodes. The RHS of Eq. (3.3.2) are known boundary values and sources as given in Eq. (3.3.1). It can be found that matrix $\mathbf{K}$ has the following properties:

(a) $\mathbf{K}$ *is a banded sparse matrix. The maximum non-zero terms of any row is 5.*
(b) *The order of matrix* $\mathbf{K}$ *is equal to the total number of unknown potentials at the interior nodes.*
(c) *The distribution and the values of the elements of* $\mathbf{K}$ *are regular. In the case of a uniform medium the value of each diagonal element is 4; all the others are* -1.
(d) *The matrix* $\mathbf{K}$ *is positive definite, but not always symmetric.* The symmetry is destroyed if the grid nodes are not coincident with the boundary.

The Gauss-elimination method is a simple and easy way to be used to solve Eq. (3.3.2). Due to the large sparsity of the matrix the Gauss-elimination method is not economical. The sparsity of the matrix should be considered during the solution process.

Compared with the direct method, iterative solution methods have many advantages for solving difference equations. Since the distribution of the elements of matrix $\mathbf{K}$ is regular, the matrix need not to be stored. So the memory requirement of the computer is considerably decreased. But the convergent speed of the iteration then becomes an important problem. The successive over-relaxation iteration is the common method to solve the finite difference equations.

3.3.2 Successive over-relaxation (SOR) method

Recall Eq. (3.2.26)

$$u_{i,j} = \tfrac{1}{4}(u_{i+1,j} + u_{i,j+1} + u_{i-1,j} + u_{i,j-1}).$$

The function value of each node is determined by the values of its four neighbouring nodes. If the initial estimate of the potential $u_{ij}^{(0)}$ is given, the first approximate values $u_{i,j}^{(1)}$ can be calculated by Eq. (3.2.26). This means each new value of the potential at the centre of the star is determined by the previous iterative values at its four adjacent nodes. While this formula is used as an iterative process, two arrays with dimensions of N (the total number of interior nodes) have to be stored. One is for storing the previous values of each node, the other is for storing the current values of the nodes. This procedure is a basic iteration method known as the *Jacobian iterative method*.

However, it is noted that when i, j are increased from 1 to N and 1 to M, the iterative sequence of Eq. (3.2.26) is a scanning of the nodes from the left-hand to the right-hand column starting from the bottom row to the top in each column. According to this procedure the iterative formulation can be modified to:

$$u_{i,j}^{n+1} = \tfrac{1}{4}(u_{i+1,j}^{n} + u_{i,j+1}^{n} + u_{i-1,j}^{n+1} + u_{i,j-1}^{n+1}) \, . \tag{3.3.4}$$

In this equation the new approximate values $u_{i-1,j}^{n+1}$ and $u_{i,j-1}^{n+1}$ are used in the $(n + 1)$th iteration as soon as they are available. This method is called the Gauss–Seidel iteration. It is more economical than the Jacobian method; only one complete array of potentials needs to be stored. It requires half the memory than in case of the Jacobian method.

Unfortunately, if the total number of the nodes is large, the convergence speed of the Gauss–Seidel iteration is still too slow. *The over-relaxation method is one of the most generally effective, stable and successful methods to accelerate the convergence speed.*

The over-relaxation method is based on the estimate of the correction required to evaluate $u_{i,j}^{n+1}$ in Eq. (3.3.4). First the residual of the function values between two iterations is defined as

$$R_{i,j} = u_{i,j}^{n+1} - u_{i,j}^{n} \, . \tag{3.3.5}$$

Then the new iterative value of the potential is determined as the sum of the old value and the residual $R_{i,j}$ multiplied by an acceleration factor α. This can be expressed by the following equation

$$u_{i,j}^{n+1} = u_{i,j}^{n} + \alpha R_{i,j} \quad 1 < \alpha < 2 \, . \tag{3.3.6}$$

The factor α is an acceleration convergence constant called a relaxation factor which determines the degree of the over-relaxation. It is greater than 1 and less than 2. If $\alpha = 1$, Eq. (3.3.6), reduces to Eq. (3.3.4). If $\alpha = 2$, the iteration becomes divergent. Therefore there is an optimum acceleration factor α_0. At this value the convergence rate is greatly increased. The difficulty is that the optimum acceleration factor is extremely problem dependent; there is no general method to estimate its value. In case of a Poissonian problem in which a rectangular region is subdivided into a square grid with $(p + 1)$ and $(q + 1)$ nodes on each side the optimum factor α_0 can be estimated by the following equation [3]

$$\alpha_0 = 2 - \pi(2)^{\tfrac{1}{2}} \left(\frac{1}{p^2} + \frac{1}{q^2} \right)^{\tfrac{1}{2}} \, . \tag{3.3.7}$$

3.3 Solution methods for difference equations

Table 3.3.1. The influence of the acceleration factor ($\varepsilon = 10^{-5}$)

Acceleration factor, α	Iterative times, N
1.00	609
1.50	228
1.70	127
1.76	96
1.763*	94
1.78	83
1.80	86
1.90	153
1.92	892

If the problem domain is a square and subdivided into a square grid, then

$$\alpha_0 = \frac{2}{1 + \sin\left(\frac{\pi}{p}\right)}. \tag{3.3.8}$$

For an area of arbitrary shape, the equivalent rectangular area can be used to estimate the optimum acceleration factor.

In order to show the effect of the relaxation factor α the Dirichlet boundary value problem shown in Fig. 3.3.1 is used as an example. Assume $p = 30$, $q = 24$, and the convergence criterion of iteration as $\varepsilon = 1.E - 5$ (the relative error of potential of each node). The iterative times and the relaxation factor are listed in Table 3.3.1.

The table shows that 1.78 is the optimum relaxation factor. 1.763* is calculated by Eq. (3.3.7), it is very close to the optimum value. With $p = 40$ and $q = 24$ the optimum acceleration factor is $\alpha_0 = 1.78$ at $N = 102$. α_0 is exactly equal to the value which is calculated by Eq. (3.3.7). Hence Eq. (3.3.7) is an approximate estimate.

In general, for a stable convergence process, the number of iterations N depends on the largest error ε at any node and the factor F [4]:

$$N = -F(\log \varepsilon). \tag{3.3.9}$$

In this equation F is known as the asymptotic rate of convergence, it is a function of the boundary conditions, the number of nodes, and the particular type of difference equations. A more detailed discussion is given in reference [8].

Even though the initial value can be given arbitrarily, it will influence the speed of convergence. The proper estimation of the initial value is helpful in obtaining good convergence, especially in solving non-linear and time-dependent problems. The result may be divergent if the initial value is not appropriate.

As the successful over-relaxation (SOR) method is simple, flexible and relatively quickly convergent, it is the most useful method. However, it is a point iterative method. Other rapidly convergence methods include line iterations, block iterations, and alternating directions implicit (ADI) methods [2, 17, 18].

Table 3.3.2 Iterations of different methods ($\varepsilon = 10^{-10}$)

Method	Time/iteration cycle	No. of iteration
Gauss–Seidel	$k \cdot 4p^2$	840
SOR	$k \cdot 7p^2$	70
ADI	$k \cdot 7p^2$	25

In the block iteration method the nodes are no longer treated seperately but in lines or blocks. In ADI method nodes are treated line by line, but the direction of the sweep of lines across the mesh is alternated (in each iteration there is a sweep in the x-direction and a sweep in the y-direction). Reference [4] gives the number of iteration by using different methods for solving the Dirichlet problem in a square region with $(p + 1)^2$ nodes. Some of the data are listed in Table 3.3.2 to show the efficiency of the different methods. k and p are constants. The table shows that the number of iterations in using SOR or ADI is much less than with the Gauss–Seidel iteration. Other effective iterative methods for specific problems are given in [19].

If the speed and the size of a computer are sufficient, direct solution methods are still faster than iterative methods.

3.3.3 Convergence criterion

In the iterative process the residual $R_{i,j}^{(n)}$ is defined by

$$R_{i,j}^{(n)} = u_{i,j}^{(n+1)} - u_{i,j}^{(n)}. \tag{3.3.10}$$

The convergence criterion gives that the residuals $R_{i,j}^{(n)}$ at every interior node become less than a predetermined error ε. Note that if the residual $R_{i,j}$ becomes small the convergence rate becomes very slow. In this case the iteration times can be used as a criterion for stopping the calculation.

On the other hand, the choice of either the absolute residual or the relative residual should also be carefully considered. To solve static problems the relative error should be taken. When solving time-dependent problems if the solution of the problem itself becomes very small the relative error becomes very large; consequently the absolute error is considered.

3.4 Difference formulations of arbitrary boundaries and interfacial boundaries between different materials

In the former sections discussions have been limited to homogeneous media which are subject to Dirichlet boundary conditions and where the grid nodes are identical with the boundaries. In more general cases there are different materials

3.4 Difference formulations of arbitrary boundaries and interfacial boundaries

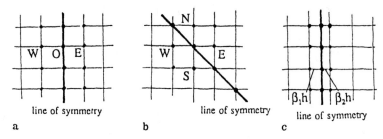

Fig. 3.4.1a–c. Different types of the lines of symmetry

and the grid nodes may not be coincident with the boundary or the interfacial boundaries. The formulations for these cases will be discussed in this section.

3.4.1 Difference formulations on the lines of symmetry

Consider the case shown in Fig. 3.4.1(a); the grid nodes are coincident with the symmetric line. The points 0, N, S lie on the line of symmetry.

Due to symmetry a fictitious point E is placed on the symmetric position of point W. With $u_E = u_W$ it is found that

$$u_0 = \tfrac{1}{4}(2u_W + u_N + u_S - h^2 F_0) . \tag{3.4.1}$$

If the line of symmetry is diagonal to the grids as shown in Fig. 3.4.1(b), then

$$u_0 = \tfrac{1}{4}[2(u_S + u_W) - h^2 F_0] . \tag{3.4.2}$$

with $u_N = u_W$ and $u_E = u_S$.

In Eqs. (3.4.1) and (3.4.2) h is the distance between the lines of the square grid.

If the line of symmetry is parallel to, but not coincident with, the grid line as shown in Fig. 3.4.1(c), then with $u_E = u_W$ and $\beta_1 + \beta_2 = 1$, Eq. (3.2.24) becomes

$$\frac{2u_E}{\beta_1 \beta_2} + u_N + u_S - 2\left(1 + \frac{1}{\beta_1 \beta_2}\right)u_0 - h^2 F_0 = 0 . \tag{3.4.3}$$

If $\beta_1 = \beta_2 = \tfrac{1}{2}$, this equation reduces to

$$u_0 = \frac{1}{10}(8u_E + u_N + u_S - h^2 F_0) . \tag{3.4.4}$$

3.4.2 Difference equation of a curved boundary

There are problems where the grid nodes are not located on the boundary of the domain, as shown in Fig. 3.4.2.

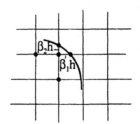

Fig. 3.4.2. Discretization model for a curved boundary

According to Eqs. (3.2.5), (3.2.6) the following equations exist:

$$u_E = u_0 + \beta_1 h \left(\frac{\partial u}{\partial x}\right)_0 + \frac{1}{2}\beta_1^2 h^2 \left(\frac{\partial^2 u}{\partial x^2}\right)_0 \tag{3.4.5}$$

$$u_W = u_0 - h \left(\frac{\partial u}{\partial x}\right)_0 + \frac{1}{2}h^2 \left(\frac{\partial^2 u}{\partial x^2}\right)_0 . \tag{3.4.6}$$

By using similar procedures as derived in Eq. (3.2.16) and (3.2.17) one obtains:

$$\frac{\partial^2 u}{\partial x^2} = \frac{1}{h^2}\left[\frac{2}{\beta_1(1+\beta_1)}u_E + \frac{2}{(1+\beta_1)}u_W - \frac{2}{\beta_1}u_0\right] \tag{3.4.7}$$

$$\frac{\partial^2 u}{\partial y^2} = \frac{1}{h^2}\left[\frac{2}{\beta_2(1+\beta_2)}u_N + \frac{2}{(1+\beta_2)}u_S - \frac{2}{\beta_2}u_0\right] \tag{3.4.8}$$

Introducing the above equations into Poisson's equation the difference equation at point 0 is

$$\frac{2u_E}{\beta_1(1+\beta_1)} + \frac{2u_N}{\beta_2(1+\beta_2)} + \frac{2u_W}{(1+\beta_1)} + \frac{2u_S}{(1+\beta_1)}$$
$$- 2\left(\frac{1}{\beta_1}+\frac{1}{\beta_2}\right)u_0 = hF_0 \tag{3.4.9}$$

with $\beta_1, \beta_2 \leq 1$. It is obvious that the program for solving problems with curved boundaries needs quite a little complicated program.

3.4.3 Difference formulations for the interface of different materials

Now the case is considered where the interface between two different materials is coincident with the grid lines shown in Fig. 3.4.3. Nodes 0, N, S are located on the interface of two regions having different permeabilities μ_a and μ_b. Let A_a and A_b denote 2-D magnetic vector potentials in the region a and b, respectively. Assume that only region a carries current distributed with uniform density J, region b is free of current. Thus the magnetic vector potential A_a in region a satisfies Poisson's equation $\nabla^2 A_a = -\mu_a J$. In region b A_b satisfies Laplace's equation $\nabla^2 A_b = 0$.

3.4 Difference formulations of arbitrary boundaries and interfacial boundaries

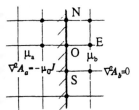

Fig. 3.4.3. Grid nodes located on the interface

If both two regions are considered as fulfilling permeabilities μ_a and μ_b, respectively, then the difference equation at node 0 is

$$A_{a_E} + A_{a_N} + A_{a_W} + A_{a_S} - 4A_{a_0} = h^2 F_0 \tag{3.4.10}$$

or

$$A_{b_E} + A_{b_N} + A_{b_W} + A_{b_S} - 4A_{b_0} = h^2 F_0 \tag{3.4.11}$$

The subscripts 'a' and 'b' represent that the region fulfills with material with permeability, μ_a and μ_b, respectively. However, the potentials A_{a_E} and A_{b_W} are fictitious, they can be eliminated by the following interface boundary conditions

$$A_{a_0} = A_{b_0} \qquad A_{a_N} = A_{b_N} \qquad A_{a_S} = A_{b_S} \tag{3.4.12}$$

and

$$\frac{1}{\mu_a}(A_{a_E} - A_{a_W}) = \frac{1}{\mu_b}(A_{b_E} - A_{b_W}). \tag{3.4.13}$$

Forming the sum of μ_a, μ_b times Eqs. (3.4.10) and (3.4.11), respectively, and considering the boundary conditions of Eqs. (3.4.12) and (3.4.13) the difference equation of node 0 is:

$$A_0 = \frac{1}{4}\left(\frac{2}{1+K}A_{b_E} + A_N + \frac{2K}{1+K}A_{a_W} + A_S - \frac{K}{1+K}h^2 F_0\right) \tag{3.4.14}$$

with

$$K = \frac{\mu_b}{\mu_a}^\dagger. \tag{3.4.15}$$

If region b is a ferromagnetic material where the lines of the magnetic flux density are orthogonal to the ferromagnetic surface, i.e. $\partial A/\partial n = 0$, then

$$A_0 = \tfrac{1}{4}(2A_W + A_N + A_S - h^2 F_0). \tag{3.4.16}$$

In accordance with the same method used before the formulations for other interfacial boundary conditions are given in Fig. 3.4.4.

If in Fig. 3.4.4(a) and (b) both regions satisfy Laplace's equation and if $\mu_a \to \infty$ then

$$A_0 = \tfrac{1}{6}(2A_E + 2A_N + A_W + A_S). \tag{3.4.17}$$

This case usually occurs in electromagnetic devices.

---

† For 2-D electrostatic fields A is replaced by φ and $K = \dfrac{\varepsilon_a}{\varepsilon_b}$.

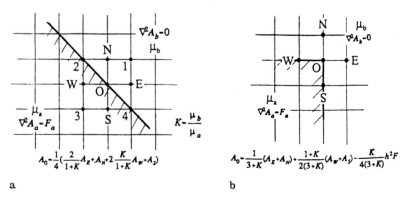

Fig. 3.4.4a, b. Different cases of interfacial boundaries

3.5 Examples

Example 3.5.1. Calculate the potential distribution in a grounded slot as shown in Fig. 3.3.1.

Due to the symmetry only half of the domain is considered. The potential distribution in the slot is given in Fig. 3.5.1.

Example 3.5.2. Assume that an infinitely long steel plate of a 2-D thickness (shown in Fig. 3.5.2) is immersed in a uniform magnetic field $\mathbf{H} = -H_m \sin \omega t \mathbf{k}$. Calculate the magnetic distribution in the ferromagnetic plate.

H component of the field satisfies the partial differential equation as follows:

$$\frac{\partial^2 H}{\partial x^2} = \mu \gamma \frac{\partial H}{\partial t} \qquad (3.5.1)$$

where μ and γ are the permeability and the conductivity of the steel. The finite difference form of the above equation was given in Eq. (3.2.51). If μ and γ are constants, the distribution of the H field may be easily calculated by Eq. (3.2.51). The iteration times of the computation are strongly dependent on the initial values. In the case of $H_m = 6000$ A/m, $\gamma = 5 \times 10^6$ S/m, $\mu = 156 \mu_0$, $\omega = 314$, $D = 2.5$ mm, assume that the time step is $\Delta t = T/360$, the step of position is $\Delta x = 0.1$ mm. The computation results of the magnetic field strength on the plane at $x = 2.5, 2.2, 1.9, 1.6, 1.3, 1.0$ (mm) are shown in Fig. 3.5.3. They still vary sinusoidally. The results show that H_{max} at these points is decaying according to the exponential function. The phases of these curves are retardant on each other, as shown in Fig. 3.5.4. In this figure the horizontal axis represents the real component of H and the vertical axis represents the image component of H. At the surface of the plate, the image component is zero. The curve shown in

3.5 Examples

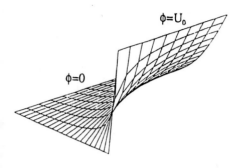

Fig. 3.5.1. Potential distribution in 2-D slot

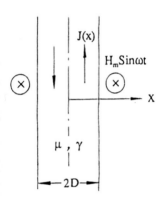

Fig. 3.5.2. A conducting plate immersed in a uniform magnetic field

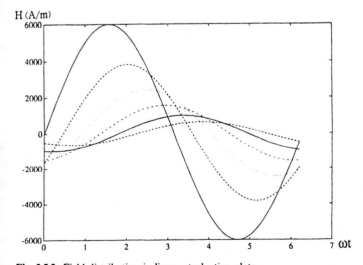

Fig. 3.5.3. Field distribution in linear conducting plate

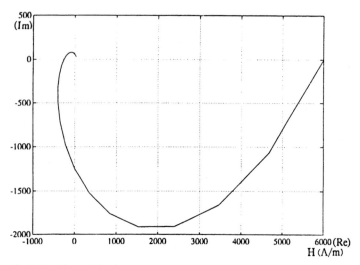

Fig. 3.5.4. Phase shift of *H* component within linear conducting plate

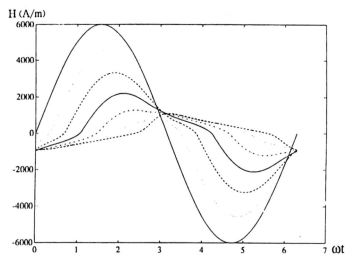

Fig. 3.5.5. Field distribution in non-linear conducting plate

Fig. 3.5.4 is the locus of the end point of the vector *H*. It represents the phase delay of the induced field in different positions of the plate.

If the permeability μ of the plate is non-linear and it varies according to the following function:

$$B = \frac{H}{156 + 0.59H} \tag{3.5.2}$$

then the magnetic field strengths on the plane of $x = 2.5, 2.1, 1.7, 1.3, 0.9, 0.5, 0.1$ (mm) are shown as the seven curves of Fig. 3.5.5.

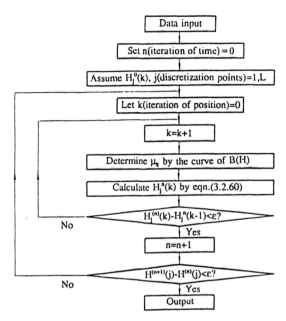

Fig. 3.5.6. Iterative procedures for non-linear diffusion equation

Due to non-linear permeability, all these curves are non-sinusoidal. The distortion is much stronger if the planes are far away from the surface. During the process of iteration, the times of iteration are extremely dependent upon the initial values of the iteration. The flow-chart of the computation is given in Fig. 3.5.6. To solve this problem, the implicit iterative formulation must be used.

3.6 Further discussions about the finite difference method

In Sect. 3.2, finite difference equations of a partial differential equation were developed using Taylor's series. The discretization equations can also be derived by particular physical principles, weighted residual approaches or variational principles. In this section these methods will be investigated. The error norm of FDM is also examined in this section.

3.6.1 Physical explanation of the finite difference method

In FDM the region being analysed is divided into a number of regular lumps shown in Fig. 3.6.1. Each of these interior lumps is assumed to have a constant value of the pertinent field variable. Figure 3.6.1 shows that the centre point of

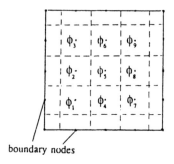
boundary nodes

Fig. 3.6.1. Finite difference discretization of a 2-D problem

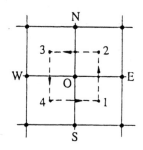

Fig. 3.6.2. Ampre's law around a point

a small area is a representation of this area. The surface lumps are one half the size of the interior lumps. The corner lumps are one-fourth of the size of interior lumps. According to this kind of discretization, the finite difference equation can be derived directly from the lumped parametric model by the physical meaning for different fields.

In consideration of a 2-D static magnetic field, the four centre points 1, 2, 3, 4 of a symmetric grid are shown in Fig. 3.6.2.

Assuming that the magnetic flux density along the lines 12, 23, 34, 41 are constants, respectively, and they can be approximated by the first order difference quotients of the magnetic potentials, i.e.

$$\begin{cases} H_{12} \cong \dfrac{(A_E - A_0)}{\mu h} & H_{23} \cong \dfrac{(A_N - A_0)}{\mu h} \\ H_{34} \cong \dfrac{(A_W - A_0)}{\mu h} & H_{41} \cong \dfrac{(A_S - A_0)}{\mu h} \end{cases} \quad (3.6.1)$$

So using Ampre's law $\oint H dl = 0$ along the contour 12341, the following equation then holds

$$A_E + A_N + A_W + A_S - 4A_0 = 0 . \quad (3.6.2)$$

For a static electric field or a steady current field, the five-point difference formulation can be derived by using the principle of the continuity of the electric flux density **D** or by the continuity of current density **J**. In heat transfer problems, the difference equation can be derived by the conservation of heat.

3.6.2 The error analysis of the finite difference method

As the high order terms of Taylor's series are neglected, the truncation error is presented. In order to analyse the error influenced by the step length of discretization, multiply Eqs. (3.2.5) and (3.2.6) by r and p, respectively, which

3.6 Further discussions about the finite difference method

results in

$$ru_E = ru_0 + rph\left(\frac{\partial u}{\partial x}\right)_0 + \frac{1}{2!}r(ph)^2\left(\frac{\partial^2 u}{\partial x^2}\right)_0 + \frac{1}{3!}r(ph)^3\left(\frac{\partial^3 u}{\partial x^3}\right)_0$$
$$+ \frac{1}{4!}r(ph)^4\left(\frac{\partial^4 u}{\partial x^4}\right)_0 \quad (3.6.3)$$

$$pu_W = pu_0 - prh\left(\frac{\partial u}{\partial x}\right)_0 + \frac{1}{2!}p(rh)^2\left(\frac{\partial^2 u}{\partial x^2}\right)_0 - \frac{1}{3!}p(rh)^3\left(\frac{\partial^3 u}{\partial x^3}\right)_0$$
$$+ \frac{1}{4!}p(rh)^4\left(\frac{\partial^4 u}{\partial x^4}\right)_0 \quad (3.6.4)$$

where $h_E = ph$, $h_W = rh$, p and $r \leq 1$. Subtraction of Eq. (3.6.4) from Eq. (3.6.3) and consideration of the assumption $p = r = 1$, yields

$$\frac{\partial u}{\partial x} = \frac{u_E - u_W}{2h} + O(h^2) \quad (3.6.5)$$

where

$$O(h^2) = \frac{1}{3!}h^2\frac{\partial^3 u}{\partial x^3} + \cdots \quad (3.6.6)$$

This equation shows that the truncation error of the difference quotient of the first order is proportional to h^2. If the partial derivatives of the higher order are neglected directly from Eq. (3.2.5), the difference formulation of $\partial u/\partial x$ is

$$\frac{\partial u}{\partial x} = \frac{u_E - u_0}{h} + O(h) \quad (3.6.7)$$

$$O(h) = \frac{1}{2!}h\frac{\partial^2 u}{\partial x^2} + \cdots \quad (3.6.8)$$

It shows that the truncation error is proportional to h. Consequently, the accuracy of the central difference quotient of the first order is higher than the forward difference quotient or the backward difference quotient of the first order.

In a similar manner, the summation of Eq. (3.6.3) and Eq. (3.6.4) is

$$r(u_E - u_0) + p(u_W - u_0) = \frac{h^2}{2}\frac{\partial^2 u}{\partial x^2}[rp(r + p)]$$
$$+ \frac{1}{3!}h^3 pr(p^2 - r^2)\left(\frac{\partial^3 u}{\partial x^3}\right)_0$$
$$+ \frac{1}{4!}h^4 pr(p^3 + r^3)\left(\frac{\partial^4 u}{\partial x^4}\right)_0 \cdot \quad (3.6.9)$$

If $p = r = 1$, the second term of RHS in Eq. (3.6.9) vanishes, then the error contained in $(\partial^2 u/\partial x^2)$ is proportional to h^2, i.e.

$$\frac{\partial^2 u}{\partial x^2} = \frac{u_E + u_W - 2u_0}{h^2} + O(h^2) . \tag{3.6.10}$$

If $p \neq r$,

$$\frac{\partial^2 u}{\partial x^2} = \frac{2[r(u_E - u_0) + p(u_W - u_0)]}{h^2[rp(r + p)]} + O(h) \tag{3.6.11}$$

where

$$O(h) = \frac{2}{3!} \frac{(p - r)}{h} \frac{\partial^3 u}{\partial x^3} . \tag{3.6.12}$$

The error contained in the second order derivative is proportional to h. Therefore, the errors introduced by using an asymmetric star is greater than those introduced by using symmetric one. Hence the merit of symmetric star is that not only can it be used more easily but more accurate results will be obtained.

Since the error contained in $\partial^2 u/\partial x^2$ is proportional to h^2, will a more accurate result be obtained if a smaller h is chosen? Equation (3.2.24) shows that the value of the potential at any point is influenced by the potentials only in its immediate neighbourhood. *As the number of grids becomes very large, the nodes become closer and closer, and the change of function between neighbouring points becomes very small, thus any further reduction of h becomes unimportant.* Reference [20] gives practical examples to explain this point of view.

The truncation error can be reduced by using the discretization formula of a nine-pointed star, as shown in Fig. 3.6.3. The nodes NE, NW, SW, SE are at the diagonal corner of the star O–E,N,W,S. By using a similar manipulation as before the following equation is obtained.

$$4(u_1 + u_3 + u_5 + u_7) + u_2 + u_4 + u_6 + u_8 - 20u_0 + 6hF_0 = 0 \tag{3.6.13}$$

3.6.3[†] Difference equation and the principle of weighted residuals [21]

It has been shown in Sect. 2.3.1 that the weighted residual approach accords to the following principle

$$\int_\Omega WRd\Omega = 0 \tag{3.6.14}$$

where W is the weighting function and R is residual of the approximation. In a point matching method, the Dirac-Delta function $\delta(x - x_0)\delta(y - y_0)$ is chosen as the weighting function. Let

$$u \cong \sum_{i=1}^{N} N_i u_i = N_E u_E + N_N u_N + N_W u_W + N_S u_S + N_0 u_0 \tag{3.6.15}$$

[†] It is advised to read this section after Chap. 4.

3.6 Further discussions about the finite difference method

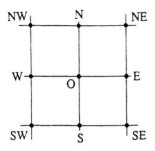

Fig. 3.6.3. A nine-pointed star of 2-D case

where N_E, N_N, N_W, N_S and N_0 are shape functions which satisfy the following relationships

$$N_E = \tfrac{1}{2}\xi(\xi + 1) \quad N_W = \tfrac{1}{2}\xi(\xi - 1)$$
$$N_N = \tfrac{1}{2}\eta(\eta + 1) \quad N_S = \tfrac{1}{2}\eta(\eta - 1) \quad (3.6.16)$$
$$N_0 = (1 - \xi)(1 + \xi) + (1 - \eta)(1 + \eta) .$$

In these equations, the local coordinates ξ, η are defined as

$$\xi = \frac{x}{h}, \quad \eta = \frac{y}{h} . \quad (3.6.17)$$

Substitution of Eq. (3.6.16) into Eq. (3.6.15) and consideration of

$$\frac{\partial^2 u}{\partial x^2} = \frac{1}{h^2}\frac{\partial^2 u}{\partial \xi^2} = \frac{1}{h^2}(u_E + u_W - 2u_0)$$

$$\frac{\partial^2 u}{\partial y^2} = \frac{1}{h^2}\frac{\partial^2 u}{\partial \eta^2} = \frac{1}{h^2}(u_N + u_S - 2u_0) \quad (3.6.18)$$

yields

$$u_0 = \frac{1}{4}(u_E + u_W + u_N + u_S) . \quad (3.6.19)$$

This is the difference equation of a five-pointed symmetric star.

3.6.4 Difference equation and the variational principle [11]

The finite difference equation can also be developed using the variational principle. For example, the corresponding functional of Laplace's equation is

$$I(u) = \sum \int_{\Omega_e} \left[\left(\frac{\partial u}{\partial x}\right)^2 + \left(\frac{\partial u}{\partial y}\right)^2 \right] dxdy \quad (3.6.20)$$

where Ω_e is a small region of the mesh shown in Fig. 3.6.4. It consists of areas 0123, 0345, 0567 and 0781. Assume that $h_E = h_W$ and $h_N = h_S$, on segments 0-1 and 0-5 $\partial u/\partial x$ are constants $(u_E - u_0)/h_E$ and $(u_W - u_0)/h_E$, respectively, and on

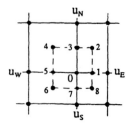

Fig. 3.6.4. A small region for deriving the difference equation

segments 0–3 and 0–7 $\partial u/\partial y$ are constants of $(u_N - u_0)/h_N$ and $(u_S - u_0)/h_N$, respectively. Thus the integrations in area 0123 are

$$\int_{0123} \left(\frac{\partial u}{\partial x}\right)^2 dxdy = \int_0^{h_N/2} dy \int_0^{h_E/2} \left(\frac{\partial u}{\partial x}\right)^2 dx = \frac{h_E}{2} \int_0^{h_N/2} \left(\frac{\partial u}{\partial x}\right)^2 dy$$

$$= \frac{h_E h_N}{4} \left(\frac{u_E - u_0}{h_E}\right)^2 \qquad (3.6.21)$$

$$\int_{0123} \left(\frac{\partial u}{\partial y}\right)^2 dxdy = \int_0^{h_E/2} dx \int_0^{h_N/2} \left(\frac{\partial u}{\partial y}\right)^2 dy = \frac{h_N}{2} \int_0^{h_E/2} \left(\frac{\partial u}{\partial y}\right)^2 dx$$

$$= \frac{h_E h_N}{4} \left(\frac{u_N - u_0}{h_N}\right)^2. \qquad (3.6.22)$$

Similarly, Eqs. (3.6.21) and (3.6.22) are to be found in areas 0345, 0567 and 0781. Substitution of these equations into Eq. (3.6.20) yields

$$I(u)|_{\Omega_r} = \frac{h_E h_N}{4} \left[\left(\frac{u_N - u_0}{h_N}\right)^2 + \left(\frac{u_E - u_0}{h_E}\right)^2 + \left(\frac{u_W - u_0}{h_E}\right)^2 \right.$$
$$+ \left(\frac{u_N - u_0}{h_N}\right)^2 + \left(\frac{u_S - u_0}{h_N}\right)^2 + \left(\frac{u_W - u_0}{h_E}\right)^2$$
$$\left. + \left(\frac{u_E - u_0}{h_E}\right)^2 + \left(\frac{u_S - u_0}{h_N}\right)^2 \right] \qquad (3.6.23)$$

By definition of the variational principle, the extremum of the functional is the solution of the original partial differential equation; i.e. let $\partial I(u)/\partial u_0 = 0$, and suppose $h_E = h_N$, the five-point difference equation as Eq. (3.2.24) is obtained.

3.7 Summary

In this chapter a number of finite difference equations are derived to satisfy various requirements (different field problems, coordinates and boundary conditions). In summary the FDM, presented in Sects. 3.1 to 3.4, is suitable for obtaining an approximate solution within a regular domain. Reference [11] indicates that the FDM is more efficient than the FEM by a factor of 2 in computer storage for calculating the propagation constants and fields of a dielectric wave guide. For solving 3-D problems within a cubic volume, the FDM is still considered to be an efficient method.

If a region containing different materials and complex shapes, then the programs are more complex. If the field contains rapid changes of the gradient, then the accuracy declines. In these cases, the finite element method is preferred.

For solving a time-dependent problem, the variation of time is approximated by a backward difference quotient. This quotient is iterated together with the iteration of the positions. The iterative methods are used to solve the difference equations. For 2-D problems, because of the large size of the grids, line iteration, block iteration or the alternative iteration techniques should be considered.

References

1. L.F. Richardson, The Approximate Arithmetical Solution by Finite Differences of Physical Problems, *Trans. R. Soc. (London)*, A210, 307–357, 1910
2. G.E. Forsythe, W.R. Wasow, *Finite Difference Methods for Partial Differential Equations*, New York, Wiley, 1960
3. G.D. Smith, *Numerical Solution of Partial Difference Equation*, London, Oxford University Press, 1965
4. K.J. Binns, P.J. Lawrenson, *Analysis and Computation of Electric and Magnetic Field Problems* (2nd edn), Pergamon Press, 1973
5. R.L. Stoll, Numerical Method of Calculating Eddy Current in Nonmagnetic Conductors, *Proc. IEE*, 114, 775–780, 1967
6. P.F. Ryff, P.P. Biringer, P.E. Burke, Calculation Methods for Current Distribution in Single Turn Coils of Arbitrary Cross Section, *IEEE Trans. on PAS*, 89(2), 228–232, 1967
7. M.S. Sarma, Potential Functions in Electromagnetic Field Problems, *IEEE Trans. on Mag.*, 6(3), 513–518, 1970
8. A.R. Mitchell, D.F. Griffiths, *The Finite Difference Method in Partial Differential Equations*, Wiley, 1980
9. J. Noye (Ed.), *Numerical Solution of Partial Difference Equations* (1981 Conference on Numerical Solution of Partial Differential Equations), North-Holland, 1982
10. W.J. Minkowycz, E.M. Sparrow *et al.*, *Handbook of Numerical Heat Transfer*, Wiley, 1988
11. E. Schweig, W.B. Bridges, Computer Analysis of Dielectric Waveguides, A Finite Difference Method, *IEEE Trans. on MTT*, 32(5), 531–541, 1984
12. Sheng Jian-ni (Ed.) *Numerical Analysis of Electromagnetic Fields* (in Chinese), Science Press, Beijing, 1984
13. M.S. Sarma, A. Yamamura, Nonlinear Analysis of Magnetic Bearings for Space Technology, *IEEE Trans. on AES*, 15(1), 134–140, 1979
14. M.S. Sarma, James C. Wilson, Accelerating the Magnetic Field Iterative Solution, *IEEE Trans. on Mag.* 12(6), 1042–1044, 1976
15. Steven C. Chapra, Raymond P. Canala, *Numerical Methods for Engineers* (2nd edn), Ch. 24, McGraw-Hill, 1988
16. William F. Ames, *Numerical Methods for Partial Differential Equations* (2nd edn), Academic Press, 1977
17. R.J. Arms, L.D. Gates, B. Zondek, A method of Block Iteration, *J. Soc. Indust. Alli. Math.*, 4, 220, 1956
18. Richard S. Verga, *Matrix Iteration Analysis*, Prentice-Hall, Inc. 1962

19. Wilhelm Muller, A New Iteration Technique for Solving Stationary Eddy Current Problem Using the Method of Finite Differences, *IEEE Trans. on Mag.*, **18**(2), 588–593, 1982
20. N.A. Demerdash, T.W. Nehl, An Evaluation of the Methods of Finite Element and Finite Differences in the Solution of Nonlinear Electromagnetic Fields in Electrical Machine, *IEEE Trans. on PAS*, **98**(1), 74–87, 1979
21. Zhou Ke-ding, *Special Topics on Engineering Electromagnetic Fields* (in Chinese), Hua-Zhong University Press, 1986

Chapter 4

Fundamentals of Finite Element Method (FEM)

4.1 Introduction

The main idea behind FDM is the approximation of the derivative operations $\partial u/\partial x$ and $\partial^2 u/\partial x^2$ by the difference quotients $\Delta u(x)/\Delta x$ and $\Delta^2 u(x)/\Delta x^2$, which reduces the partial differential equation to a set of algebraic equations. The application of FDM has two serious limitations. First, the regular steps of h_x, h_y, h_z which construct an array of grid nodes in the x, y, z directions are not suitable for a field with a rapidly changing gradient or for problems having a curved boundary. Second, different formulae must be derived for specific interfaces between different media and for the various shapes of boundaries.

The concept of finite element analysis can be dated back to the 1930s [1] and was originally developed to handle problems in structural mechanics. The technology of finite element analysis has advanced rapidly since the middle of the 1950s. The term 'finite element' was first used by Clough in 1960 [2]. The method has grown exponentially since then, expanding into almost all areas of scientific and engineering disciplines including solid and fluid mechanics, structural analysis, heat transfer, electromagnetics and even medical sciences. The earliest and the most comprehensive book introducing FEM was contributed by Zienkiewicz [3].

In the area of electromagnetic fields, the earliest application of FEM was used by Winslow to calculate the magnetic field in 1967 [4]. In 1970s, Silvester and Chari published the first paper [5] to expound FEM by use of the variational principle for electromagnetic field problems and in 1980 wrote the first book [6] to introduce the finite element method for the analysis of electromagnetic field problems. Now, many sophisticated software packages, such as MAGNET [7], ANSYS [8], TOSCA and Elektra [9], are available for analysis of electrical engineering problems of static and time varying fields. FEM has become a powerful and necessary tool for CAE (computer aided engineering). The method is now presented in text books for graduate and undergraduate students in many countries (see references [10–15]).

The basic idea of FEM is to divide the solution domain into a number of small interconnected subregions, called 'elements', as shown in Fig. 4.1.1. This shows that the discretization form of FEM is quite different from FDM (dashed line

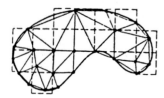

Fig. 4.1.1. Discretization form of FEM

represents the mesh used in FDM). The shape and the size of elements are arbitrary, so it is flexible to fit the different shape of the boundary. The density of the elements may easily be adjusted according to the problem. *The behaviour of the unknown function within the element is approximated by an interpolation function. By using the weighted residuals approach or the variational principle, the partial differential equation is reduced to a sparse, banded, symmetric and positive definite matrix equation. The values of the unknown function are represented by all of the nodal values of the elements which are solved from the matrix equation. The remarkable advantage of FEM is the flexibility of the method. It is particularly well suited to problems with complicated geometries and complicated distribution of media.* This method can be used for time-dependent, linear or non-linear and two-dimensional or three-dimensional problems.

The contents of FEM are discussed in the three chapters. In this chapter, in order to avoid variational concepts, the matrix equation of finite elements is derived using the weighted residual approach, and a general survey of the methods is illustrated by solving a static potential problem in the following sections. The variational finite element method and the problem of high order approximation are discussed in Chaps. 5 and 6.

4.2 General procedures of the finite element method

The method contains the following steps.

(1) Discretize the solution domain into sub-regions by elements. Within the elements, select an appropriate approximate function in terms of the unknown function included in the partial differential equation. If the element is a triangle, then the unknown function u within an element is approximated by the nodal values u_i, u_j, u_m, and the shape functions N_k^e, i.e.

$$u_e = \sum_k N_k^e u_k \quad (k = i, j, m) \tag{4.2.1}$$

where i, j, m are vertices of the triangle shown in Fig. 4.2.1 and u_e is the function within the element.

(2) Derive the element matrix equation using the principle of weighted residual or the variational principle. For a triangular element where is no charge

4.2 General procedures of the finite element method

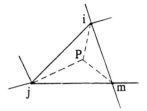

Fig. 4.2.1. A triangle element

density, the element matrix equation is presented as

$$\begin{bmatrix} k_{ii} & k_{ij} & k_{im} \\ k_{ji} & k_{jj} & k_{jm} \\ k_{mi} & k_{mj} & k_{mm} \end{bmatrix} \begin{Bmatrix} u_i \\ u_j \\ u_m \end{Bmatrix} = 0 \qquad (4.2.2)$$

(3) Assemble the element matrix equation at every node of the domain to form a system matrix equation

$$\mathbf{Ku} = \mathbf{B} \qquad (4.2.3)$$

where $\mathbf{u}$ is a column matrix with order N (N is the total number of nodes) containing all the nodal values of the domain. The global matrix $\mathbf{K}$ is a sparse, symmetric and positive definite matrix with order $N \times N$. It is also called an assembled matrix, system matrix or stiffness matrix. The name 'stiffness' comes from mechanics. The column matrix $\mathbf{B}$ includes the source term included in Poisson's equation and the known boundary conditions.

(4) Solve Eq. (4.2.3) to obtain the discretized function values on every node.

(5) Make additional computation if desired. For example, in potential field problems, the field strength, forces, parameters (resistance, capacitance, inductance) are normally of interest.

As an example of the method, a two-dimensional Poisson's equation subject to mixed boundary conditions as shown in Eq. (4.2.4) is considered.

$$\begin{cases} \nabla^2 u = -f(x, y) & \text{in } \Omega \\ u|_{\Gamma_1} = g_1(\Gamma) & \text{on } \Gamma_1 \\ \dfrac{\partial u}{\partial n}\bigg|_{\Gamma_2} = g_2(\Gamma) & \text{on } \Gamma_2 \end{cases} \qquad (4.2.4)$$

where Ω is the domain of the problem and Γ_1, Γ_2 are two parts of the boundary Γ ($\Gamma = \Gamma_1 + \Gamma_2$). The matrix equation will be derived in following sub-sections.

4.2.1 Domain discretization and shape functions

To discretize the domain into subregions (finite elements) is the first step in FEM. This is to replace the solution domain with infinite degrees of freedom by a system having a finite number of degrees of freedom. The shapes, sizes and the configurations of the elements may be of very different types and these will be discussed in

Chap. 6. Here a three-node triangular element is chosen as an example to illustrate the complete procedures of the method.

The second step is to choose a trial function to approximate the behaviour of the unknown function within the element. Usually the polynomial is adaptable, because it is easy to differentiate and integrate and it can approximate any function if the polynomial contains enough terms. For example, a linear polynomial is used,

$$u(x, y) = \tilde{u}(x, y) = \alpha_1 + \alpha_2 x + \alpha_3 y \tag{4.2.5}$$

where $\tilde{u}^{\dagger}$ represents the unknown function u within the element. It is assumed that the potential varies linearly depending on the coordinates x, y and the field strength is uniform in a small element. The unknown parameters α_1, α_2, α_3 in Eq. (4.2.5) will be determined by the nodal parameters u_k, x_k, y_k ($k = i, j, m$) of the element.

Substitution of the nodal values u_i, u_j, u_m and the nodal coordinates (x_i, y_i), (x_j, y_j), (x_m, y_m) into Eq. (4.2.5), yields

$$\begin{cases} u_i = \alpha_1 + \alpha_2 x_i + \alpha_3 y_i \\ u_j = \alpha_1 + \alpha_2 x_j + \alpha_3 y_j \\ u_m = \alpha_1 + \alpha_2 x_m + \alpha_3 y_m \end{cases} \tag{4.2.6}$$

Thus

$$\begin{Bmatrix} \alpha_1 \\ \alpha_2 \\ \alpha_3 \end{Bmatrix} = \begin{bmatrix} 1 & x_i & y_i \\ 1 & x_j & y_j \\ 1 & x_m & y_m \end{bmatrix}^{-1} \begin{Bmatrix} u_i \\ u_j \\ u_m \end{Bmatrix}. \tag{4.2.7}$$

The expansion of the above equation is

$$\begin{cases} \alpha_1 = \dfrac{1}{2S}[u_i(x_j y_m - x_m y_j) + u_j(x_m y_i - x_i y_m) + u_m(x_i y_j - x_j y_i)] \\ = \dfrac{1}{2S}[a_i u_i + a_j u_j + a_m u_m] \\ \alpha_2 = \dfrac{1}{2S}[u_i(y_j - y_m) + u_j(y_m - y_i) + u_m(y_i - y_j)] \\ = \dfrac{1}{2S}[b_i u_i + b_j u_j + b_m u_m] \\ \alpha_3 = \dfrac{1}{2S}[u_i(x_m - x_j) + u_j(x_i - x_m) + u_m(x_j - x_i)] \\ = \dfrac{1}{2S}[c_i u_i + c_j u_j + c_m u_m] \end{cases} \tag{4.2.8}$$

$\dagger$ In order to simplify the symbol, the ' ~ ' is omitted in the following text, u is used to express the approximate solution.

4.2 General procedures of the finite element method

where S is the area of the triangle having vertices i, j, m, i.e.

$$S = \frac{1}{2} \begin{vmatrix} 1 & x_i & y_i \\ 1 & x_j & y_j \\ 1 & x_m & y_m \end{vmatrix} = \frac{1}{2}(b_i c_j - b_j c_i) . \qquad (4.2.9)$$

In Eq. (4.2.8)

$$\begin{cases} a_i = \begin{vmatrix} x_j & y_j \\ x_m & y_m \end{vmatrix} = x_j y_m - x_m y_j \quad b_i = -\begin{vmatrix} 1 & y_j \\ 1 & y_m \end{vmatrix} = y_j - y_m \\ c_i = \begin{vmatrix} 1 & y_j \\ 1 & y_m \end{vmatrix} = x_m - x_j \\ a_j = x_m y_i - x_i y_m \quad b_j = y_m - y_i \quad c_j = x_i - x_m \\ a_m = x_i y_j - x_j y_i \quad b_m = y_i - y_j \quad c_m = x_j - x_i \end{cases} \qquad (4.2.10)$$

Substitution of Eq. (4.2.10) into Eq. (4.2.5) results in:

$$u(x, y) = \frac{1}{2S}[(a_i + b_i x + c_i y)u_i + (a_j + b_j x + c_j y)u_j$$
$$+ (a_m + b_m x + c_m y)u_m] . \qquad (4.2.11)$$

The above equation can be written to

$$u(x, y) = \sum_k N_k^e u_k \quad (k = i, j, m)$$

or

$$u(x, y) = \sum_k N_k^e u_k = [N_i^e \quad N_j^e \quad N_m^e] \begin{Bmatrix} u_i \\ u_j \\ u_m \end{Bmatrix} = [N_k^e]\{u_k^e\} \qquad (4.2.12)$$

where the functions N_k^e are called shape functions. The superscript e denotes the 'element', the subscript k denotes the vertices of the element. For a three-node triangle, $k = i, j, m$, there are three shape functions, N_i, N_j, N_m. The function u within the element is a linear combination of the shape functions and the three nodal values of the triangle. In addition, the linear interpolation polynomial (Eq. 4.2.5)) is now expressed by the shape functions as Eq. (4.2.12). These shape functions are expressed as:

$$\begin{cases} N_i(x, y) = \frac{1}{2S}(a_i + b_i x + c_i y) \\ N_j(x, y) = \frac{1}{2S}(a_j + b_j x + c_j y) \\ N_m(x, y) = \frac{1}{2S}(a_m + b_m x + c_m y) . \end{cases} \qquad (4.2.13)$$

Notice that

$$a_i + b_i x + c_i y = \begin{vmatrix} x_j & y_j \\ x_m & y_m \end{vmatrix} - x \begin{vmatrix} 1 & y_j \\ 1 & y_m \end{vmatrix} + y \begin{vmatrix} 1 & y_j \\ 1 & y_m \end{vmatrix}$$

$$= \begin{vmatrix} 1 & x & y \\ 1 & x_j & y_j \\ 1 & x_m & y_m \end{vmatrix} = 2\Delta_{Pjm}$$

where Δ_{Pjm} is the area of triangle Pjm, P is any point inside the triangle shown in Fig. 4.2.1, hence $N_i = \Delta_{pjm}/S$. In a similar manner, $N_j = \Delta_{Pim}/S$, $N_m = \Delta_{Pij}/S$, thus,

$$N_i(x, y) + N_j(x, y) + N_m(x, y) = 1 . \qquad (4.2.14)$$

Consequently the shape functions of the vertices of 3-node triangle are:

node i	node j	node m
$N_i(x_i, y_i) = 1,$	$N_i(x_j, y_j) = 0,$	$N_i(x_m, y_m) = 0$
$N_j(x_i, y_i) = 0,$	$N_j(x_j, y_j) = 1,$	$N_j(x_m, y_m) = 0$
$N_m(x_i, y_i) = 0,$	$N_m(x_j, y_j) = 0,$	$N_m(x_m, y_m) = 1.$

These results can be denoted by the Kronecker function, e.g.

$$N_k^e(x, y) = \delta_{ij} = \begin{cases} 1 & i = j \\ 0 & i \neq j \end{cases}. \qquad (4.2.15)$$

On the vertices of the triangle, the shape functions are 1 or zero, within the element, the shape function varies linearly.

4.2.2 Method using Galerkin residuals

When the unknown function of Poisson's equation (4.2.4) is substituted by an approximate function $u = \sum_k N_k^e u_k$ in each element, the residual R^e is unavoidable, i.e.

$$R^e = \nabla^2 u + f(x, y) \qquad (4.2.16)$$

If the residual R in the whole domain tends to zero, then u can be regarded as an acceptable approximate solution. As mentioned in Sect. 2.2, because of the broad choices of the principle of error minimization, there are many different methods to derive the finite element equation. *The most often used principle of error distribution in deriving the finite element equation is known as Galerkin's criterion. According to Galerkin's method, the weighting functions are chosen to be the same as the shape functions,* i.e.

$$\int_\Omega WR \, d\Omega = \sum_{e=1}^M \int_{\Omega_e} W_k R^e \, d\Omega = 0 \qquad (4.2.17)$$

4.2 General procedures of the finite element method

where

$$W_k = N_k. \quad (4.2.18)$$

M is the total number of elements of the problem domain. In the next subsection, the finite element equation is derived using this principle.

4.2.2.1 Element matrix equations

Combination of Eqs. (4.2.17) and (4.2.16), leads to

$$\int_\Omega W^e R^e d\Omega = \sum \int_{\Omega_e} [N_k^e]^T \left[\frac{\partial^2 u}{\partial x^2} + \frac{\partial^2 u}{\partial y^2} + f(x, y) \right] d\Omega = 0. \quad (4.2.19)$$

Because

$$\frac{\partial}{\partial x}\left([N_k^e]^T \frac{\partial u}{\partial x}\right) = [N_k^e]^T \frac{\partial^2 u}{\partial x^2} + \frac{\partial}{\partial x}[N_k^e]^T \frac{\partial u}{\partial x} \quad (4.2.20)$$

the two terms in the volume integral of the middle term of Eq. (4.2.19) can be written as

$$\begin{cases} \int_{\Omega_e} [N_k^e]^T \frac{\partial^2 u}{\partial x^2} dxdy = \int_{\Omega_e} \frac{\partial}{\partial x}\left([N_k^e]^T \frac{\partial u}{\partial x}\right) dxdy \\ \qquad\qquad - \int_{\Omega_e} \frac{\partial}{\partial x}[N_k^e]^T \frac{\partial u}{\partial x} dxdy \\ \int_{\Omega_e} [N_k^e]^T \frac{\partial^2 u}{\partial y^2} dxdy = \int_{\Omega_e} \frac{\partial}{\partial y}\left([N_k^e]^T \frac{\partial u}{\partial y}\right) dxdy \\ \qquad\qquad - \int_{\Omega_e} \frac{\partial}{\partial y}[N_k^e]^T \frac{\partial u}{\partial y} dxdy . \end{cases} \quad (4.2.21)$$

Substitution of Eq. (4.2.21) into Eq. (4.2.19) and consideration of parameter β (reluctivity or permittivity) of the material appearing in the LHS of the differential equation, Eq. (4.2.19) becomes

$$\sum \left\{ -\int_{\Omega_e} \beta \left[\frac{\partial}{\partial x}[N_k^e]^T \frac{\partial u}{\partial x} + \frac{\partial}{\partial y}[N_k^e]^T \frac{\partial u}{\partial y} \right] dxdy \right.$$

$$+ \int_{\Omega_e} \left[\frac{\partial}{\partial x}\left([N_k^e]^T \beta \frac{\partial u}{\partial x}\right) + \frac{\partial}{\partial y}\left([N_k^e]^T \beta \frac{\partial u}{\partial y}\right) \right] dxdy$$

$$\left. + \int_{\Omega_e} [N_k^e]^T f(x, y) dxdy \right\} = 0. \quad (4.2.22)$$

Remember the column vector $[N_k^e]$ consists of three terms; the above equation implies three equations at each node. By using 2-D Green's theory

$$\int_{\Omega_e} \left(\frac{\partial P}{\partial x} + \frac{\partial Q}{\partial y} \right) dxdy = \oint_{\Gamma} [P \cos(\mathbf{n}, \mathbf{i}) + Q \cos(\mathbf{n}, \mathbf{j})] d\Gamma \qquad (4.2.23)$$

and letting $\beta N_k \frac{\partial u}{\partial x} = P$, $\beta N_k \frac{\partial u}{\partial y} = Q$ $(k = i, j, m)$, the second integral of the LHS of Eq. (4.2.22) is transferred to a boundary integral, i.e.

$$\int_{\Omega_e} \left[\frac{\partial}{\partial x} \left(N_k \beta \frac{\partial u}{\partial x} \right) + \frac{\partial}{\partial y} \left(N_k \beta \frac{\partial u}{\partial y} \right) \right] dxdy$$

$$= \oint_{\Gamma} N_k \beta \left[\frac{\partial u}{\partial x} \cos(n, x) + \frac{\partial u}{\partial y} \cos(n, y) \right] d\Gamma = \oint_{\Gamma} N_k \beta \frac{\partial u}{\partial n} d\Gamma . \qquad (4.2.24)$$

Then Eq. (4.2.19) becomes

$$\int_{\Omega_e} W_k^e R_k^e d\Omega = \oint_{\Gamma} \beta \left[N_k \frac{\partial u}{\partial n} \right] d\Gamma - \int_{\Omega_e} \beta \left[\frac{\partial N_k}{\partial x} \frac{\partial u}{\partial x} + \frac{\partial N_k}{\partial y} \frac{\partial u}{\partial y} \right] dxdy$$

$$+ \int_{\Omega_e} N_k f(x, y) dxdy . \qquad (4.2.25)$$

Recalling Eqs. (4.2.13) and (4.2.11), the partial derivatives of N_k and u with respect to x, y are

$$\begin{cases} \dfrac{\partial N_k}{\partial x} = \dfrac{1}{2S} b_k \\ \dfrac{\partial N_k}{\partial y} = \dfrac{1}{2S} c_k \end{cases} (k = i, j, m) \qquad (4.2.26)$$

and

$$\begin{cases} \dfrac{\partial u}{\partial x} = \dfrac{1}{2S}(b_i u_i + b_j u_j + b_m u_m) = \dfrac{1}{2S}[b_i \ b_j \ b_m] \begin{Bmatrix} u_i \\ u_j \\ u_k \end{Bmatrix} \\ \\ \dfrac{\partial u}{\partial y} = \dfrac{1}{2S}(c_i u_i + c_j u_j + c_m u_m) = \dfrac{1}{2S}[c_i \ c_j \ c_m] \begin{Bmatrix} u_i \\ u_j \\ u_k \end{Bmatrix} . \end{cases} \qquad (4.2.27)$$

4.2 General procedures of the finite element method

Substitute Eqs. (4.2.26), (4.2.27) into Eq. (4.2.25), and assume that in each elements $f(x, y)$ is a constant, and consider that

$$\begin{cases} \int_\Delta dxdy = S \\ \int_\Delta x\, dxdy = \dfrac{S}{3}(x_i + x_j + x_m) = \bar{x}S \\ \int_\Delta y\, dxdy = \dfrac{S}{3}(y_i + y_j + y_m) = \bar{y}S \end{cases} \tag{4.2.28}$$

where $\bar{x} = \tfrac{1}{3}(x_i + x_j + x_m)$, $\bar{y} = \tfrac{1}{3}(y_i + y_j + y_m)$. Then at node i, the result is:

$$-(k_{ii}u_i + k_{ij}u_j + k_{im}u_m) + \frac{1}{2}(a_i + b_i\bar{x} + c_i\bar{y})f(x, y)$$

$$+ \oint_\Gamma \beta N_i \frac{\partial u}{\partial n} d\Gamma = 0 \tag{4.2.29}$$

where

$$\begin{cases} k_{ii} = \dfrac{\beta}{4S}(b_i^2 + c_i^2) \\ K_{ij} = \dfrac{\beta}{4S}(b_i b_j + c_i c_j) \\ k_{im} = \dfrac{\beta}{4S}(b_i b_m + c_i c_m) \end{cases} \tag{4.2.30}$$

Consider that

$$a_i + b_i\bar{x} + c_i\bar{y} = a_j + b_j\bar{x} + c_j\bar{y} = a_m + b_m\bar{x} + c_m\bar{y} = \tfrac{2}{3}S \tag{4.2.31}$$

and apply the same procedure to the nodes j and m, the final element equation of this triangle is:

$$\begin{cases} -(k_{ii}u_i + k_{ij}u_j + k_{im}u_m) + \dfrac{S}{3}f(x, y) + \oint_\Gamma \beta N_i \dfrac{\partial u}{\partial n} d\Gamma = 0 \\ -(k_{ji}u_i + k_{jj}u_j + k_{jm}u_m) + \dfrac{S}{3}f(x, y) + \oint_\Gamma \beta N_j \dfrac{\partial u}{\partial n} d\Gamma = 0 \\ -(k_{mi}u_i + k_{mj}u_j + k_{mm}u_m) + \dfrac{S}{3}f(x, y) + \oint_\Gamma \beta N_m \dfrac{\partial u}{\partial n} d\Gamma = 0 \end{cases} \tag{4.2.32}$$

In these equations, the term $(S/3)f(x, y)$ is the effect of the forcing function presented in the RHS of Eq. (4.2.4). The boundary integral terms in Eq. (4.2.32)

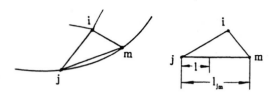

Fig. 4.2.2. An element along the boundary

represent the non-homogeneous boundary conditions of the second kind. If one edge of the element (for instance, the edge jm) is located on the boundary, as shown in Fig. 4.2.2, the shape functions along the edge jm are:

$$N_i = 0, \qquad N_j = (1 - l/l_{jm}), \qquad N_m = l/l_{jm} . \qquad (4.2.33)$$

Then

$$\int_{l_{jm}} N_j \frac{\partial u}{\partial n} dl = \int_0^{l_{jm}} \left(1 - \frac{l}{l_{jm}}\right) g_2(s) \, dl = \frac{1}{2} g_2(s) l_{jm} \qquad (4.2.34)$$

and

$$\int_{l_{jm}} N_m \frac{\partial u}{\partial n} dl = \int_0^{l_{jm}} \frac{l}{l_{jm}} g_2(s) \, dl = \frac{1}{2} g_2(s) l_{jm} \qquad (4.2.35)$$

where $g_2(s) = \left.\frac{\partial u}{\partial n}\right|_s$ is the given boundary condition of the second kind. Then Eq. (4.2.32) can be written as

$$-\begin{bmatrix} k_{ii} & k_{ij} & k_{im} \\ k_{ji} & k_{jj} & k_{jm} \\ k_{im} & k_{jm} & k_{mm} \end{bmatrix} \times \begin{Bmatrix} u_i \\ u_j \\ u_m \end{Bmatrix} + \begin{Bmatrix} P_i \\ P_j \\ P_m \end{Bmatrix} = \begin{Bmatrix} 0 \end{Bmatrix} \qquad (4.2.36)$$

where

$$P_i = \frac{S}{3} f(x, y) \qquad P_j = P_m = \frac{1}{3} f(x, y) S + \frac{1}{2} \beta l_{jm} g_2(s) . \qquad (4.2.37)$$

For Laplace's equation ($f(x, y) = 0$) and homogeneous boundary conditions of the second kind ($g_2(s) = 0$), $P_i = P_j = P_m = 0$, Eq. (4.2.36) is simplified to:

$$\begin{bmatrix} k_{ii} & k_{ij} & k_{im} \\ k_{ji} & k_{jj} & k_{jm} \\ k_{mi} & k_{mj} & k_{mm} \end{bmatrix} \times \begin{Bmatrix} u_i \\ u_j \\ u_m \end{Bmatrix} =$$

$$\frac{\beta}{4S} \begin{bmatrix} b_i^2 + c_i^2 & b_i b_j + c_i c_j & b_i b_m + c_i c_m \\ b_j b_i + c_j c_i & b_j^2 + c_j^2 & b_j b_m + c_j c_m \\ b_m b_i + c_m c_i & b_m b_j + c_m c_j & b_m^2 + c_m^2 \end{bmatrix} \times \begin{Bmatrix} u_i \\ u_j \\ u_m \end{Bmatrix} = \begin{Bmatrix} 0 \end{Bmatrix} \qquad (4.2.38)$$

or

$$\mathbf{K}_e \mathbf{u}^e = 0 \qquad (4.2.39)$$

4.2 General procedures of the finite element method

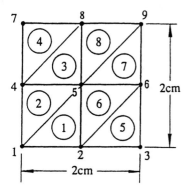

Fig. 4.2.3. Discretization of a 2-D domain

where

$$\mathbf{K}_e = \frac{\beta}{4S} \begin{bmatrix} b_i^2 + c_i^2 & b_i b_j + c_i c_j & b_i b_m + c_i c_m \\ b_j b_i + c_j c_i & b_j^2 + c_j^2 & b_j b_m + c_j c_m \\ b_m b_i + c_m c_i & b_m b_j + c_m c_j & b_m^2 + c_m^2 \end{bmatrix}. \quad (4.2.40)$$

Equation (4.2.40) is the expression of the element matrix, it shows that $\mathbf{K}_e$ is a symmetric matrix of the order of 3. The diagonal elements are always positive. The value of each element of $\mathbf{K}_e$ is dependent on the coordinates of the vertices of the element and the material parameters β (for the electric field, $\beta = \varepsilon$, for the magnetic field, $\beta = 1/\mu$). The formulation for evaluating the coefficients of the element matrix can be synthesized as

$$k_{rs} = k_{sr} = \frac{\beta}{4S}(b_r b_s + c_r c_s) \quad r, s = i, j, m \quad (4.2.41)$$

Example 4.2.1. The domain of a 2-D Laplacian problem is shown in Fig. 4.2.3. Subdivide the domain into eight triangular elements. The coordinates of the nodes are obtained with the dimension shown in Fig. 4.2.3. Using Eq. (4.2.40), the coefficients of the element matrix equation for elements 1 and 2 are:

$$\beta \begin{bmatrix} 0.5 & -0.5 & 0.0 \\ -0.5 & 1.0 & -0.5 \\ 0.0 & -0.5 & 0.5 \end{bmatrix} \times \begin{Bmatrix} u_1 \\ u_2 \\ u_5 \end{Bmatrix} = 0 \quad (4.2.42)$$

and

$$\beta \begin{bmatrix} 0.5 & 0.0 & -0.5 \\ 0.0 & 0.5 & -0.5 \\ -0.5 & -0.5 & 1.0 \end{bmatrix} \times \begin{Bmatrix} u_1 \\ u_5 \\ u_4 \end{Bmatrix} = 0 \quad (4.2.43)$$

During evaluation of the element matrix, the sequence of i, j, m in each element should be taken counterclockwise so that the area of each triangular element is positive.

4.2.2.2 System matrix equation

The system matrix is the combination of the element matrices. The principle for assembling the element matrices into a system matrix is based on the property of compatibility. It means that at every node the nodal value of the unknown function of each element is the same no matter the node belongs to which element. This requirement has a simple physical meaning. In scalar potential field, for example, the potential value is unique at any point. Hence the nodal value is related with all the elements connected at one node. In other words, the nodal value of each node is the assembly of values contributed by all the connected elements. The procedures for assembling the system matrix of the problem shown in Fig. 4.2.3 are as follows.

Assume the element matrix is of the order $n \times n$ (for a three-node triangle element, $n = 3$) and the global matrix is of the order $N \times N$ (N is the number of the total nodes of the domain). First expand the element matrices to the global matrix of the order $N \times N$ by adding zeros in the remaining locations. Then add the coefficients in the corresponding positions of the global matrix according to the address denoted by subscripts i, j. For instance, first the elements of system matrix of the order $N \times N$ are set to zero. Second, put the coefficients of the matrix of element 1 at suitable positions according to the subscripts; the result is

$$\begin{bmatrix} 0.5 & -0.5 & 0 & 0 & 0.0 & 0 & 0 & 0 & 0 \\ -0.5 & 1.0 & 0 & 0 & -0.5 & 0 & 0 & 0 & 0 \\ 0 & 0 & 0 & 0 & 0 & 0 & 0 & 0 & 0 \\ 0 & 0 & 0 & 0 & 0 & 0 & 0 & 0 & 0 \\ 0.0 & -0.5 & 0 & 0 & 0.5 & 0 & 0 & 0 & 0 \\ 0 & 0 & 0 & 0 & 0 & 0 & 0 & 0 & 0 \\ 0 & 0 & 0 & 0 & 0 & 0 & 0 & 0 & 0 \\ 0 & 0 & 0 & 0 & 0 & 0 & 0 & 0 & 0 \\ 0 & 0 & 0 & 0 & 0 & 0 & 0 & 0 & 0 \end{bmatrix}$$

In this matrix, except for the coefficients of element 1, all the elements are zeros. Third, the matrix of element 2 is added to the above matrix according to the corresponding address of the coefficients. After the elements matrix of 1 and 2 is added, the system matrix is

$$\begin{bmatrix} 1.0 & -0.5 & 0 & -0.5 & 0.0 & 0 & 0 & 0 & 0 \\ -0.5 & 1.0 & 0 & 0 & -0.5 & 0 & 0 & 0 & 0 \\ 0 & 0 & 0 & 0 & 0 & 0 & 0 & 0 & 0 \\ -0.5 & 0 & 0 & 1.0 & -0.5 & 0 & 0 & 0 & 0 \\ 0.0 & -0.5 & 0 & -0.5 & 1.0 & 0 & 0 & 0 & 0 \\ 0 & 0 & 0 & 0 & 0 & 0 & 0 & 0 & 0 \\ 0 & 0 & 0 & 0 & 0 & 0 & 0 & 0 & 0 \\ 0 & 0 & 0 & 0 & 0 & 0 & 0 & 0 & 0 \\ 0 & 0 & 0 & 0 & 0 & 0 & 0 & 0 & 0 \end{bmatrix}$$

4.2 General procedures of the finite element method

It is obvious that the matrix is still symmetric and the diagonal elements are dominant. The generic formulation of the assembly is

$$\mathbf{K} = \sum_{e=1}^{M} \mathbf{K}_e \tag{4.2.44}$$

where M is the total number of elements. For Example 4.2.1, the global matrix is

$$\begin{bmatrix}
k_{11} & k_{12} & & k_{14} & k_{15} & & & & \\
k_{21} & k_{22} & k_{23} & & k_{25} & k_{26} & & & \\
 & k_{32} & k_{33} & & & k_{36} & & & \\
k_{41} & & & k_{44} & k_{45} & & k_{47} & k_{48} & \\
k_{51} & k_{52} & & k_{54} & k_{55} & k_{56} & & k_{58} & k_{59} \\
 & k_{62} & k_{63} & & k_{65} & k_{66} & & & k_{69} \\
 & & & k_{74} & & & k_{77} & k_{78} & \\
 & & & k_{84} & k_{85} & & k_{87} & k_{88} & k_{89} \\
 & & & & k_{95} & k_{96} & & k_{98} & k_{99}
\end{bmatrix}. \tag{4.2.45}$$

Each element of the matrix is assembled by the following rules:

$$\begin{cases} k_{ii} = \sum_L k_{ii}^e & L, \text{ is the total number of triangles connected to node } i \\ k_{ij} = \sum_P k_{ij}^e & P \text{ is the number of elements which contain edge } ij \end{cases} \tag{4.2.46}$$

For instance, in Fig. 4.2.3 six triangles (the elements 1, 2, 3, 6, 7, 8) are joined at node 5. This means that these six elements have contributions to node 5. Hence the element k_{55} is added by five terms. On the other hand, two elements, 2 and 3, contain edge $\overline{45}$, hence k_{45} consists of two terms. These are given in Eq. (4.2.47) as follows

$$\begin{cases} k_{55} = k_{55}^{(1)} + k_{55}^{(2)} + k_{55}^{(3)} + k_{55}^{(6)} + k_{55}^{(7)} + k_{55}^{(8)} \\ k_{45} = k_{45}^{(2)} + k_{45}^{(3)} \end{cases} \tag{4.2.47}$$

where the superscripts $^{(1)}, ^{(2)}, \ldots$ denote the numbers of triangles. The subscripts denote the number of nodes. This kind of assembly is based on the sequence of elements. It is executed in computer programs. The program is a *recurrent* procedure according to the sequence of element ($e = 1, M$). For each element calculate S, a_i, b_i, c_i, K_{ss}, K_{rs} and sum them as in Eq. (4.2.47). The calculation of the element matrix and the stiffness matrix is therefore carried out at once. If there are different media in the solution domain, the numbering of the elements should be sequential according to the different materials one by one.

Each node connects with a very limited number of nodes of the whole domain. *The system matrix is not only a symmetric but also a sparse and banded matrix. It contains a great number of zero elements and the diagonal elements of* the matrix are positive ($k_{ii} > 0$) and dominant, and all the values of the sequent subdeterminant are positive. *Hence the system matrix is a positive definite matrix.*

In contrast to FDM, the maximum non-zero element of each row is indefinite. Usually the non-zero elements of each row are less than 9. However, the range occupied by the non-zero elements may be a large number as it depends upon the sequence in which the nodes are numbered. For example, in Fig. 4.2.4(b), the largest range of non-zero elements occupied has nine columns. If the subdivision is increased in the direction of the wide side of the domain, then the range of the non-zero elements occupied will also be increased. The range of non-zero elements occupied is described using the term 'bandwidth' (*BW*). *BW* is defined as:

$$BW = D + 1 \qquad (4.2.48)$$

where D is the maximum difference between the numbers of the two vertices of a certain element. For instance, in Fig. 4.2.4(a), the maximum difference of the nodal number is $D = 7 - 1 = 8 - 2 = 13 - 7 = \ldots = 6$, then the bandwidth is $BW = 6 + 1 = 7$. In Fig. 4.2.4(b), $BW = 9$. Consequently, the format of the numbering of nodes strongly influences the bandwidth and the computer

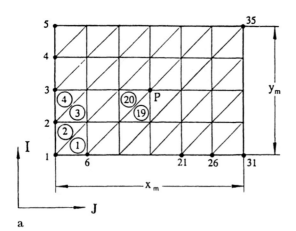

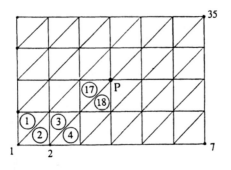

Fig. 4.2.4a, b. Different sequence of the nodal number

memory directly. Hence, in practice, a technique used to optimize the ordering of nodes in order to minimize the bandwidth is necessary [15].

Incidentally, the assembly of the system matrix can also be executed according to the sequence of nodes [13].

4.2.2.3 Storage of the system matrix

According to the different methods used to solve the matrix equation, usually two strategies are employed to store the stiffness matrix: skyline (envelope) storage and non-zero elements storage.

When the stiffness matrix is symmetric, banded and non-zero terms are clustered around the main diagonal. Only the terms within the bandwidth of the upper or lower triangle matrix need to be stored. *In the envelope method, the terms between the first non-zero element and the diagonal element of each row are stored.* This storage scheme results in a 'ragged' edge profile, called the envelope or skyline of the matrix, as shown in Fig. 4.2.5. This is the corresponding system matrix of the problem shown in Fig. 4.2.2. In this model, certain zero terms are still contained in the array $AK(IS)$ which occupy space in the computer memory. This storage is needed when using the Gaussian elimination method to solve matrix equations. The reason will be understood in next section.

In the envelope method, IS is the total dimension of the array AK which stores the coefficients of matrix $\mathbf{K}$. *IS equals the summation of the numbers between the first non-zero term to the diagonal element of every row.* For instance, in Fig. 4.2.5, $IS = 1 + 2 + 2 + \ldots + 5 = 33$. As the dimension of the array AK is counted, the address of the diagonal elements are memorized by an additional one-dimensional array $ND(LO)$. In Fig. 4.2.5, $ND(9) = 1, 3, 5, 9, 14, 19, 23, 28, 33$. The dimension of the array ND is the total number of nodes. Using this method, a considerable amount of computer memory is saved. In Example 4.3.1, the total amount of nodes is 870 and the elements of the matrix is $870 \times 870 = 756900$. When the envelope method is used, the order of array AK is 25287; it is only 3.34% of the whole matrix. In Example 4.5.2, the total number of nodes is 1066, but $IS = 28137$, so it is 2.48% of the whole size of the matrix.

However, when solving the matrix equation, the positions of k_{ii}, k_{ij} should be recovered by using the array ND first. For example, the diagonal element in the fourth row (k_{44}) is stored in the ninth position of the array AK. In general, the position of k_{ii} in AK is denoted by p, then $p = ND(i)$. The position of k_{ij} in array AK is also to be found after the position of diagonal elements have been determined. Assume the lower triangular elements are stored, the position of a non-diagonal element k_{ij} (if $i > j$) in AK is denoted by q; first to find the location p of k_{ii}, then $q = p - (i - j)$. If $j > i$, it indicates k_{ij} is located in the upper triangle. According to $k_{ij} = k_{ji}$ and the rule of storage, only k_{ij} is stored. Hence the location of k_{jj} is determined first, i.e. $p = ND(j)$, then the address of k_{jj} is $q = p - (j - i)$. These results can be proved as shown in Fig. 4.2.5.

Therefore in any program used to solve problems, the coefficients of the matrix must be recovered first.

If the iterative method is used to solve the matrix equation, all the zero elements have no relation to the calculation of iteration, hence no zero elements need to be stored. In the computation and storage of the system matrix, the first non-zero element of each row is removed to the left side of the matrix. For instance, the matrix shown in Eq. (4.2.45) is altered to

$$\mathbf{K}' = \begin{bmatrix} k_{11} & k_{12} & k_{14} & k_{15} & 0 & 0 & 0 \\ k_{21} & k_{22} & k_{23} & k_{25} & k_{26} & 0 & 0 \\ k_{32} & k_{33} & k_{36} & 0 & 0 & 0 & 0 \\ k_{41} & k_{44} & k_{45} & k_{47} & k_{48} & 0 & 0 \\ k_{51} & k_{52} & k_{54} & k_{55} & k_{56} & k_{58} & k_{59} \\ k_{62} & k_{63} & k_{65} & k_{66} & k_{69} & 0 & 0 \\ k_{74} & k_{77} & k_{78} & k_{74} & 0 & 0 & 0 \\ k_{84} & k_{85} & k_{87} & k_{88} & k_{89} & 0 & 0 \\ k_{95} & k_{96} & k_{98} & k_{99} & 0 & 0 & 0 \end{bmatrix}. \qquad (4.2.49)$$

In the alternative matrix $\mathbf{K}'$, the maximum number of columns depends on the maximum number of the neighbouring nodes connected to a certain node. For instance, in Fig. 4.2.3, node 5 connects with other six nodes, hence the maximum column of $\mathbf{K}'$ is 7. Because the zero elements are not considered, the positions of the elements in the new matrix $\mathbf{K}'$ are confused. Therefore an additional 2-D array is used to record the number of columns which the elements located in the original matrix $\mathbf{K}$. This matrix is shown as below.

$$\mathbf{AD} = \begin{bmatrix} 1 & 2 & 4 & 5 & 0 & 0 & 0 \\ 1 & 2 & 3 & 5 & 6 & 0 & 0 \\ 2 & 3 & 6 & 0 & 0 & 0 & 0 \\ 1 & 4 & 5 & 7 & 8 & 0 & 0 \\ 1 & 2 & 4 & 5 & 6 & 8 & 9 \\ 2 & 3 & 5 & 6 & 9 & 0 & 0 \\ 4 & 7 & 8 & 0 & 0 & 0 & 0 \\ 4 & 5 & 7 & 8 & 9 & 0 & 0 \\ 5 & 6 & 8 & 9 & 0 & 0 & 0 \end{bmatrix} \qquad (4.2.50)$$

The evaluation of these two matrices is carried out as follows. If k_{ij} is non-zero, then it is put in column p, i.e. $k'_{ip} = k_{ij}$ and a_{ip} (element of $\mathbf{AD} = j$. In this method, although an additional matrix is required, the column of matrix $\mathbf{AD}$ (usually less than 9) is much smaller than the column of the original matrix, especially for problems having a large number of nodes.

4.2 General procedures of the finite element method

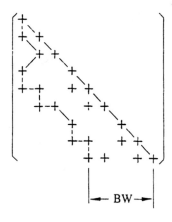

Fig. 4.2.5. Skyline storage of the system matrix

Consequently, the real non-zero elements storage is much more economical than skyline storage for saving the computer memory. But the disadvantage is that more complex programming is required to keep track of the terms stored and more computational time is needed when using the iterative method. These two methods to store the matrix coefficients are chosen depending upon the size of the problem to be solved.

4.2.2.4 Treatment of the Dirichlet boundary condition

For the problem subject to the Dirichlet boundary condition, the nodal value of those nodes located on the boundary are known. The known value must be removed to the RHS of the matrix equation. The system matrix and the column vector of the RHS are modified by the following way.

Assume that the node m is located on boundary Γ_1 and its potential value is U_0. It means that the equation of m row in the simultaneous equations is

$$u_m = U_0$$

Hence let $b_m = U_0$, b_m be the element of column matrix in the RHS of the matrix equation. For the LHS of the matrix equation, the diagonal element of the mth row is replaced by 1 and other elements in the mth row and mth column are filled by zero. The other terms in the RHS of the matrix equation become

$$b'_i = b_i - k_{im} U_0 \qquad (4.2.51)$$

Then the matrix equation

$$\mathbf{Ku} = \mathbf{B}$$

is changed to

$$\begin{bmatrix} k_{11} & k_{12} & \cdots & 0 & \cdots & k_{1N} \\ k_{21} & k_{22} & \cdots & 0 & \cdots & k_{2N} \\ \vdots & \vdots & & \vdots & & \vdots \\ 0 & 0 & \cdots & 1 & \cdots & 0 \\ \vdots & \vdots & & \vdots & & \vdots \\ k_{N1} & k_{N2} & \cdots & 0 & \cdots & k_{NN} \end{bmatrix} \times \begin{Bmatrix} u_1 \\ \vdots \\ u_m \\ \vdots \\ u_N \end{Bmatrix} = \begin{Bmatrix} b_1 - k_{1m}U_0 \\ b_2 - k_{2m}U_0 \\ \vdots \\ U_0 \\ \vdots \\ b_N - k_{Nm}U_0 \end{Bmatrix}. \quad (4.2.52)$$

If the number of nodes subject to known boundary values of the first kind is denoted by L01, then repeat the above procedure L01 times to yield the equation

$$\tilde{\mathbf{K}}\mathbf{u} = \tilde{\mathbf{B}} \qquad (4.2.53)$$

In this approach, the dimension of the column matrix **u** is not reduced by introducing the known potential values of the boundary and the size of matrix **K** is unchanged. The advantage of this method is that the construction of the system matrix is unchanged and the computer program is independent of the specific geometry. If there are homogeneous boundary conditions of the first kind, in order to deduct the equations $u_{m+i} = 0$, the best way is to number these nodes at the bottom of the nodal sequence, then Eq. (4.2.52) becomes

$$\begin{bmatrix} k_{11} & \cdots & k_{1m} & 0 & \cdots & 0 \\ \vdots & & \vdots & & & \vdots \\ k_{m1} & \cdots & k_{mm} & 0 & & 0 \\ 0 & & 0 & 1 & \cdots & 0 \\ \vdots & & \vdots & & & \vdots \\ 0 & \cdots & 0 & 0 & & 1 \end{bmatrix} \times \begin{Bmatrix} u_1 \\ \vdots \\ u_m \\ \vdots \\ u_N \end{Bmatrix} = \begin{Bmatrix} b_1 \\ \vdots \\ b_m \\ 0 \\ 0 \\ 0 \end{Bmatrix} \qquad (4.2.54)$$

Then the dimension of the matrix equation to be solved is reduced by the order of $N - (m + 1)$.

Similarly, if all the nodes are numbered sequentially according to the interior nodes and boundary nodes, for example, the number of interior nodes are $1, \ldots, N$, and the boundary nodes having known potentials are $N + 1, \ldots, N_0$, then the finite element equation for this Laplacian problem can be altered to

$$\sum_{j=1}^{N} k_{ij} u_j = -\sum_{j=N+1}^{N_0} k_{ij} u_j \qquad (4.2.55)$$

Thus the order of simultaneous equations is according to the number of interior nodes.

4.3 Solution methods of finite element equations

The matrix equation derived by FEM is symmetric and positive definite. All the main sub-determinants have positive values. Therefore the solution of the finite element equation exists and is unique. The matrix equation to be solved is

$$KX = B \qquad (4.3.1)$$

The solution algorithms of linear matrix equations can generally be classified as direct methods and iterative schemes. The over-relaxation iteration and the conjugate gradient method are two well known schemes using iterative methods. The results obtained from the iterative method are the limiting value of a sequence of approximations. The direct methods include Gaussian elimination and Cholesky's decomposition. The direct methods are considered as an exact solution from the theoretical point of view. Due to the rounding off errors produced by computation, the results are still approximate. A brief introduction of the various methods is given in this section. More detailed information is found in references [13–17].

4.3.1 Direct methods

4.3.1.1 Gaussian elimination method

The Gaussian elimination method is a common method used to solve symmetric and unsymmetric matrix equations. It consists of two steps: forward elimination and backward substitution. Equation (4.3.2) is used as an example to describe the process.

$$\begin{bmatrix} 4 & 3 & 2 \\ 3 & 5 & 1 \\ 2 & 1 & 8 \end{bmatrix} \times \begin{Bmatrix} x_1 \\ x_2 \\ x_3 \end{Bmatrix} = \begin{Bmatrix} 12 \\ 14 \\ 12 \end{Bmatrix} \qquad (4.3.2)$$

The first step is to transfer the matrix into an upper triangular matrix; it is called forward elimination. Divide the elements of the first row by the first element (including the column matrix of the right-hand side) so that the first element of the diagonal is 1. Then subtract a multiple of the first equation from the second and the third to obtain zeroes below the diagonal in the first column. This procedure is diagrammed as:

$$\begin{matrix} \text{row } a1 \\ \text{row } a2 \\ \text{row } a3 \end{matrix} \begin{bmatrix} 4 & 3 & 2 & 12 \\ 3 & 5 & 1 & 14 \\ 2 & 1 & 8 & 12 \end{bmatrix}$$

$$\begin{matrix} a1/4 & \to \text{row } b1 \\ a2 - 3 \times (b1) & \to \text{row } b2 \\ a3 - 2 \times (b1) & \to \text{row } b3 \end{matrix} \begin{bmatrix} 1 & 3/4 & 1/2 & 3 \\ 0 & 11/4 & -1/2 & 5 \\ 0 & -1/2 & 7 & 6 \end{bmatrix}.$$

The above calculation can be formulated by

$$\begin{cases} k^1_{1j} = k_{1j}/k_{11} \\ b^1_1 = b_1/k_{11} \end{cases} \quad (j = 1, 2, \ldots, N) \tag{4.3.3}$$

$$\begin{cases} k_{ij} = k_{ij} - k^1_{1j} \times k_{i1} & (i, j = 2, 3, \ldots, N) \\ b^1_i = b_i - k_{i1}b^1_1 & (i = 2, 3, \ldots, N) \end{cases} \tag{4.3.4}$$

N is the order of the matrix, b_i are the elements of the column matrix of the RHS of Eq. (4.3.1). After this step, all the elements in the first column are zeroes except the first element.

Repeat the same procedure, until an upper triangular matrix and a new column matrix are obtained.

$$\begin{matrix} \text{row } b1 & \to \text{row } c1 \\ \text{row } b2/\tfrac{4}{11} & \to \text{row } c2 \\ \text{row } b3 + c2/2 & \to \text{row } c3 \end{matrix} \begin{bmatrix} 1 & 3/4 & 1/2 \\ 0 & 1 & -2/11 \\ 0 & 0 & 76/11 \end{bmatrix}, \begin{Bmatrix} 3 \\ 20/11 \\ 76/11 \end{Bmatrix}.$$

Then the matrix equation of (4.3.2) becomes

$$\begin{bmatrix} 1 & 3/4 & 1/2 \\ 0 & 1 & -2/11 \\ 0 & 0 & 76/11 \end{bmatrix} \times \begin{Bmatrix} x_1 \\ x_2 \\ x_3 \end{Bmatrix} = \begin{Bmatrix} 3 \\ 20/11 \\ 76/11 \end{Bmatrix}. \tag{4.3.5}$$

It is obvious from Eq. (4.3.5) that x_3, x_2, x_1 are obtained by the back-substitution from the last equation to the first. The results of Eq. (4.3.5) are:

$x_3 = 1$

$x_2 = \tfrac{20}{11} + \tfrac{2}{11}x_3 = 2$

$x_1 = 3 - \tfrac{3}{4}x_2 - \tfrac{1}{2}x_3 = 1$.

The general formulation of the back-substitution is:

$$x_i = b^{(i)}_i - \sum k^{(i)}_{ij} x_j \quad (i = 1, \ldots, N). \tag{4.3.6}$$

Generally, in order to overcome the problem where the pivot element is zero or a very small number, a rearrangement of the elements is necessary. At the kth stage in the elimination, the rows are rearranged to ensure that the coefficient with the largest absolute value in the kth column in the lower triangle is on the leading diagonal. Thus a search is made in the kth column of the coefficient matrix, beginning at row k and ending with the last row N.

However, in solving the matrix equation of FEM, the diagonal elements are pivotal and it is not necessary to search for a pivot in the elimination. Because of the symmetry of the matrix, only the lower or upper triangular elements need to be calculated.

During the process of elimination, the coefficients of the matrix are changed in each step. Some zero elements may be evaluated as non-zeroes. Only the zeroes before the first and after the last non-zero element of each row and column are always zeroes. Hence only the zeroes before the first and after the last non-zero elements need not be stored. Therefore the skyline storage of non-zero elements fits the Gaussian elimination method.

4.3.1.2 Cholesky's decomposition (triangular decomposition)

The Gaussian elimination method transforms the original matrix into a triangular matrix. It has been proved that if the matrix **K** is a symmetric and positive definite matrix, then **K** can be uniquely decomposed into two triangular matrices, e.g.

$$\mathbf{K} = \mathbf{U}\mathbf{U}^T \tag{4.3.7}$$

U is a lower triangular matrix, the transposition of **U** is an upper triangular matrix.

$$\mathbf{U} = \begin{bmatrix} u_{11} & & & \\ u_{21} & u_{22} & & \\ \vdots & & & \\ u_{n1} & & & u_{nn} \end{bmatrix}, \quad \mathbf{U}^T = \begin{bmatrix} u_{11} & u_{21} & \cdots & u_{n1} \\ & u_{22} & & \\ & & \ddots & \\ & & & u_{nn} \end{bmatrix}. \tag{4.3.8}$$

The elements of **U** are obtained by the following equations:

$$\sum_{m=1}^{n} u_{im} u_{jm} = k_{ij} \qquad (j = 1, 2, \ldots, i;\ i = 1, 2, \ldots, n) \tag{4.3.9}$$

$$u_{ij} = \left(k_{ij} - \sum_{m=1}^{j-1} u_{im} u_{jm}\right)\bigg/ u_{jj} \qquad (i = j+1, \ldots, n) \tag{4.3.10}$$

$$u_{jj} = \left(k_{jj} - \sum_{m=1}^{j-1} u_{jm}^2\right)^{\frac{1}{2}} \qquad (j = 1, 2, \ldots, n) \tag{4.3.11}$$

Let $\mathbf{U}^T\mathbf{X} = \mathbf{g}$. Once the equation $\mathbf{Ug} = \mathbf{B}$ is solved, the solution is obtained by solving the equation:

$$\mathbf{U}^T\mathbf{X} = \mathbf{g}. \tag{4.3.12}$$

In order to avoid the computation of a root square, Eq. (4.3.7) is decomposed by

$$\mathbf{K} = \mathbf{L}\mathbf{D}\mathbf{L}^T. \tag{4.3.13}$$

4 Fundamentals of finite element method (FEM)

Since **D** is a diagonal matrix, then **LD** is

$$\mathbf{LD} = \begin{bmatrix} 1 & & & \\ l_{21} & 1 & & \\ \vdots & l_{32} & 1 & \\ l_{n1} & & & 1 \end{bmatrix} \begin{bmatrix} d_1 & & & \\ & d_2 & & \\ & & \ddots & \\ & & & d_n \end{bmatrix}. \quad (4.3.14)$$

The elements of **D** and **L** are

$$\begin{cases} d_1 = k_{11} \\ l_{21} = k_{21}/d_1 \\ l_{n1} = k_{n1}/d_1 \\ d_2 = k_{22} - l_{21}^2 d_1 \\ l_{32} = (k_{32} - l_{31}d_1 l_{21})/d_2 \end{cases} \quad (4.3.15)$$

In general,

$$d_i = k_{ii} - \sum_{m=1}^{i-1} l_{im} d_m l_{jm} \quad (i = 1, 2, \ldots, n) \quad (4.3.16)$$

$$l_{ij} = (k_{ij} - \sum_{m=1}^{i-1} l_{im} d_m l_{jm})/d_j$$

$$(j = 1, 2, \ldots, i-1; i = 1, 2, \ldots, n). \quad (4.3.17)$$

Using Eqs. (4.3.16) and (4.3.17), the elements of **L** and **D** may be calculated. Then

$$\mathbf{KX} = \mathbf{LDL}^T\mathbf{X} = \mathbf{B}. \quad (4.3.18)$$

Let

$$\tilde{\mathbf{X}} \cong \mathbf{DL}^T\mathbf{X} \quad (4.3.19)$$

then

$$\mathbf{L}\tilde{\mathbf{X}} = \mathbf{B}. \quad (4.3.20)$$

X is easy to obtain from Eq. (4.3.19). In the procedure of Cholesky's decomposition, the vector **B** on the RHS is unchanged, it is useful to decrease the error. Using this method to solve Eq. (4.3.2), the procedures are as follows:

$$d_1 = k_{11} = 4$$

$$l_{21} = k_{21}/d_1 = 3/4$$

$$l_{31} = k_{31}/d_1 = 1/2$$

$$d_2 = k_{22} - l_{21}^2 d_1 = 11/4$$

$$l_{32} = (k_{32} - l_{31}d_2 l_{21})/d_2 = -2/11$$

$$d_3 = k_{33} - l_{31}d_1 l_{31} - l_{32}d_2 l_{32} = 76/11$$

4.3 Solution methods of finite element equations

thus

$$\begin{bmatrix} 1 & & \\ 3/4 & 1 & \\ 1/2 & -2/11 & 1 \end{bmatrix} \times \{x\} = \begin{Bmatrix} 12 \\ 14 \\ 12 \end{Bmatrix}$$

$$\tilde{X} = \{12 \quad 5 \quad 76/11\}^T$$

then X is obtained by Eq (4.3.19), i.e.

$$\begin{bmatrix} 4 & & \\ & 11/4 & \\ & & 76/11 \end{bmatrix} \begin{bmatrix} 1 & 3/4 & 1/2 \\ & 1 & -2/11 \\ & & 1 \end{bmatrix} \{x\} = \begin{Bmatrix} 12 \\ 5 \\ 76/11 \end{Bmatrix}.$$

The result is:

$$X = \{1 \quad 2 \quad 1\}^T$$

Appendix 3 of references [15] and [17] provides the computer program of Cholesky's decomposition.

A great advantage of this method is that if the problem to be solved has different values of the RHS, then only the solver of the matrix equation is repeated.

4.3.2 Iterative methods

Iterative methods have the advantage that the sparseness of the coefficient matrix is utilized. Only the nonzero elements are stored and which leave them unchanged during iteration. At each step of the iteration, a single row of the system matrix is used in the calculation. The total system matrix need not to be stored. The structure of the matrix K plays no influence in iterative methods. Thus the optimal ordering of nodes is not necessary. Consequently, for a large system, the iterative method is more suitable.

A simple example of the iterative method is shown as below:

$$\begin{cases} 5x_1 - 2x_2 = 8 \\ 3x_1 - 20x_2 = 26 \end{cases} \tag{4.3.21}$$

Equation (4.3.21) can be written as

$$\begin{Bmatrix} x_1 \\ x_2 \end{Bmatrix} = \begin{bmatrix} 0 & 0.4 \\ 0.15 & 0 \end{bmatrix} \begin{Bmatrix} x_1 \\ x_2 \end{Bmatrix} + \begin{Bmatrix} 1.6 \\ -1.3 \end{Bmatrix}. \tag{4.3.22}$$

This method proceeds from some initial 'guess' $\{x\}^0$ and defines a sequence of successive approximations $\{x\}^1$, $\{x\}^2$ .... which converge to the exact solution. Suppose the initial value of $\{x\}^1$ is $\{x\}^{(0)} = \{0, 0\}^T$. Substitution of

Eq. (4.3.22) into Eq. (4.3.21) yields

$$\begin{Bmatrix} x_1 \\ x_2 \end{Bmatrix}^{(1)} = \begin{bmatrix} 0 & 0.4 \\ 0.15 & 0 \end{bmatrix} \begin{Bmatrix} 0 \\ 0 \end{Bmatrix} + \begin{Bmatrix} 1.6 \\ -1.3 \end{Bmatrix} = \begin{Bmatrix} 1.6 \\ -1.3 \end{Bmatrix}$$

Subsequently

$$\begin{Bmatrix} x_1 \\ x_2 \end{Bmatrix}^{(k+1)} = \begin{bmatrix} \mathbf{M} \end{bmatrix} \begin{Bmatrix} x \end{Bmatrix}^{(k)} + \begin{Bmatrix} 1.6 \\ -1.3 \end{Bmatrix} \tag{4.3.23}$$

The matrix **M** is called the iterative matrix; it is unchanged in each iteration. When $k \to \infty$ $\lim_{k \to \infty} (x)^{(k)} = (x)^*$, the iterative procedure is convergent. For the above example, the approximate values of each step are:

k	1	2	3	4	5	6
x_1	0.	1.6	2.12	1.9928	1.99856	2.00432
x_2	0.	−1.3	−1.06	−0.9964	−1.00108	−1.000226

The differences of the approximate solution between adjacent steps are reduced as the number of iterations, k, is increased. When $k = 6$, $x_1^{(6)} - x_1^{(5)} = 0.001872$, and $x_2^{(6)} - x_2^{(5)} = 0.000854$. If these differences are less than a predetermined criterion ε, then $\{x\}^{(k)}$ is accepted as the solution of Eq. (4.3.18).

4.3.2.1 Method of over-relaxation iteration [16]

The difference of the iterative method used in FDM and FEM is that in FDM the coefficients are recognized by double subscripts; in FEM, the coefficients are recognized by a single subscript. In FEM, the nth equation of a finite element equation is

$$\sum_{j=1}^{i-1} k_{ij} u_j + k_{ii} u_i + \sum_{j=i+1}^{N} k_{ij} u_j = b_i .$$

In iterative formulation, the above equation is altered to

$$u_i^{(m+1)} = - \left[\sum_{j=1}^{i-1} k_{ij} u_j^{(m+1)} + \sum_{j=i+1}^{N} k_{ij} u_j^{(m)} - b_i \right] \Big/ k_{ii}$$

$$i = 1, \ldots, N \tag{4.3.24}$$

where N is the order of the matrix equation. When $m \to \infty$, $u_i^{(m)}$ must be convergent to the real solution. Usually a permissible error ε is given as the convergence criterion. If the adjacent iterative value of each node is less than ε, then the iteration is stopped. In order to increase the iterative speed, the over-relaxation iterative formula is used:

$$u_i^{(m+1)} = \alpha \tilde{u}_i^{(m+1)} + (1 - \alpha) u_i^{(m)} \tag{4.3.25}$$

4.3 Solution methods of finite element equations

where

$$\tilde{u}_i^{(m+1)} = -\left[\sum_{j=1}^{i-1} k_{ij}u_j^{(m+1)} + \sum_{j=i+1}^{N} k_{ij}u_j^{(m)} - b_i\right]\bigg/k_{ii}$$

$$i = 1, \ldots, N \tag{4.3.26}$$

and α is an accelerative factor, $1 < \alpha < 2$. The factor α is very problem dependent, as discussed in FDM. Usually the iterative methods take more computing time than direct methods.

4.3.2.2 Conjugate–gradient method (CGM)

The conjugate gradient scheme is used for the solution of sparse positive definite symmetric matrix equations of the type $\mathbf{KX} = \mathbf{B}$. Detailed concepts of the conjugate gradient method will be explained in Sect. 11.4.2. or can be found in references [15, 16]. Only the formulae are listed here.

To solve equation $\mathbf{KX} = \mathbf{B}$, assume an initial vector $\mathbf{X}_0$, then the residual is:

$$\mathbf{r}_0 = \mathbf{KX}_0 - \mathbf{B}. \tag{4.3.27}$$

The initial direction $\mathbf{p}_0$ to search the minimum of the function $F(\mathbf{X}) = \mathbf{KX} - \mathbf{B}$ (the minimizer of $F(\mathbf{X})$ is the solution of $\mathbf{KX} = \mathbf{B}$) is chosen to coincide with $\mathbf{r}_0$, i.e.

$$\mathbf{p}_0 = \mathbf{r}_0. \tag{4.3.28}$$

Then the successive estimate of the next approximation of $\mathbf{X}$ is:

$$\mathbf{X}_{i+1} = \mathbf{X}_i + \alpha_i \mathbf{p}_i \tag{4.3.29}$$

where

$$\alpha_i = -\frac{\mathbf{p}_i^T \mathbf{r}_i}{\mathbf{p}_i^T \mathbf{K} \mathbf{p}_i}. \tag{4.3.30}$$

Then

$$\mathbf{r}_{i+1} = \mathbf{r}_i + \alpha_i \mathbf{K} \mathbf{p}_i \tag{4.3.31}$$

$$\mathbf{p}_{i+1} = \mathbf{r}_{i+1} - \frac{\mathbf{p}_i^T \mathbf{K} \mathbf{r}_{i+1}}{\mathbf{p}_i^T \mathbf{K} \mathbf{p}_i} \mathbf{p}_i. \tag{4.3.32}$$

The directions $\mathbf{p}_i$ are selected so as to make successive residuals orthogonal to each other, i.e.

$$\mathbf{r}_i^T \mathbf{r}_{i+1} = 0. \tag{4.3.33}$$

In this process, error is removed in one independent search direction at a time and not reintroduced subsequently. After N steps there is no direction left in which correction is required. The resulting solution is therefore the exact solution. The iteration is stopped when $\mathbf{KX}_{i+1} - \mathbf{B} < \varepsilon$.

Another efficient iterative method is the Preconditioned Conjugate Gradient Method (PCGM). It can improve the condition number of the matrix equation and is regarded as a good method for solving large systems. Detailed procedures are given in references [15, 16, 18, 19].

4.4 Mesh generation

In mesh generation, the following principles should be satisfied:

(1) Nodes are placed within the region and on the boundary wherever the field distribution is located. The density of nodes should be high in those areas where the function varies rapidly; the elements could be large while the field is uniform.

(2) The elements cannot overlap or overspill, or leave empty spaces. All of the elements must be well proportioned. For instance, it is important that there is no great disparity between the edge lengths in one element, i.e. sharp corners in each element must be avoided. In other words, equilateral triangles are better than long narrow triangles.

(3) Nodes may not be placed on the side of adjacent elements, e.g. Fig. 4.4.1(a) is improper, (b) is proper.

(4) For a region composed of different materials, the parting lines of the materials should be represented by the boundaries of elements, as shown in Fig. 4.4.1(c).

(5) In order to reduce computer storage, the bandwidth of the system matrix should be as small as possible. Thus the numbering scheme of the nodes should be circulated from the narrow side to the wide side. For example, the ordering of nodes in Fig. 4.2.4(a) is better than that in Fig. 4.2.4(b). The general technique of the optimized numbering of nodes is explained in references [15, 19].

(6) The local numbering of the vertices of each element must be counterclockwise, as shown in Fig. 4.4.2, otherwise the area of the triangle will be negative.

Mesh generation and data construction of mesh information are the most annoying problems encountered when using FEM. They consist of a topological description of the meshes and the coordinates of the nodes. For instance, in

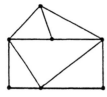 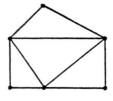 

a b c

Fig. 4.4.1a–c. Proper and improper nodes

4.4 Mesh generation

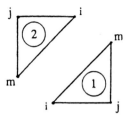

Fig. 4.4.2. The sequence of local nodes of a pair triangle

calculating the coefficients of the matrix, the coordinates of the vertices of each node is necessary. The coordinates of nodes is given by $x(NO), y(NO)$. NO is the global numbering of nodes. Hence the relationship between the number of global nodes $(1, 2, \ldots, NO)$ and the ordering of local number (i, j, m) is necessary information when calculating the element coefficients. The input data is a set of arrays $I(EO), J(EO), M(EO)$ and $x(NO), y(NO)$. $I(EO)$ denotes the node i of every element. $J(EO), M(EO)$ are nodes J, M of every element. In Fig. 4.2.3, the arrays of $I(EO), J(EO)$ and $M(EO)$ are

I: /1, 5, 4, 8, 2, 6, 5, 9/ J: /2, 4, 5, 7, 3, 5, 6, 8/

M: /5, 1, 8, 4, 6, 2, 9, 5/ .

The arrays of $x(NO), y(NO)$ are:

x: /0., 1., 2., 0., 1., 2., 0., 1., 2./ y: /0., 0., 0., 1., 1.,1., 2., 2., 2./ .

The arrays of $I(EO), J(EO), M(EO)$ can also be expressed by a two-dimensional array $IJM(k, EO)$. The first number in the parentheses denotes the number of i, j, m. The second number in the parentheses denotes the number of elements. For example, in Fig. 4.2.3, $IJM(1, 1) = 1$, $IJM(2, 1) = 2$, $IJM(3, 1) = 5$ and so on. If there are hundreds or thousands of nodes, the preparation of the data input is very tedious and time consuming. If the input data contains errors, it will result in a waste of labour and computer resources. Consequently, automatic mesh generation is very important. It is a special technique including graphics. In this section only a simple method of mesh generation is described.

4.4.1 Mesh generation of a triangular element

Two steps are included in generating a mesh module: a logical step and a geometric step. The logical step describes the relationship between the global number and the local number of the nodes of each element. A geometric step gives the geometric coordinates of every node in the module. For convenience, the logical step and the geometric step associated with a module of a structure are set up together within a major logical step.

In order to explain the basic algorithm, a rectangular domain (Fig. 4.2.4(a)) is chosen as an example. Assume that the lengths of the two edges of a rectangular area are X_m and Y_m, N_x and N_y are defined as the number of divisions in x and y directions, respectively. Depending upon the node density, the lengths of the segments in both the x and y directions are selected. Then the total number of nodes (*LO*) and elements (*LEO*) are

$$LO = (N_x + 1) \times (N_y + 1) \tag{4.4.1}$$

$$LEO = 2N_x \times N_y. \tag{4.4.2}$$

If the nodal number is arranged column by column, as shown in Fig. 4.2.4(a), then the nodal number of point P at the intersection of N_i row and N_j column is

$$N_P = (N_j - 1) \times (N_y + 1) + N_i. \tag{4.4.3}$$

If the nodal number is arranged row by row, as in Fig. 4.2.4(b), then

$$N_P = (N_i - 1) \times (N_x + 1) + N_j. \tag{4.4.4}$$

The global coordinates of N_P are

$$\begin{aligned} x_P &= x_1 + (N_j - 1) \times (X_M - x_1)/N_x \\ y_P &= y_1 + (N_i - 1) \times (Y_M - y_1)/N_y \end{aligned} \tag{4.4.5}$$

where x_1, y_1 are the coordinates of an initial point. In Fig. 4.2.4(a), assume $X_m = 6$ cm, $Y_m = 4$ cm, let $N_x = 6$, $N_y = 4$, and $x_1 = 0$, $y_1 = 0$, for node P, $N_i = 3$, $N_j = 4$, then

$$N_p = (4-1) \times (4+1) + 3 = 18, \qquad x(18) = 3 \text{ cm}, \qquad y(18) = 2 \text{ cm}.$$

To define the logical code of the elements and nodes, the elements are divided into two types, as shown in Fig. 4.4.2: an upper triangle and a lower triangle. The nodal code of each element is memorized by a two-dimensional array $IJM(k, E)$. The following formulations are used to construct the global code of a pair of triangular elements. If the codes are numbered column by column, then:

$$\begin{cases} E = 2N_y \times (N_j - 1) + 2N_i - 1 \\ IJM(I, E) = IJM((N_j - 1) \times (N_y + 1) + N_i, E) \\ IJM(J, E) = IJM(I + N_y + 1, E) \\ IJM(M, E) = IJM(J + 1, E) \\ E = E + 1 \\ IJM(I, E) = IJM(M(E - 1), E) \\ IJM(J, E) = IJM(I(E - 1) + 1, E) \\ IJM(M, E) = IJM(I(E - 1), E). \end{cases} \tag{4.4.6}$$

4.4 Mesh generation

If the codes are numbered row by row, then

$$\begin{cases} E = 2N_x \times (N_i - 1) + 2N_j - 1 \\ IJM(I, E) = IJM((N_i - 1) \times (N_x + 1) + N_j, E) \\ IJM(J, E) = IJM(I + N_x + 1 + 1, E) \\ IJM(M, E) = IJM(J - 1, E) \\ E = E + 1 \\ IJM(I, E) = IJM(I(E - 1), E) \\ IJM(J, E) = IJM(I + 1, E) \\ IJM(M, E) = IJM(J + N_x + 1, E) \end{cases} \quad (4.4.7)$$

For instance, in Fig. 4.2.4(a), $IJM(I, 19) = (12, 19)$, $IJM(J, 19) = (17, 19)$.

4.4.2 Automatic mesh generation

The former section gives a simple idea of mesh generation in a rectangular domain. Actually, automatic mesh generation should be capable of producing a valid finite element mesh for any geometry without user intervention. It should be able to change the position of nodes automatically, delete elements and rearrange the topological relation in an attempt to improve the element geometry. As an example, in the package MAGNET 2-D, when the coordinates (x, y) of the vertices of the polygon, shown in Fig. 4.4.3(a) and the division numbers on each edge are input, then the meshed model is displayed on the screen automatically as shown in Fig. 4.4.4(b). During the process of the connection of nodes, the obtuse angle must be avoided. Many schemes exist for optimizing the connection. All these schemes try to make connections yielding triangles as close to equilateral as possible. The well known Delaunary criterion [20] is one of the most commonly used principles in mesh generation. Detailed methods may be found in references [18] and [20-22].

Generally, automatic mesh generation consists of the following steps:

(1) Subdivide the domain into several sub-regions in which the material is homogeneous and determine the density of the elements in each region.

(2) Subdivide each sub-region into triangles, as shown in Fig. 4.4.3(b), join all these subregions into a whole. It is imperative that the nodes at the common sides of each sub-region be coincident.

In order to obtain faster and more accurate results, the more advanced method such as adaptive mesh generation has been developed ([23-28]). Improvement of the mesh is not only concerned with geometrical aspects of the mesh but also with the errors of the results. The size of element depends on the different variations of the field. In the process of adaptive mesh generation,

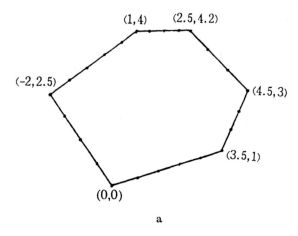

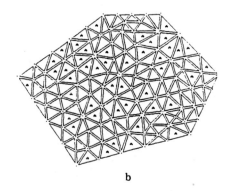

Fig. 4.4.3a, b. Automatic mesh generation by using MAGNET 2-D

the finite element model is generated iteratively. Potential distribution is calculated sequentially beginning with a coarse mesh, the mesh is then refined in locations where the greatest error exists. The error estimation may be based on the complementary variational principle [25], the energy minimum [26] or on computing the residual in the finite element solution directly [27].

4.5 Examples

Example 4.5.1. A 2-D Laplacian problem subject to Dirichlet boundary condition in a rectangular domain is chosen as an example to illustrate the essential feature of the finite element method. Due to its symmetric property, half of the

4.5 Examples

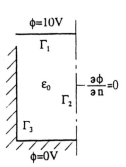

Fig. 4.5.1. A 2-D Laplacian problem

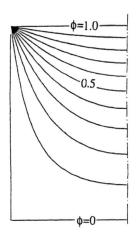

Fig. 4.5.2. Equipotential lines of Fig. 4.5.1

domain, shown in Fig. 4.5.1, is considered. The problem is then stated by

$$\begin{cases} \nabla^2 \varphi = 0 \\ \varphi|_{\Gamma_1} = 1 \quad \varphi|_{\Gamma_3} = 0 \\ \dfrac{\partial \varphi}{\partial n}\bigg|_{\Gamma_2} = 0 \end{cases} \qquad (4.5.1)$$

This is a problem having mixed boundary conditions. Boundaries Γ_1, Γ_3 satisfy boundary conditions of the first kind and boundary Γ_2 satisfies a homogeneous boundary condition of the second kind. The associated finite element equation is:

$$\mathbf{K}\{\varphi\} = 0 \qquad (4.5.2)$$

The domain is subdivided into a non-uniform triangular element; the steps in x and y directions are 8 * 2.0, 16 * 0.5, 5 * 0.2 and 4 * 4.0, 9 * 2.0, 10 * 0.5, 5 * 0.2, respectively. The total number of nodes is $LO = 30 \times 29 = 870$. Among these 87 nodes are boundary nodes with known potentials of $1V$ and $0V$. These data are given by arrays LUO and UO. The total number of elements is $LEO = 2 \times 28 \times 29 = 1624$.

The homogeneous boundary condition of the second kind on Γ_2 is satisfied automatically, hence the boundary nodes on Γ_2 are not dealt with specifically. After the matrix equation is solved, the potential value at each node is obtained. If the distribution of the equipotential line is of interest, the coordinates of equal potential lines are obtained by the linear interpolation or Lagrange interpolation. The potential distribution of Fig. 4.5.1 is shown in Fig. 4.5.2.

If the domain consists of different materials as shown in Fig. 4.5.3, the edges of the element must be coincident with the interfacial line. The result shows that

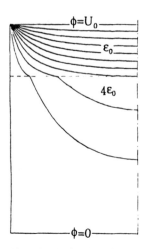 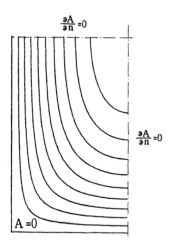

Fig. 4.5.3. Refractive phenomenon at the interface of different materials

Fig. 4.5.4. B lines within the ferromagnetic conductor

the equipotential lines are condensed into the upper part with smaller permittivity. At the interface, the field is refracted according to the ratio of the permittivities.

Example 4.5.2. Plot the distribution of B lines of a long current carrying ferromagnetic conductor ($\mu = 400\mu_0$) with a rectangular cross-section.

Solution. Due to the symmetry, one quarter of the domain, shown in Fig. 4.5.4, is subdivided into 2000 triangular elements and 1066 nodes. Because $\mu \gg \mu_0$. the magnetic field is assumed within the conductor. The boundary of the conductor is assumed as an equipotential line of $A = 0 (A = A_z)$, the two symmetric lines satisfy $\partial A/\partial n = 0$. In each element, the current density is assumed as being constant. For a 2-D case, the equipotential lines of A consist of the B-lines. By using the program as before, the B lines are drawn in Fi.g 4.5.4.

Example 4.5.3. Consider the influence of a ferromagnetic plate near a pair of rectangular current carrying conductor with infinite length.

Solution: This is an open boundary problem. It means the field may extend to infinity. To simplify the calculation, an artificial boundary, shown by the contour a, b, c, d in Fig. 4.5.5, is assumed as the boundary at infinity where $A = 0$. Due to symmetry, half of the domain is to be analysed. Using translational symmetry, the scalar potential A_z is chosen as the variable to be calculated. The boundary conditions of the solution domain are given in Fig. 4.5.5, and the solution of the B lines are plotted in the same diagram. The result shows the tangential component of the flux density along the surface of the steel plate is almost zero. This is because the permeability of the plate is $10\mu_0$. Another

4.6 Summary

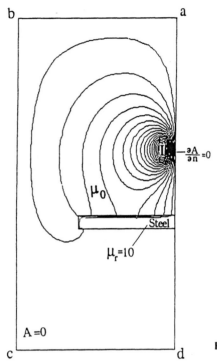

Fig. 4.5.5. Solution of an open boundary problem

conclusion is that the artificial boundary will influence the result of calculation. In other words, the accuracy of the solution is limited by the assumed zero boundary. In general, if the artificial boundary is sufficiently large, the accuracy is enough. This can be defined as when the solution considered does not change when the zero boundary is removed. A more advanced method to deal with the open boundary problem is introduced in the next chapter.

4.6 Summary

In this chapter, general procedures of the finite element method to solve a potential problem are illustrated. In order to avoid using the functional and its variations, a matrix equation of finite elements is derived by the principle of weighted residuals. To illustrate the basic idea of the method, a 3-node triangular element is used as discretization elements. Formulations for calculating the coefficients of a matrix equation of a 2-D translational symmetric problem are given. The coefficients of this element matrix depends on the parameter of the material and the coordinates of the vertices of the element, i.e.

$$k_{rs} = k_{sr} = \frac{\beta}{4S}(b_r b_s + c_r c_s) \quad r, s = i, j, m$$

in the electrical field, $\beta = \varepsilon$ (permittivity of dielectric).

The formulations for calculating an axisymmetric problem will be given in the next chapter. The solution methods for solving algebraic equations are introduced.

The disadvantages of FEM are:

(1) Like FDM, if the problem to be solved is unbounded, then an artificial boundary must be assumed or a special process is required (this will be introduced in the next chapter).

(2) The pre- and post-data processing are more complex than for the finite difference method and integral equation methods (these will be introduced in part three). Because the whole region has to be discretized, a tremendous number of nodes and elements are required.

The variational finite element method and matrix equations of other electromagnetic field problems are discussed in Chap. 5. The high order elements will be introduced in Chap. 6. If the suitable mesh discretization is adapted, accuracy of the method is high.

References

1. Richard H. Gallagher, *Finite Element Analysis*, Prentice-Hall, 1975, Ch. 1, p. 33
2. R.W. Clough, The Finite Element Method in Plane Stress Analysis, *J. Struct. Div., ASCE, Proc. 2nd Conf. Electronic Computation*, pp. 345–378, 1960
3. O.C. Zienkiewicz, Y.K. Cheung, *The Finite Element Method in Structural and Continuum Mechanics*, McGraw-Hill, London, 1967
4. A.M. Winslon, Numerical Solution of the Quasilinear Poisson's Equation in a Nonuniform Triangle Mesh, *J. Comp. Phys.*, 2, 149–172, 1967
5. P.P. Silvester, M.V.K. Chari, Finite Element Solution of Saturable Magnetic Field Problem *IEEE Trans. on PAS-89*, pp. 1642–1651, 1970
6. M.V.K. Chari, P.P. Silvester (ed.), *Finite Elements in Electrical and Magnetic Field Problems*, John Wiley, 1980
7. *Magnet 2D*, Infolytica Limited, 1988
8. *ANSYS, Engineering Analysis System*, Swanson Analysis System Inc. 1989
9. *TOSCA, Elektra*, Vector Fields Limited, Oxford, England
10. P. Tong. John N. Rossettos, *Finite Element Method*. The MIT Press, 1977
11. Sheng Jian-ni (ed.), *Numerical Analysis of Electromagnetic Fields* (in Chinese), Science Press, Beijing, 1984
12. Frank L. Stasa, *Applied Finite Element Analysis for Engineers*, Holt, Rinehart and Winston, 1985
13. R.E. White, *An Introduction to the Finite Element Method with Application to Nonlinear Problems*, John Wiley, 1985
14. S. Ratnajeevan, H. Hoole, *Computer-Aided Analysis and Design of Electromagnetic Devices*, Elsevier, 1989
15. P.P. Silvester, R.L. Ferrari, *Finite Elements for Electrical Engineers* (2nd. ed), Cambridge University Press, 1990

References

16. H.R. Schwarz, *Finite Element Methods*, Academic Press, 1988
17. S.S. Rao, *The Finite Element Method in Engineering* (2nd ed.), Pergamon Press, 1989
18. Z.J. Cendes (ed.), *Computational Electromagnetics* (section one), Elsevier, 1986
19. Alan Jennings, *Matrix Computation for Engineers and Scientists*, John Wiley, 1977
20. Z.S. Cendes, D. Shenton, H. Shahnasser, Magnetic Field Computation Using Delaunay Triangulation and Complementary Finite Element Methods, *IEEE Trans. on Mag.*, **19**(6), 2551–2554, 1983
21. O.W. Anderson, Laplacian Electrostatic Field Calculation by Finite Elements with Automatic Grid Generation, *IEEE Trans. on PAS-92*, No. 5, pp. 1485–1492, 1973
22. Kenneth Baldwin (ed.), *Modern Methods for Automating Finite Element Mesh Generation*, ASCE, 1986
23. Z.J. Cendes, D.N. Shenton, Adaptive Mesh Refinement in the Finite Element Computation of Magnetic Field, *IEEE Trans.*, **Mag-21**(5), 1811–1816, 1985
24. A.R. Pinchuk, P.P. Silvester, Error Estimation for Automatic Adaptive Finite Element Mesh Generation, *IEEE Trans.*, **Mag-21**(6), 2551–2553, 1985
25. C.S. Biddlecombe, J. Simkin, C.W. Trowbridge, Error Analysis in Finite Element Models of Electromagnetic Field, *IEEE Trans.*, **Mag-22**(5), 811–813, 1986
26. M. Fujita, M. Yamana, Two-Dimensional Automatically Adaptive Finite Element Mesh Genertion, *IEEE Trans. on Mag.*, **24**(1) 303–305, 1988
27. P. Fernandes, Local Error Estimates for Adaptive Mesh Refinement, *IEEE Trans. on Mag.*, **24**(1), 299–302, 1988
28. David Shenton, Zoltan Cendes, MAX – An Expert System for Automatic Adaptive Magnetics Modeling, *IEEE Trans. on Mag.*, **22**(5), 805–807, 1986

Chapter 5

Variational Finite Element Method

5.1 Introduction

In general, most of the problems in engineering and science can be described by variational principles. For instance, the principle of least action exists in mechanics and electrodynamics [1]. In electrostatic fields, Thomson's theory [2] states that the electric energy is minimum if the system is in equilibrium. In classical thermodynamics, the entropy remains at maximum for any equilibrated isolated system.

Variational expressions succinctly summarize the governing equations of these problems and provide a means for an approximate solution. Therefore the solution of any boundary value problem is characterized by a function which yields an extremum (minimum, maximum) value or is stationary to a related functional $I(u)$[†]. *From a historical point of view, variational problems (to find the extremum function of a functional) are solved by the solution of their equivalent Euler's equations* (differential equation).

However, the development of high-speed digital computers has enabled the numerical solution of many variational problems. *Hence a partial differential equation can be solved using the approximate method of its equivalent variational problem. One of the most prominent methods is the variational finite element method.* The classical variational formulation of a continuum problem has advantages over the differential formulation for obtaining an approximate solution. The reasons are as follows.

First, as indicated in Chap. 2, the extremum function of a functional is the solution of the corresponding operator equations. In potential problems, *the unknown function contained in the equivalent functional of the problem has lower order derivatives than those contained in differential equations and consequently an approximate solution can be sought in a large number of functions.* For example, Poisson's equation subject to Dirichlet boundary conditions is mathematically expressed as:

$$\begin{cases} \nabla^2 \varphi = -\rho/\varepsilon & \text{in Domain } \Omega \\ \varphi|_\Gamma = U_0 & \text{on Boundary } \Gamma \end{cases} \quad (5.1.1)$$

[†] In this book, symbol I represents functional.

The equivalent functional of Poisson's equation, subject to the Dirichlet boundary condition, is equivalent to a constrained functional expressed by:

$$\begin{cases} I(\varphi) = \int_\Omega \tfrac{1}{2}\varepsilon |\nabla \varphi|^2 \, d\Omega - \int_\Omega \rho\varphi \, d\Omega \\ \varphi|_\Gamma = U_0 \, . \end{cases} \quad (5.1.2)$$

The second equation of Eq. (5.1.2) is the constrained condition of the equivalent functional $I(\varphi)$. It shows that $I(\varphi)$ contains only the first order derivative of φ [the equivalence of Eqs. (5.1.1) and (5.1.2) was proved in Sect. 2.4.2], while the second order partial derivative of φ is contained in the partial differential equation.

Second, *some problems may possess reciprocal variational formulations. This means that when describing a physical problem one functional has to be minimized and another functional of a different form has to be maximized.* In such cases one may find the upper and lower bounds on the functional. This has important engineering significance. For instance, it can be used to calculate the parameters of electromagnetic fields [3, 4].

Third, *in variational formulations, it is possible to treat complicated interfacial boundary conditions as natural boundary conditions.* (This property will be proved in Sect. 5.3.)

Finally, *from the mathematical point of view, with variational formulations, it is easy to prove the existence of the solution.* It is proved, if the operator $\mathscr{L}$ of the operator equation ($\mathscr{L}u = f$) is symmetric and positive definite, then this equation has only one solution [5].

The variational finite element method is based on the principle of the variations. The process in the variational finite element method is to find equivalent variations of the physical problem first, then to minimize the equivalent functional approximately to obtain a set of algebraic equations. The solution of these simultaneous equations is the approximate solution of the problem to be solved.

The basic concepts of the functional and its variations are reviewed in Sect. 5.2. The equivalent functionals for electromagnetic field problems are derived in Sect. 5.3. During the derivation of the functionals, it is shown that the boundary conditions of the second and third kind are included in the equivalent functional. The discretized finite element equations of various electromagnetic field problems are derived in Sect. 5.4. Finally, some special problems such as the solution of open boundary problems and problems containing conductors with free potentials are discussed in the last section.

5.2 Basic concepts of the functional and its variations

The theory and calculus of the functional and its variations are beyond the scope of this book. Without becoming too involved with mathematical difficulties, some basic concepts of functionals and their variations are reviewed briefly. For the complete theory of functionals and variations, the reader may refer to any of the references [5–9].

5.2.1 Definition of the functional and its variations

5.2.1.1 The functional

The function $y = f(x)$ or $y = f(x_1, x_2, \ldots, x_n)$ expresses the relationship between a set of variables $(x_1, \ldots, x_n)$ and a set of numbers. Therefore, *a function is a mapping connecting one space of numbers to another space of numbers*.

A functional is a different kind of mapping. It relates a set of functions to a set of numbers. For instance, what is the shortest length between the two points A and B of a curve shown in Fig. 5.2.1(a)? What is the minimum surface suspended between two circular wire loops as shown in Fig. 5.2.1(b)?

Obviously, the length is a number. It is determined by the shape of the curve which is a function of the variable x, i.e. the length of the curve is expressed by:

$$L(y) = \int_A^B dl = \int_A^B \sqrt{dx^2 + dy^2} = \int_A^B \sqrt{1 + (dy/dx)^2}\,dx$$

$$= \int_{x_A}^{x_B} \sqrt{1 + (y')^2}\,dx \ . \tag{5.2.1}$$

This equation means that the length of the curve depends on the function $y = f(x)$ and its derivative of the first order. The value of $L(y)$ depends on the argument of function $y(x)$. Hence $L(y)$ is called a functional. *Therefore, the functional is a real or complex value, it is the function of a function, not a function of variable.* In electrostatic fields, the field intensity $E(r)$ is a function of coordinates, the potential difference U between two points and the electrostatic energy W_e included in volume Ω are functions of the field strength, i.e.

$$U = \int_A^B \mathbf{E}(\mathbf{r})\,dl \tag{5.2.2}$$

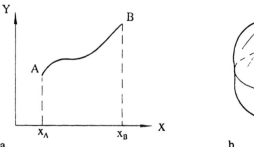
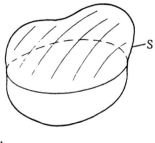

a   b

Fig. 5.2.1a, b. A planar curve and a curved surface

5.2 Basic concepts of the functional and its variations

$$W_e = \tfrac{1}{2}\int_\Omega \varepsilon E^2 d\Omega \ . \tag{5.2.3}$$

Consequently, the U and W_e are functionals which are determined by a vector function $\mathbf{E}(\mathbf{r})$. Once the function $\mathbf{E}(\mathbf{r})$ is specified, the functional $W(\mathbf{E})$ is evaluated. Equation (5.2.1) can be expressed in a generic form:

$$I(y) = \int_{x_A}^{x_B} F(x, y, y') dx \ . \tag{5.2.4}$$

Usually, the functional is expressed by an integral. In Eq. (5.2.4), F is the integrand of the functional.

In two-dimensional cases, the functional is expressed as:

$$I(u) = \int_\Omega F(x, y, u, u_x, u_y, u_{xx}, u_{yy}) dxdy \tag{5.2.5}$$

where

$$u = f(x, y), \quad u_x = \partial u/\partial x, \quad u_y = \partial u/\partial y, \quad u_{xx} = \partial^2 u/\partial x^2$$

$$u_{yy} = \partial^2 u/\partial y^2 \ . \tag{5.2.6}$$

In the notation of operators, the functional is a special operator. It is defined as that the operator $\int_\Omega F d\Omega$ which maps its domain Ω onto a set of real or complex numbers called functionals. The domain of a functional is the space of admissible functions which may be restricted to satisfy certain continuity restrictions or boundary conditions. Thus, a functional assigns every element $u \in \Omega_F$ to a certain number $I(u)$. Hence *all concepts and results of an operator are valid for the functional.* For example, the stationary property of a functional is analogous to the stationary property of a function.

It is known that the point x_0 is a stationary point, if the function $y(x)$ is stationary at point x_0, i.e. $dy(x)/dx|_{x=x_0} = 0$ or

$$\lim_{\alpha \to 0} \frac{y(x_0 + \alpha) - y(x_0)}{\alpha} = 0 \tag{5.2.7}$$

α is a predetermined infinitesimal value.

Similarly, the derivative of a functional is defined as:

$$\lim_{\alpha \to 0} \frac{I(y + \alpha\eta) - I(y)}{\alpha} = 0 \tag{5.2.8}$$

where

$$\eta = \eta(x) = dY(x, \alpha)/d\alpha|_{\alpha=0} \tag{5.2.9}$$

η is zero at two endpoints A and B in Fig. 5.2.1(a). The concept of the variation of a functional will be illustrated by comparison of the variation of a function.

5.2.1.2 The differentiation and variation of a function

The differentiation of a simple function $y = f(x)$ is defined as:

$$y'(x) = \lim_{\Delta x \to 0} \frac{\Delta y}{\Delta x} \qquad (5.2.10)$$

i.e.

$$\frac{\Delta y}{\Delta x} = y'(x) + \alpha \quad (\text{when } \Delta x \to 0, \alpha \to 0) \,.$$

Thus

$$\Delta y = y'(x)\Delta x + \alpha \Delta x = dy + \alpha dx = dy + 0(\Delta x) \qquad (5.2.11)$$

where $\alpha \Delta x$ is an infinitesimal of a high order. When $|\Delta x|$ is small enough dy is the approximation of Δy. This means that *the differentiation of a function is the principal value of the increment of the function*. The high order derivatives of $y(x)$ are obtained by expansion of Taylor's series. The incremental of the function y caused by the variable x is derived as

$$y(x + dx) - y(x) = y'(x)\,dx + \frac{1}{2!}y''(x)\,dx^2 + \ldots + \frac{1}{n!}y^n(x)\,dx^n \qquad (5.2.12)$$

where neglecting the high order derivatives yields:

$$dy(x) = y'(x)\,dx \,. \qquad (5.2.13)$$

In Eq. (5.2.13) $dy(x)$ is the principal value of the increment of the function $y = f(x)$, it is the first order derivative of function $y(x)$. Similarly, the second order derivative and the nth order derivative are:

$$d^2 y(x) = y''(x)\,dx^2 \qquad (5.2.14)$$

$$d^n y(x) = y^n(x)\,dx^n \,. \qquad (5\,2.15)$$

The definition of the variation of function is:

$$\delta y = Y(x) - y_0(x) = \alpha \eta(x) \qquad (5.2.16)$$

where δy is the increment of $y(x)$ in the functional $I(y)$, i.e. $I(y') = I(y + \delta y)$. This equation represents that if $y(x)$ is changed from $y_0(x)$ to $Y(x)$, then $Y(x) - y_0(x)$ is called the variation of $y(x)$ at $y_0(x)$, where $\alpha \ll 1$, $\eta(x) \in [x_A, x_B]$ is any acceptable function defined in the domain under consideration. The symbol δ (which reads 'the variation of') is used in calculus of the variations to indicate the various functions. The difference between the derivation and variation is that the variation of the function is concerned with the parameter α not the variable x.

One important property of the variation of function is that the sequence of the variation and the derivation can be alternated with each other, i.e.

$$d(\delta y) = \delta(dy) \,. \qquad (5.2.17)$$

5.2 Basic concepts of the functional and its variations

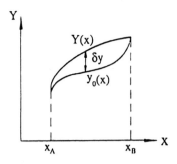

Fig. 5.2.2. Variation of a function

This is because

$$\delta y_x = \delta \frac{d}{dx} y = \frac{dY}{dx} - \frac{dy}{dx} = \frac{d}{dx}(Y - y) = \frac{d}{dx}\delta y. \qquad (5.2.18)$$

This property is useful in deriving the equivalent functional in field problems in the sections to follow.

5.2.1.3 Variation of the functional

The definition of the variation of a functional is similar to the derivation of a function y. The variation of a functional $I[y(x)]$ caused by the variation of function y is expressed by:

$$I(y + \delta y) - I(y) = \int_{x_1}^{x_2} F(x, y + \delta y) dx - \int_{x_1}^{x_2} F(x, y) dx = \int_{x_1}^{x_2} \frac{\partial F}{\partial y} \delta y \, dx$$

$$+ \frac{1}{2!} \int_{x_1}^{x_2} \left(\frac{\partial^2 F}{\partial y^2} (\delta y)^2 \right) dx + \ldots$$

$$= \delta I(y) + \frac{1}{2!} \delta^2 I(y) + \ldots \qquad (5.2.19)$$

Therefore the first order variation (usually simplified as a variation) of the functional is:

$$\delta I(y) = \int_{x_1}^{x_2} \frac{\partial F}{\partial y} \delta y \, dx \qquad (5.2.20)$$

where $\delta I(y)$ is the principal value of the increment of a functional. The variation of the first order of functional $I(y)$ depends on δy linearly. Similarly, the second order variation and the nth order variation of the functional are:

$$\delta^2 I(y) = \int_{x_1}^{x_2} \frac{\partial^2 F}{\partial y^2} (\delta y)^2 \, dx \quad (5.2.21)$$

and

$$\delta^n I(y) = \int_{x_1}^{x_2} \frac{\partial^n F}{\partial y^n} (\delta y)^n \, dx \quad (5.2.22)$$

The second order variation of functional $\delta^2 I(y)$ depends on the quadratic variation of the function. Comparing Eq. (5.2.21) with Eq. (5.2.14), the operation for the variation is similar to the operation for the derivative. *The variation of the first and second order of a functional are extensions of the derivatives of functions.* For example, if the integrand of a functional is $F(x) = y^2(x)$, then the increment of a functional is:

$$\Delta I = \int_a^b [F(x) + \delta y] \, dx - \int_a^b F(x) \, dx$$

$$= \int_a^b [y^2(x) + 2y(x)\delta y + (\delta y)^2] \, dx - \int_a^b y^2(x) \, dx$$

$$= \int_a^b 2y(x)\delta y \, dx + \int_a^b (\delta y)^2 \, dx = \delta I(y) + 0(\delta y) \quad (5.2.23)$$

where

$$\delta I(y) = \int_a^b \frac{\partial F}{\partial y} \delta y \, dx \quad (5.2.24)$$

The first term of the RHS of Eq. (5.2.23) is a linear functional. The second term of the RHS of Eq. (5.2.23) is an infinitesimal of a high order. Thus the variation of a functional is also the principal value of the increment of the functional. According to the extremum condition of a function, the extremum condition or the stationary condition of a functional is:

$$\delta I(y) = 0 \quad (5.2.25)$$

i.e. *the first order variation of the functional being equal to zero is the necessary condition under which the functional has its extremum function.*

If $\delta^2 I(y) < 0$, the function is maximum, if $\delta^2 I(y) > 0$, the function is minimum.

5.2.2 Calculus of variations and Euler's equation

In the conventional method of variations, the extremal function of the functional is solved by the equivalent Euler's equation. In numerical approaches, the

5.2 Basic concepts of the functional and its variations

differential equation is solved by the equivalent variational problem. Hence the relations between the differential equation and the variation are discussed in this section.

5.2.2.1 Euler's equation [10]

Consideration of the functional consists of the first derivative of a function y, e.g.

$$I(y) = \int_{x_1}^{x_2} F(x, y, y')dx . \tag{5.2.26}$$

Let $\delta y = \alpha \eta(x)$, $\delta y' = \alpha \eta'(x)$. Then, the variety of the integrand of the functional is:

$$\delta F = F(x, y + \delta y, y' + \delta y') - F(x, y, y') = \frac{\partial F}{\partial y}\delta y + \frac{\partial F}{\partial y'}\delta y' + \frac{\partial^2 F}{\partial y \partial y'}\delta y \delta y'$$

$$+ \frac{1}{2!}\left[\frac{\partial^2 F}{\partial y^2}(\delta y)^2 + \frac{\partial^2 F}{\partial y'^2}(\delta y')^2\right] + \cdots$$

$$= \alpha\left[\frac{\partial F}{\partial y}\eta(x) + \frac{\partial F}{\partial y'}\eta'(x)\right] + \frac{\alpha^2}{2!}\left[\frac{\partial^2 F}{\partial y^2}\eta^2(x) + \frac{\partial^2 F}{\partial y'^2}\eta'^2(x)\right] + \cdots \tag{5.2.27}$$

Comparing Eq. (5.2.27) with Eq. (5.2.20), the variation of the first order of the functional is:

$$\delta I(y) = \int_{x_1}^{x_2}\alpha\left[\frac{\partial F}{\partial y}\eta(x) + \frac{\partial F}{\partial y'}\eta'(x)\right]dx = \alpha\int_{x_1}^{x_2}\eta\left[\frac{\partial F}{\partial y} - \frac{d}{dx}\left(\frac{\partial F}{\partial y'}\right)\right]dx$$

$$+ \alpha\eta\frac{\partial F}{\partial y'}\bigg|_{x=x_1}^{x=x_2} . \tag{5.2.28}$$

By using the extremum condition $\delta I(y) = 0$, the extremal function $y(x)$ of the functional $I(y)$ is thus obtained. In Eq. (5.2.28), as α is a constant, $\eta(x)$ is an arbitrary function. Instead of $\delta I(y) = 0$, the following equations must be satisfied:

$$\frac{\partial F(x, y, y')}{\partial y} = \frac{d}{dx}\left(\frac{\partial F(x, y, y')}{\partial y'}\right) \tag{5.2.29}$$

$$\eta(x)\frac{\partial F(x, y, y')}{\partial y'}\bigg|_{x=x_1} = 0 \quad \eta(x)\frac{\partial F(x, y, y')}{\partial y'}\bigg|_{x=x_2} = 0 . \tag{5.2.30}$$

These equations are equivalent to Eq. (5.2.25). Hence the solution of the variational equation equals the solution of the differential equation (5.2.29) subject to the additional condition (5.2.30). Equation (5.2.29) can be written as:

$$F_y - \frac{d}{dx}F_{y'} = 0 \tag{5.2.31}$$

This is called the Euler–Lagrange Equation. It was proved by Euler in 1755 and independently by Lagrange, also in 1755. Therefore, *any variational problem of the first order is equivalent to solving Euler's equation subject to the additional conditions of Eq. (5.2.30)*. Equation (5.2.30) includes two cases: if $\eta(x_1) = 0$, $\eta(x_2) = 0$, e.g. $y(x_1) = A$, $y(x_2) = B$ (Fig. 5.2.1(a)), then Eq. (5.2.29) and Eq. (5.2.30) are equivalent to the boundary conditions of the variational problem of the first kind. If $\eta(x_1) \neq 0$, $\eta(x_2) \neq 0$, e.g. $y(x_1)$, $y(x_2) \neq \text{const}$, then the variational problem corresponds to:

$$\frac{\partial F}{\partial y} - \frac{d}{dx}\left(\frac{\partial F}{\partial y'}\right) = 0 \tag{5.2.32}$$

and

$$\left.\frac{\partial F}{\partial y'}\right|_{x=x_1} = 0 \qquad \left.\frac{\partial F}{\partial y'}\right|_{x=x_2} = 0 . \tag{5.2.33}$$

These two equations represent the variational problem of the second kind.

Therefore, in association with $\eta = 0$ or $\eta \neq 0$, the fixed boundary value problem of variation equals the boundary value conditions of the first kind of Euler's equation. The free boundary value problem of variation is equal to the boundary value conditions of the second kind of Euler's equation.

If the functional contains the second derivatives of the function, i.e.

$$I(y) = \int_{x_A}^{x_B} F(x, y, y', y'') \, dx \tag{5.2.34}$$

then the corresponding Euler's equation is:

$$F_y - \frac{d}{dx} F_{y'} + \frac{d}{dx} F_{y''} = 0 . \tag{5.2.35}$$

5.2.2.2 Euler's equation for multivariable functions

Consider that the functional consists of the function $u = f(x, y, z)$ and its partial derivatives of the first order, e.g.

$$I(u) = \int_\Omega F(x, y, z, u, u_x, u_y, u_z) \, dxdydz . \tag{5.2.36}$$

Let

$$\delta u = \alpha \eta(x, y, z) \tag{5.2.37}$$

then

$$\delta I(u(x, y, z)) = \int_\Omega \left[\frac{\partial F}{\partial u}\delta u + \frac{\partial F}{\partial u_x}\delta u_x + \frac{\partial F}{\partial u_y}\delta u_y + \frac{\partial F}{\partial u_z}\delta u_z\right] d\Omega$$

5.2 Basic concepts of the functional and its variations

$$= \alpha \int_\Omega \left[\frac{\partial F}{\partial u}\eta + \frac{\partial F}{\partial u_x}\eta_x + \frac{\partial F}{\partial u_y}\eta_y + \frac{\partial F}{\partial u_z}\eta_z \right] d\Omega$$

$$= \alpha \int_\Omega \left[\frac{\partial F}{\partial u}\eta + \left(\frac{\partial F}{\partial u_x}\mathbf{i} + \frac{\partial F}{\partial u_y}\mathbf{j} + \frac{\partial F}{\partial u_z}\mathbf{k}\right) \cdot \nabla\eta \right] dxdydz \quad (5.2.38)$$

Using the vector identity:

$$\mathbf{A} \cdot \nabla u = \nabla \cdot (u\mathbf{A}) - u\nabla \cdot \mathbf{A} \quad (5.2.39)$$

and Gauss's theorem:

$$\int_\Omega \nabla \cdot \mathbf{A} d\Omega = \oint_s \mathbf{A} \cdot d\mathbf{S} \quad (5.2.40)$$

Eq. (5.2.38) results in:

$$\delta I(u) = \alpha \int_\Omega \eta \left[\frac{\partial F}{\partial u} - \frac{\partial}{\partial x}\left(\frac{\partial F}{\partial u_x}\right) - \frac{\partial}{\partial y}\left(\frac{\partial F}{\partial u_y}\right) - \frac{\partial}{\partial z}\left(\frac{\partial F}{\partial u_z}\right) \right] d\Omega$$

$$+ \alpha \oint_s \eta \left[\frac{\partial F}{\partial u_x}(\mathbf{n}\cdot\mathbf{i}) + \frac{\partial F}{\partial u_y}(\mathbf{n}\cdot\mathbf{j}) + \frac{\partial F}{\partial u_z}(\mathbf{n}\cdot\mathbf{k}) \right] ds . \quad (5.2.41)$$

Thus the variational equation $\delta I(u) = 0$ is equivalent to Euler's equation:

$$\frac{\partial F}{\partial u} = \frac{\partial}{\partial x}\left(\frac{\partial F}{\partial u_x}\right) + \frac{\partial}{\partial y}\left(\frac{\partial F}{\partial u_y}\right) + \frac{\partial}{\partial z}\left(\frac{\partial F}{\partial u_z}\right) \quad (5.2.42)$$

subject to the additional condition:

$$\eta \left[\frac{\partial F}{\partial u_x}(\mathbf{n}\cdot\mathbf{i}) + \frac{\partial F}{\partial u_y}(\mathbf{n}\cdot\mathbf{j}) + \frac{\partial F}{\partial u_z}(\mathbf{n}\cdot\mathbf{k}) \right]\bigg|_{x,y,z \in \Omega} = 0 . \quad (5.2.43)$$

If the functional consists of the partial derivative of the second order, then Euler's equation is:

$$\frac{\partial F}{\partial u} + \frac{\partial^2}{\partial x^2}\left(\frac{\partial F}{\partial u_{xx}}\right) + \frac{\partial^2}{\partial y^2}\left(\frac{\partial F}{\partial u_{yy}}\right) + \frac{\partial^2}{\partial z^2}\left(\frac{\partial F}{\partial u_{zz}}\right) = 0 \quad (5.2.44)$$

subject to the additional condition:

$$\mathbf{n} \cdot \left\{ \mathbf{i}\left[\eta_x \frac{\partial F}{\partial u_{xx}} - \eta \frac{\partial}{\partial x}\left(\frac{\partial F}{\partial u_{xx}}\right) \right] + \mathbf{j}\left[\eta_y \frac{\partial F}{\partial u_{yy}} - \eta \frac{\partial}{\partial y}\left(\frac{\partial F}{\partial u_{yy}}\right) \right] \right.$$

$$\left. + \mathbf{k}\left[\eta_z \frac{\partial F}{\partial u_{zz}} - \eta \frac{\partial}{\partial z}\left(\frac{\partial F}{\partial u_{zz}}\right) \right] \right\} = 0 . \quad (5.2.45)$$

5.2.2.3 The shortest length of a curve

The aim of this section is to find an extremum function $y(x)$ which minimizes the following functional:

$$I(y) = \int_{x_A}^{x_B} \sqrt{1 + (Y')^2}\, dx \qquad (5.2.46)$$

by using Euler's equation.

Let $\eta(x)$ represent a function of x which is zero at the end points x_A and x_B and has a continuous derivative to the first order in the interval of x_A and x_B. The function $Y(x)$ is defined by:

$$Y(x) = y(x) + \alpha\eta(x) \qquad (5.2.47)$$

where $y(x)$ is the desired extremum and α ($\ll 1$) is a parameter. Because of the arbitrariness of $\eta(x)$, $Y(x)$ represents any curve drawn between the points A and B as shown in Fig. 5.2.2. The purpose is to pick one curve out of all these curves $Y(x)$, which satisfies Eq. (5.2.46). Based on the definition of the variation of the function, I is a function of parameter α. If $\alpha = 0$, then $Y(x) = y(x)$ is the desired extremum, i.e. the result is determined by:

$$dI/d\alpha|_{\alpha=0} = 0. \qquad (5.2.48)$$

Substituting Eqs. (5.2.46) into (5.2.48) leads to:

$$\left(\frac{dI}{d\alpha}\right)_{\alpha=0} = \int_{x_A}^{x_B} \frac{1}{2}\frac{1}{(1 + (Y')^2)^{1/2}} 2Y' \left(\frac{dY'}{d\alpha}\right) dx. \qquad (5.2.49)$$

Since

$$Y'(x) = y'(x) + \alpha\eta'(x) \qquad (5.2.50)$$

and

$$dY'/d\alpha = \eta'(x). \qquad (5.2.51)$$

Substituting Eq. (5.2.50), (5.2.51), into Eq. (5.2.49) and letting $dI/d\alpha$ equal zero at $\alpha = 0$, leads to:

$$\left(\frac{dI}{d\alpha}\right)_{\alpha=0} = \int_{x_A}^{x_B} \frac{y'(x)\eta'(x)}{\sqrt{1 + (y')^2}}\, dx = 0. \qquad (5.2.52)$$

Integration of Eq. (5.2.52) by parts results in:

$$\left(\frac{dI}{d\alpha}\right)_{\alpha=0} = \frac{y'}{\sqrt{1 + (y')^2}}\eta(x)\bigg|_{x_A}^{x_B} - \int_{x_A}^{x_B} \eta(x)\frac{d}{dx}\left(\frac{y'}{\sqrt{1 + (y')^2}}\right) dx. \qquad (5.2.53)$$

The first term on the RHS of Eq. (5.2.53) is zero because at the end points $\eta(x) = 0$. Consider the second term on the RHS of eq. (5.2.53) and recall that $\eta(x)$

5.2 Basic concepts of the functional and its variations

is an arbitrary function. The integral $\int_{x_A}^{x_B} \eta(x) f(x) \, dx$ will be zero only if the function $f(x)$ is zero, i.e.

$$\frac{d}{dx}\left(\frac{y'}{\sqrt{1+(y')^2}}\right) = 0. \tag{5.2.54}$$

Integration of Eq. (5.2.54) with respect to x, results in

$$\frac{y'}{\sqrt{1+(y')^2}} = \text{const} \tag{5.2.55}$$

or

$$y' = \text{const} \tag{5.2.56}$$

then

$$y = Ax + B. \tag{5.2.57}$$

The above equation shows that $y(x)$ is a straight line as expected.

This process shows that the solution of a variational problem is equivalent to solving the differential equation (Eq. (5.2.54)) which is Euler's equation of the variational problem (Eq. (5.2.46)). In FEM, the problem is dealt with in different way: the solution of a partial differential equation is obtained by finding the extremum function of an equivalent functional. The first step is to find the equivalent functional of the operator equation. Then the extremum solution of the functional is also the solution of the operator equation.

5.2.3 Relationship between the operator equation and the functional

Variational expressions can be established by the physical problem directly, or can be derived by a differential or an integral equation. For an operator equation

$$\mathscr{L}(\mathbf{u}) = \mathbf{f} \tag{5.2.58}$$

only if the operator $\mathscr{L}$ is positive definite and self-adjoint may the equivalent functional exist and can be determined by:

$$I(\mathbf{u}) = \langle \mathscr{L}(\mathbf{u}), \mathbf{u} \rangle - \langle \mathbf{u}, \mathbf{f} \rangle - \langle \mathbf{f}, \mathbf{u} \rangle. \tag{5.2.59}$$

This can be proved in the following ways [8, 11]:
(a) If $\mathbf{v}$ ($\mathbf{v} = \mathbf{u} + \eta$) is the solution of Eq. (5.2.58), then $I(\mathbf{v}) > I(\mathbf{u})$.
(b) If $\mathbf{u}$ satisfies Eq. (5.2.59), then $\mathbf{u}$ is the solution of Eq. (5.2.58).

Proof (a):

$$\begin{aligned} I(\mathbf{v}) &= \langle \mathscr{L}\mathbf{v}, \mathbf{v} \rangle - \langle \mathbf{v}, \mathbf{f} \rangle - \langle \mathbf{f}, \mathbf{v} \rangle \\ &= [\langle \mathscr{L}\mathbf{u}, \mathbf{u} \rangle - \langle \mathbf{u}, \mathbf{f} \rangle - \langle \mathbf{f}, \mathbf{u} \rangle] + \langle \mathscr{L}\eta, \eta \rangle \\ &\quad + [\langle \mathscr{L}\mathbf{u}, \eta \rangle - \langle \mathbf{f}, \eta \rangle] + [\langle \mathscr{L}\eta, \mathbf{u} \rangle - \langle \eta, \mathbf{f} \rangle]. \end{aligned} \tag{5.2.60}$$

In consideration of the symmetry and the positive definite of the operator $\mathscr{L}$, the following conditions exist

$$\langle \mathscr{L}\eta, \eta \rangle > 0, \qquad \langle \mathscr{L}\eta, \mathbf{u} \rangle = \langle \eta, \mathscr{L}\mathbf{u} \rangle$$

Then

$$I(\mathbf{v}) - I(\mathbf{u}) = \langle \mathscr{L}\eta, \eta \rangle + \langle \mathscr{L}\mathbf{u} - \mathbf{f}, \eta \rangle + \langle \eta, \mathscr{L}\mathbf{u} - \mathbf{f} \rangle$$
$$= \langle \mathscr{L}\eta, \eta \rangle > 0. \tag{5.2.61}$$

Hence $I(\mathbf{u}) = \min$.

Proof (b):

Let $\mathbf{v} = \mathbf{u} + \alpha\eta$ (α is a complex number), by assumption

$$\Delta I = I(\mathbf{v}) - I(\mathbf{u}) = \langle \mathscr{L}\alpha\eta, \alpha\eta \rangle + \langle \mathscr{L}\mathbf{u} - \mathbf{f}, \alpha\eta \rangle + \langle \alpha\eta, \mathscr{L}\mathbf{u} - \mathbf{f} \rangle \geq 0. \tag{5.2.62}$$

If $\alpha \to 0$, then $\Delta I = \min$. If α is a real number, then

$$\Delta I = \alpha^2 \langle \mathscr{L}\eta, \eta \rangle + \alpha \langle \mathscr{L}\mathbf{u} - \mathbf{f}, \eta \rangle + \alpha \langle \eta, \mathscr{L}\mathbf{u} - \mathbf{f} \rangle$$
$$= \alpha^2 \langle \mathscr{L}\eta, \eta \rangle + \alpha \mathrm{Re} \langle \mathscr{L}\mathbf{u} - \mathbf{f}, \eta \rangle. \tag{5.2.63}$$

If $\alpha = j\alpha$, then

$$\Delta I = \alpha^2 \langle \mathscr{L}\eta, \eta \rangle - j\alpha \langle \mathscr{L}\mathbf{u} - \mathbf{f}, \eta \rangle + j\alpha \langle \eta, \mathscr{L}\mathbf{u} - \mathbf{f} \rangle$$
$$= \alpha^2 \langle \mathscr{L}\eta, \eta \rangle + \alpha \mathrm{Im} \langle \mathscr{L}\mathbf{u} - \mathbf{f}, \eta \rangle. \tag{5.2.64}$$

Based on $\lim_{\alpha \to 0} \partial \Delta I / \partial \alpha = 0$, only if $\mathscr{L}$ is bounded operator, then $\langle \mathscr{L}\eta, \eta \rangle$ is also bounded and $\langle \mathscr{L}\mathbf{u} - \mathbf{f}, \eta \rangle = 0$, $\alpha^2 \langle \mathscr{L}\eta, \eta \rangle \geq 0$. Then $\mathscr{L}\mathbf{u} = \mathbf{f}$. Hence $\mathscr{L}$ must be positive definite.

If $\mathscr{L}$ is a real operator, u and f are real functions, then

$$I(u) = \langle \mathscr{L}(u), u \rangle - 2\langle u, f \rangle. \tag{5.2.65}$$

Consequently, the equivalent functional $I(u)$ of an operator equation can be derived by Eq. (5.2.59).

5.3 Variational expressions for electromagnetic field problems [12]

In electromagnetic field theory, the differential equations of the continuum problems are more well known than the variational expressions. In general, with electromagnetic field problems it is possible to find a corresponding functional. A variational equation can be derived by the differential equation or be set up on the basis of energy balance without knowing the differential operator. In this

5.3.1 Variational expressions for Poisson's equation

5.3.1.1 Mathematical manipulation [13]

The boundary value problem of Poisson's equation subject to different boundary conditions is expressed by:

$$\begin{cases} \mathscr{L}(\varphi) - f = 0 & \text{in } \Omega \\ \varphi|_{\Gamma_1} = U_0 & \text{boundary condition of the first kind} \\ f_1(\Gamma)\varphi + \dfrac{\partial \varphi}{\partial n} = f_2(\Gamma) & \text{boundary condition of the third kind} \end{cases} \quad (5.3.1)$$

where $\mathscr{L} = \nabla^2$. It was proved that if $\mathscr{L}$ is a definite self-adjoint operator, its functional exists.

Multiplying the first equation of Eq. (5.3.1) by the first order variation of φ, ($\delta\varphi$), and integrating the result over the domain yields:

$$\int_\Omega \delta\varphi [\mathscr{L}(\varphi) - f] d\Omega = 0 . \quad (5.3.2)$$

The purpose of this step is to manipulate the resulting expression (Eq. (5.3.2)) into a form that allows the variational operator δ to be moved out of the integral, i.e.

$$\delta \int_\Omega \varphi [\mathscr{L}(\varphi) - f] d\Omega = 0 . \quad (5.3.3)$$

Then, the integral is a functional written by:

$$I(\varphi) = \int_\Omega \varphi [\mathscr{L}(\varphi) - f] d\Omega . \quad (5.3.4)$$

Apply the vector identity:

$$\nabla \cdot (u\mathbf{v}) = \nabla u \cdot \mathbf{v} + u \nabla \cdot \mathbf{v}$$

letting $u = \delta\varphi$ and $\mathbf{v} = \nabla\varphi$, and noting that $\nabla^2\varphi = \nabla \cdot \nabla\varphi$, one obtains:

$$\delta\varphi \nabla \cdot \nabla\varphi = \nabla \cdot (\delta\varphi \nabla\varphi) - \nabla(\delta\varphi) \cdot \nabla\varphi . \quad (5.3.5)$$

Substitution of Eq. (5.3.5) into Eq. (5.3.2) and consideration of the assumption $\mathscr{L}(\varphi) = \nabla^2\varphi$, leads to:

$$\int_\Omega [\nabla \cdot (\delta\varphi \nabla\varphi) - \nabla(\delta\varphi) \cdot \nabla\varphi - \delta f \varphi] d\Omega = 0 . \quad (5.3.6)$$

The first term of Eq. (5.3.6) can be transformed into a surface integral using Green's theorem. Then Eq. (5.3.6) becomes

$$\oint_s \delta\varphi \nabla\varphi \mathbf{n} ds - \int_\Omega \nabla(\delta\varphi) \cdot \nabla\varphi d\Omega - \int_\Omega \delta f \varphi d\Omega = 0 . \qquad (5.3.7)$$

where $\mathbf{n}$ is a unit vector normal to the surface S of domain Ω. Since $\varphi|_\Gamma = U_0$ then $\delta\varphi = 0$ on the surface and Eq. (5.3.7) is reduced to:

$$\int_\Omega [\nabla(\delta\varphi) \cdot \nabla\varphi + \delta f \varphi] d\Omega = 0 . \qquad (5.3.8)$$

By noting that $\nabla(\delta\varphi) \cdot \nabla\varphi = \frac{1}{2} \delta(\nabla\varphi \cdot \nabla\varphi)$ Eq. (5.3.8) leads to:

$$\delta \int_\Omega [\frac{1}{2}(\nabla\varphi \cdot \nabla\varphi) + f\varphi] d\Omega = 0 . \qquad (5.3.9)$$

Hence the equivalent functional of Eq. (5.3.1) is:

$$I(\varphi) = \int_\Omega [\frac{1}{2}(\nabla\varphi)^2 + f\varphi] d\Omega \qquad (5.3.10)$$

subject to $\varphi|_\Gamma = U_0$.

This is the equivalent functional of Poisson's equation subject to Dirichlet boundary conditions. In comparison of this equation with Poisson's equation, it is clear that the order of the derivative of function φ contained in the equivalent functional is reduced by one.

If the boundary condition of the third kind ($\partial\varphi/\partial n = f_2 - f_1\varphi$) is substituted into Eq. (5.3.7), one obtains:

$$\oint_s \delta\varphi(f_2 - f_1\varphi) dS - \delta\int_\Omega [\frac{1}{2}\nabla\varphi \cdot \nabla\varphi) - f\varphi] d\Omega = 0 . \qquad (5.3.11)$$

Consideration of that $f = -\rho/\varepsilon$, thus the corresponding functonal is:

$$I(\varphi) = \frac{1}{2}\varepsilon\int_\Omega |\nabla\varphi|^2 d\Omega - \int_\Omega \rho\varphi d\Omega + \oint_s \varepsilon \frac{1}{2}(f_1\varphi^2 - f_2\varphi) dS \qquad (5.3.12)$$

where the boundary condition of the third kind is included in the functional. For inhomogeneous boundary conditions of the second kind $\partial\varphi/\partial n = f_2$, $f_1 = 0$, then:

$$I(\varphi) = \int_\Omega \frac{1}{2}\varepsilon|\nabla\varphi|^2 d\Omega - \int_\Omega \rho\varphi d\Omega - \int_S \varepsilon f_2 \varphi dS . \qquad (5.3.13)$$

If $\partial\varphi/\partial n = 0$, then

$$I(\varphi) = \int_\Omega \frac{1}{2}\varepsilon |\nabla\varphi|^2 d\Omega - \int_\Omega \rho\varphi d\Omega . \qquad (5.3.14)$$

5.3 Variational expressions for electromagentic field problems

Equations (5.2.10), (5.3.12), (5.3.13) and (5.3.14) are functionals of Poisson's equation subject to different boundary conditions. These equations can be written in a unified form, i.e.

$$\begin{cases} I(\varphi) = \frac{1}{2}\int_\Omega \varepsilon |\nabla \varphi|^2 d\Omega + \oint_s \varepsilon (\frac{1}{2} f_1 \varphi^2 - f_2 \varphi) dS - \int_\Omega \rho \varphi d\Omega = \min \\ \varphi|_{\Gamma_1} = U_0 . \end{cases} \quad (5.3.15)$$

This equation represents a constraint variational problem. Hence, *the problem subject to Dirichlet boundary conditions corresponds to a constrained functional. All other boundary conditions (Neumann and Robin) are contained in the equivalent functionals.*

Alternatively, Euler's equation of Eq. (5.3.15) can be derived from the following procedures. Rewrite the functional in Eq. (5.3.15) in the form of

$$I(\varphi) = \frac{1}{2}\int_\Omega \varepsilon \left[\left(\frac{\partial \varphi}{\partial x}\right)^2 + \left(\frac{\partial \varphi}{\partial y}\right)^2 + \left(\frac{\partial \varphi}{\partial z}\right)^2 \right] d\Omega$$

$$+ \oint_s \varepsilon \left(\frac{1}{2} f_1 \varphi^2 - f_2 \varphi \right) dS - \int_\Omega \rho \varphi d\Omega$$

according to the definition of $\delta I(\varphi) = \lim_{\alpha \to 0} \frac{F(\varphi + \delta \varphi) - F(\varphi)}{\alpha}$, $\delta \varphi = \alpha \eta$, and in consideration of $\left(\frac{\partial \varphi}{\partial x} + \alpha \eta_x\right)^2 = \left(\frac{\partial \varphi}{\partial x}\right)^2 + 2\left(\frac{\partial \varphi}{\partial x}\right) \alpha \eta_x + \alpha^2 \eta_x^2$, then

$$\delta I(\varphi) = \int_\Omega \left[\varepsilon \left(\frac{\partial \varphi}{\partial x} \eta_x + \frac{\partial \varphi}{\partial y} \eta_y + \frac{\partial \varphi}{\partial z} \eta_z \right) - \rho \eta \right] d\Omega$$

$$+ \oint_s \varepsilon f_1 \varphi \eta dS - \oint_s \varepsilon f_2 \eta dS$$

$$= \int_\Omega (\varepsilon \nabla \varphi \cdot \nabla \eta - \rho \eta) d\Omega + \oint_s \varepsilon f_1 \varphi \eta dS - \oint_s \varepsilon f_2 \eta dS$$

$$= \varepsilon \int_\Omega (\nabla \cdot (\eta \nabla \varphi) - \eta \nabla^2 \varphi) d\Omega - \int_\Omega \rho \eta d\Omega + \oint_s \varepsilon (f_1 \eta \varphi - f_2 \eta) dS .$$

If $\delta I(\varphi) = 0$, the following equations must be satisfied

$$\begin{cases} -\varepsilon \nabla^2 \varphi = \rho \\ \dfrac{\partial \varphi}{\partial n} + f_1 \varphi = f_2 . \end{cases} \quad (5.3.16)$$

Thus, it is proved that Eq. (5.3.15) is the equivalent functional of Poisson's equation with inhomogeneous boundary conditions of a different kind. It is obvious that the homogeneous or inhomogeneous boundary conditions of the second and the third kind are automatically satisfied in the process of variations where the divergence theorem can be used. *Hence the boundary conditions of the second and third kind, called natural boundary conditions, do not need to be dealt with in finite element methods. This is an important advantage of FEM.* In other words, the boundary conditions of the second and third kind are automatically satisfied using the variational approach. The only exception is where boundary conditions of the first kind are considered as a constrained condition of the variational equation.

For axisymmetric fields, the functional of Poisson's equation is

$$\begin{cases} I(\varphi) = 2\pi \int_\Omega \frac{1}{2}\left[\varepsilon\left[\left(\frac{\partial \varphi}{\partial r}\right)^2 + \left(\frac{\partial \varphi}{\partial z}\right)^2\right] - 2\rho\varphi\right] r\,dr\,dz \\ \qquad + 2\pi \int_\Gamma \varepsilon\left(\frac{1}{2}f_1\varphi^2 - f_2\varphi\right) r\,d\Gamma \\ \varphi|_{\Gamma_2} = g(r, z) \,. \end{cases} \qquad (5.3.17)$$

5.3.1.2 Physical manipulation

In electrostatic fields, the energy is a functional of potential φ, i.e.

$$W_e(\varphi) = \frac{1}{2}\int_\Omega \mathbf{D}\cdot\mathbf{E}\,d\Omega = \frac{\varepsilon}{2}\int_\Omega |\nabla\varphi|^2 d\Omega \,. \qquad (5.3.18)$$

According to Eq. (5.2.18)

$$\delta W_e(\varphi) = W_e(\varphi + \delta\varphi) - W_e(\varphi) = \varepsilon \int_\Omega \nabla\varphi \cdot \nabla\delta\varphi\, d\Omega \qquad (5.3.19)$$

By using vector identity and then

$$\nabla\varphi \cdot \nabla\delta\varphi = \nabla\cdot(\delta\varphi\nabla\varphi) - \delta\varphi\nabla\cdot\nabla\varphi \,. \qquad (5.3.20)$$

Substitution of Eq. (5.3.20) into Eq. (5.3.19), leads to:

$$\delta W_e(\varphi) = \varepsilon\int_\Omega \nabla\cdot(\delta\varphi\nabla\varphi)\,d\Omega + \int_\Omega \varepsilon\delta\varphi\nabla\cdot\mathbf{E}\,d\Omega$$

$$= \oint_s \varepsilon\delta\varphi\frac{\partial\varphi}{\partial n}\,dS + \int_\Omega \delta\varphi\rho\,d\Omega \qquad (5.3.21)$$

5.3 Variational expressions for electromagentic field problems

where S is the boundary surface of the domain and n is the unit length of the normal direction of surface. Equation (5.3.21) demonstrates that the incremental of energy δW_e in the domain equals the incremental of energy due to the volume charge ρ and the surface charge σ. Substitution of the third kind of boundary condition into Eq. (5.3.21), yields:

$$\delta W_e(\varphi) = \oint_s \varepsilon \delta \varphi (f_2 - f_1 \varphi) \, dS + \int_\Omega \rho \delta \varphi \, d\Omega . \tag{5.3.22}$$

Combining Eqs. (5.3.19) and (5.3.22), the result is:

$$\varepsilon \int_\Omega \nabla \varphi \delta \nabla \varphi \, d\Omega - \oint_s \varepsilon \delta \varphi (f_2 - f_1 \varphi) \, dS - \int_\Omega \rho \delta \varphi \, d\Omega = 0 . \tag{5.3.23}$$

This is the variational expression of Poisson's equation with boundary conditions of the third kind. In other words, the equivalent functional is:

$$I(\varphi) = \tfrac{1}{2} \int_\Omega \varepsilon |\nabla \varphi|^2 d\Omega + \oint_s \varepsilon (\tfrac{1}{2} f_1 \varphi^2 - f_2 \varphi) \, dS - \int_\Omega \rho \varphi \, d\Omega . \tag{5.3.24}$$

Of course this equation is exactly the same as Eq. (5.3.12)

5.3.2 Variational expressions for Poisson's equation in piece-wise homogeneous materials

Since Gauss's theorem is used in proving the self-adjoint property of the Laplacian operator $\mathscr{L}$, the variational equation (5.3.24) cannot be extended to the case where the permeabilities are discontinuous at the interface, S, shown in Fig. 5.3.1.

Using Eq. (5.3.24) in both areas of Ω_1 and Ω_2 respectively leads to

$$I_1(\varphi) = \tfrac{1}{2} \int_{\Omega_1} \varepsilon |\nabla \varphi|^2 d\Omega + \oint_{\Gamma_1 + s} \varepsilon (\tfrac{1}{2} f_1 \varphi^2 - f_2 \varphi) \, d\Gamma - \int_{\Omega_1} \rho \varphi \, d\Omega \tag{5.3.25}$$

and

$$I_2(\varphi) = \tfrac{1}{2} \int_{\Omega_2} \varepsilon |\nabla \varphi|^2 d\Omega + \oint_{\Gamma_2 + s} \varepsilon (\tfrac{1}{2} f_1 \varphi^2 - f_2 \varphi) \, d\Gamma - \int_{\Omega_2} \rho \varphi \, d\Omega . \tag{5.3.26}$$

Fortunately, at the interface S, the normal directions of the boundary are

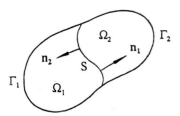

Fig. 5.3.1. Piece-wise uniform domain

opposite to each other, and which satisfies:

$$\varphi|_{s_-} = \varphi|_{s_+}, \qquad \varepsilon_1\left(\frac{\partial\varphi}{\partial n}\right)\bigg|_{s_-} = \varepsilon_2\left(\frac{\partial\varphi}{\partial n}\right)\bigg|_{s_+}. \tag{5.3.27}$$

Combining Eqs. (5.3.25) and (5.3.26) to Eq. (5.3.27), and considering the edge of the discretized elements along the interface, then the variational expression of Poisson's equation in piece-wise uniform materials are:

$$\delta I(\varphi) = \tfrac{1}{2} \int_{\Omega_1+\Omega_2} \varepsilon |\nabla\varphi|^2 d\Omega - \int_{\Omega_1+\Omega_1} \rho\varphi d\Omega - \oint_{\Gamma_1+\Gamma_2} \varepsilon(\tfrac{1}{2}f_1\varphi^2 - f_2\varphi)d\Gamma$$

$$= \min. \tag{5.3.28}$$

where Γ_1 and Γ_2 are boundaries of region 1 and 2, respectively. It is obvious that the interfacial boundary conditions at the interface are automatically satisfied in the variational approach.

5.3.3 Variational expression for the scalar Helmholtz equation

Neglecting the displacement current, Maxwell's equations are reduced to:

$$\begin{cases} \nabla \times \mathbf{H} = \mathbf{J}_s + \mathbf{J}_e \\ \nabla \times \mathbf{E} = -\dfrac{\partial \mathbf{B}}{\partial t} \\ \nabla \cdot \mathbf{H} = 0 \\ \nabla \cdot \mathbf{E} = 0. \end{cases} \tag{5.3.29}$$

By introducing the vector potential $\mathbf{A}$ and considering in the case of sinusoidal excitation, the Helmholtz equation is:

$$(\nabla^2 + \beta^2)\mathbf{A} = \mu\mathbf{J}_s \tag{5.3.30}$$

where

$$\beta^2 = -j\omega\mu\gamma \tag{5.3.31}$$

and $\mathbf{J}_s$ is the imposed current density.

Using formula (5.2.59), the functional corresponding to the Helmholtz equation is:

$$I(\mathbf{A}) = \langle \nabla^2 \mathbf{A}, \mathbf{A} \rangle + \beta^2 \langle \mathbf{A}, \mathbf{A} \rangle + 2\langle \mu\mathbf{J}_s, \mathbf{A} \rangle$$

$$= \int_\Omega \nabla^2 \mathbf{A} \cdot \mathbf{A} d\Omega + \int_\Omega \beta^2 \mathbf{A} \cdot \mathbf{A} d\Omega + 2\int_\Omega \mu\mathbf{J}_s \mathbf{A} d\Omega. \tag{5.3.32}$$

Applying the vector identity: $\nabla^2 \mathbf{A} = \nabla(\nabla \cdot \mathbf{A}) - \nabla \times \nabla \times \mathbf{A}$ and Green's theorem (1.2.15):

$$\int_\Omega (\nabla \times \mathbf{P} \cdot \nabla \times \mathbf{Q} - \mathbf{P} \cdot \nabla \times \nabla \times \mathbf{Q}) d\Omega = \oint_s \mathbf{P} \times \nabla \times \mathbf{Q} d\mathbf{S}$$

5.3 Variational expressions for electromagentic field problems

Eq. (5.3.32) becomes:

$$I(\mathbf{A}) = \int_\Omega \mathbf{A} \cdot \nabla(\nabla \cdot \mathbf{A}) d\Omega - \int_\Omega \nabla \times \mathbf{A} \cdot \nabla \times \mathbf{A} d\Omega + \int_\Omega \beta^2 \mathbf{A} \cdot \mathbf{A} d\Omega$$

$$+ \int_\Omega 2\mu \mathbf{J}_s \mathbf{A} d\Omega + \oint_s \mathbf{A} \times \nabla \times \mathbf{A} d\mathbf{S} \ . \qquad (5.3.33)$$

Considering a 2-D case, A_z is replaced by A and the Coulomb gauge $\nabla \cdot \mathbf{A} = 0$ is used, then Eq. (5.3.33) reduces to:

$$I(A) = \int_\Omega \left[\left(\frac{\partial A}{\partial x}\right)^2 + \left(\frac{\partial A}{\partial y}\right)^2 \right] dxdy + j\omega\mu\gamma \int_\Omega A^2 dxdy - \int_\Omega 2\mu J_s A dxdy$$

$$+ \oint_s A \frac{\partial A}{\partial n} dS$$

$$= \int_\Omega F(x, y, A, A_x, A_y) dxdy + \oint_s A \frac{\partial A}{\partial n} dS \qquad (5.3.34)$$

where A_x, A_y are partial derivatives with respect to x and y, respectively.

If the problem is subject to Dirichlet boundary conditions, then the functional is a constrained functional:

$$I(A) = \int_\Omega \left[\left(\frac{\partial A}{\partial x}\right)^2 + \left(\frac{\partial A}{\partial y}\right)^2 \right] dxdy + j\omega\mu\gamma \int_\Omega A^2 dxdy - \int_\Omega 2\mu J_s A dxdy$$

$$+ \oint_s A \frac{\partial A}{\partial n} dS$$

$$A|_\Gamma = g(s) \ . \qquad (5.3.35)$$

If the problem is axisymmetric, then the equivalent functional is [12]:

$$I(A) = \int_\Omega \left\{ r\left[\left(\frac{\partial A}{\partial r}\right)^2 + \left(\frac{\partial A}{\partial z}\right)^2 - \beta^2 A^2 - 2\mu J A \right] \right.$$

$$\left. + 2A \frac{\partial A}{\partial r} + \frac{A^2}{r} \right\} d\Omega \qquad (5.3.36)$$

where $A = A_\vartheta$, $J = J_\vartheta$.

The variational expression of the Helmholtz equation in linear media can also be derived by the extremum principle of energy.

5.3.4 Variational expression for the magnetic field in a non-linear medium [14]

In this section, the discussion is treated in a different way. Assume that the corresponding functional of a specific partial differential equation is given, then proving the Euler's equation of the known functional is just the partial differential equation which is to be solved.

A 2-D non-linear boundary value problem is represented by:

$$\begin{cases} \dfrac{\partial}{\partial x}\left(\dfrac{1}{\mu}\cdot\dfrac{\partial A}{\partial x}\right) + \dfrac{\partial}{\partial y}\left(\dfrac{1}{\mu}\cdot\dfrac{\partial A}{\partial y}\right) = -J & \text{in } \Omega \\ \dfrac{\partial A}{\partial n} + f_1 A = f_2 & \text{on } \Gamma_1 \\ A = \text{const} & \text{on } \Gamma_2 \end{cases} \qquad (5.3.37)$$

where $A = A_z$. The equivalent variational problem of Eq. (5.3.37) is:

$$\begin{cases} I(A) = \int_\Omega \left(\int_0^B \dfrac{1}{\mu} B dB\right) dxdy - \int_\Omega JA_z \, dxdy + \oint_{\Gamma_1} \dfrac{1}{\mu}\left(\dfrac{1}{2} f_1 A^2 - f_2 A\right) d\Gamma \\ = \min \\ A|_{\Gamma_2} = \text{const}. \end{cases} \qquad (5.3.38)$$

Now, the problem is to prove that Euler's equation of (5.3.38) is the first equation of (5.3.37). Based on the definition of variation of a functional:

$$\delta I(A) = \dfrac{\partial}{\partial \alpha} I[A(x,y) + \alpha \delta A]|_{\alpha=0}$$

$$= \left\{\dfrac{\partial}{\partial \alpha}\left[\int_\Omega \left(\int_0^B \dfrac{1}{\mu} B dB\right) dxdy - \int_\Omega JA \, dxdy \right.\right.$$

$$\left.\left. + \int_{\Gamma_1} \dfrac{1}{\mu}\left(\dfrac{1}{2} f_1 A^2 - f_2 A\right) d\Gamma\right]\right\}\bigg|_{\alpha=0} = 0. \qquad (5.3.39)$$

Because $\dfrac{\partial F}{\partial \alpha} = \dfrac{\partial F}{\partial B}\dfrac{\partial B}{\partial A}\dfrac{\partial A}{\partial \alpha} = \dfrac{\partial F}{\partial B}\dfrac{\partial B}{\partial A}\delta A,$ then

$$\delta I(A) = \int_\Omega \dfrac{\partial}{\partial B}\left(\int_0^B \dfrac{1}{\mu} B dB\right)\dfrac{\partial B}{\partial A} \delta A \, dxdy - \int_\Omega J\delta A \, dxdy$$

5.3 Variational expressions for electromagentic field problems

$$+ \int_\Gamma \frac{1}{\mu}(f_1 A - f_2)\delta A d\Gamma$$

$$= \int_\Omega \frac{1}{\mu} B \frac{\partial B}{\partial A} \delta A dxdy - \int_\Omega J\delta A dxdy + \int_{\Gamma_1} \frac{1}{\mu}(f_1 A - f_2)\delta A d\Gamma = 0.$$

(5.3.40)

By considering the relationship between $\mathbf{B}$ and A, the following equations are obtained.

$$B^2 = B_x^2 + B_y^2, \quad B_x = \frac{\partial A}{\partial y}, \quad B_y = -\frac{\partial A}{\partial x} \quad (5.3.41)$$

then

$$B\frac{\partial B}{\partial A} = B\left(\frac{\partial B}{\partial B_x}\frac{\partial B_x}{\partial A} + \frac{\partial B}{\partial B_y}\frac{\partial B_y}{\partial A}\right) = B_x\frac{\partial B_x}{\partial A} + B_y\frac{\partial B_y}{\partial A} \quad (5.3.42)$$

and thus

$$\delta I = \int_\Omega \left(\frac{1}{\mu}\frac{\partial A}{\partial x}\frac{\partial(\delta A)}{\partial x} + \frac{1}{\mu}\frac{\partial A}{\partial y}\frac{\partial(\delta A)}{\partial y}\right)dxdy - \int_\Omega J\delta A dxdy$$

$$+ \int_{\Gamma_1} \frac{1}{\mu}(f_1 A - f_2)\delta A d\Gamma = 0. \quad (5.3.43)$$

The first integral term of the above equation can be expressed by:

$$\int_\Omega \left[\frac{\partial}{\partial x}\left(\frac{1}{\mu}\frac{\partial A}{\partial x}\delta A\right) + \frac{\partial}{\partial y}\left(\frac{1}{\mu}\frac{\partial A}{\partial y}\delta A\right)\right]dxdy$$

$$= \int_\Omega \left[\frac{\partial}{\partial x}\left(\frac{1}{\mu}\frac{\partial A}{\partial x}\right) + \frac{\partial}{\partial y}\left(\frac{1}{\mu}\frac{\partial A}{\partial y}\right)\right]\delta A dxdy$$

$$+ \int_\Omega \left(\frac{1}{\mu}\frac{\partial A}{\partial x}\frac{\partial(\delta A)}{\partial x} + \frac{1}{\mu}\frac{\partial A}{\partial y}\frac{\partial(\delta A)}{\partial y}\right)dxdy. \quad (5.3.44)$$

Using Green's formula

$$\int_\Omega \left(\frac{\partial N}{\partial x} + \frac{\partial M}{\partial y}\right)dxdy = \oint_\Gamma [N\cos(\mathbf{n}, \mathbf{i}) + M\cos(\mathbf{n}, \mathbf{j})]d\Gamma \quad (5.3.45)$$

the LHS of Eq. (5.3.44) is transformed to a surface integral, i.e.

$$\int_\Omega \left[\frac{\partial}{\partial x}\left(\frac{1}{\mu}\frac{\partial A}{\partial x}\delta A\right) + \frac{\partial}{\partial y}\left(\frac{1}{\mu}\frac{\partial A}{\partial y}\delta A\right) \right] dxdy$$

$$= \int_\Gamma \left[\frac{1}{\mu}\frac{\partial A}{\partial x}\delta A \cdot \cos(\mathbf{n}, \mathbf{i}) + \frac{1}{\mu}\frac{\partial A}{\partial y}\delta A \cdot \cos(\mathbf{n}, \mathbf{j}) \right] d\Gamma$$

$$= \oint_\Gamma \frac{1}{\mu}\frac{\partial A}{\partial n}\delta A d\Gamma . \qquad (5.3.46)$$

If $A = $ const on Γ_2, then $\delta A = 0$. Equation (5.3.46) then reduces to:

$$\oint_\Gamma \frac{1}{\mu}\frac{\partial A}{\partial n}\delta A d\Gamma = \oint_{\Gamma_1} \frac{1}{\mu}\frac{\partial A}{\partial n}\delta A d\Gamma . \qquad (5.3.47)$$

Combining Eqs. (5.3.43), (5.3.44) and (5.3.47) results in:

$$\delta I = -\int_\Omega \left[\frac{\partial}{\partial x}\left(\frac{1}{\mu}\frac{\partial A}{\partial x}\right) + \frac{\partial}{\partial y}\left(\frac{1}{\mu}\frac{\partial A}{\partial y}\right) + J \right]\delta A dxdy$$

$$+ \int_{\Gamma_1} \frac{1}{\mu}\left(\frac{\partial A}{\partial n} + f_1 A - f_2\right)\delta A d\Gamma = 0 . \qquad (5.3.48)$$

Therefore, the following equations must be satisfied:

$$\begin{cases} \dfrac{\partial}{\partial x}\left(\dfrac{1}{\mu}\cdot\dfrac{\partial A}{\partial x}\right) + \dfrac{\partial}{\partial y}\left(\dfrac{1}{\mu}\cdot\dfrac{\partial A}{\partial y}\right) = -J & \text{in } \Omega \\ \dfrac{\partial A}{\partial n} + f_1 A = f_2 & \text{on } \Gamma_1 \\ A|_{\Gamma_2} = \text{const} & \text{on } \Gamma_2 \end{cases} \qquad (5.3.49)$$

This proves that Eq. (5.3.37) is Euler's equation of the functional given in Eq. (5.3.38).

The variational expressions for the integral operators are discussed in reference [12].

5.4 Variational finite element method

It has been shown in Sect. 5.2 *that to solve a partial differential equation subject to the given boundary conditions is equivalent to solving a variational problem. This problem is to find an extremum function of the functional subject to the given constraints. This equivalence is based on the fact that the functional is maximum, minimum or stationary only when the corresponding Euler's equations and the corresponding boundary conditions are satisfied.* Before the development of digital computers, only simple variational problems could be solved by the Ritz method, the Galerkin method or the orthogonal series method. With digital computers, numerical methods are developed rapidly. The extremum of a functional can be found by using numerical methods.

5.4.1 Ritz method

The Ritz method is used for the variational approach. The unknown solution is assumed in terms of a summation of series which is called trial function. The trial function consists of unknown parameters, C_i, and basis functions, ψ_i. For example, the approximate solution is expressed as:

$$u = \sum_{i=1}^{n} C_i \psi_i \quad (i = 1, \ldots, n). \tag{5.4.1}$$

Substituting the trial function into the functional allows the functional to be expressed in terms of unknown parameters. Differentiating the functional with respect to these parameters and then setting to zero gives:

$$\frac{\partial I(C_i \psi_i)}{\partial C_i} = 0. \tag{5.4.2}$$

If there are N unknown parameters, there will be N simultaneous equations to be solved to obtain these unknown parameters. In this method, *the trial function is defined over the whole solution domain and satisfies at least some and usually all the boundary conditions*. The accuracy of the approximate solution depends on the choice of the trial function. The approximation improves with the size of the trial function family and the number of adjustable parameters. If the trial functions are part of an infinite set of functions they are capable of representing the unknown function to any degree of accuracy. If the form of the solution can be guessed, the trial function should be close to that imaged solution.

Example 5.4.1. Find the function $u(x)$ which satisfies the differential equation

$$\frac{d^2 u}{dx^2} = -f(x) = -2 \tag{5.4.3}$$

subject to

$$u(a) = A, \quad u(b) = B.$$

Solution. $u(x)$ is a continuous function in the closed region $[a, b]$. This problem is equivalent to the problem of finding the function $u(x)$ which minimizes the functional

$$I(u) = \int_a^b \left[\frac{1}{2} \left(\frac{du}{dx} \right)^2 - f(x) u(x) \right] dx. \qquad (5.4.4)$$

According to the Ritz method, the solution of Eq. (5.4.4) is approximately represented by a trial function of the form:

$$u(x) = C_1 \psi_1(x) + C_2 \psi_2(x) + \ldots + C_n \psi_n(x) = \sum_{i=1}^n C_i \psi_i(x) \qquad (5.4.5)$$

where C_i are unknown parameters to be determined. ψ_i should be selected in such a way that $u(x)$ satisfies the boundary conditions. If $A = B = 0$, the simplest trial function is chosen as:

$$u(x) \cong (x - a)(x - b)(C_1 + C_2 x + C_3 x^2 + \ldots + C_n x^{n-1}). \qquad (5.4.6)$$

To avoid the tedious calculation, let

$$u(x) = C_1 (x - a)(x - b). \qquad (5.4.7)$$

Substituting Eq. (5.4.7) into Eq. (5.4.4) and minimizing the functional $I(u)$ with respect to the parameter C_1, leads to:

$$\frac{\partial I(u)}{\partial C_1} = \frac{\partial}{\partial C_1} \int_a^b \left\{ \frac{1}{2} C_1^2 \left[2x - (a+b) \right]^2 - 2 C_1 (x - a)(x - b) \right\} dx = 0$$

$$C_1 \left[\tfrac{4}{3} (a^2 + ab + b^2) - (a+b)^2 \right] = 2 \left[\tfrac{1}{3} (a^2 + ab + b^2) \right.$$
$$\left. - \tfrac{1}{2} (a+b)^2 + ab \right]$$

hence

$$C_1 = -1$$

thus

$$u(x) = -(x - a)(x - b).$$

This result is exactly the same as the analytic solution of Eq. (5.4.3). For this problem, the lower order approximation is sufficient to obtain a satisfactory result.

The advantage of the Ritz method is that the order of the derivatives of the function contained in the functional is reduced by one compared to the differential equation. However, if the shape of the boundary is complicated, it is hard to choose a trial function which satisfies the boundary conditions. The finite element method is an improvement of the Ritz method. It subdivides the whole domain into a large number of elements, then the trial functions only need to be satisfied in each element. Hence it is much easier to choose the approximate function than in the Ritz method.

5.4.2 Finite element method (FEM)

The FEM and the Ritz method are essentially the same. Both use a trial function as an approximate solution and determine the parameters included in the trial functions by making the functional minimum, maximum or stationary. The main difference between these two methods is that the assumed trial functons in the Ritz method are defined over the whole domain which satisfy the boundary conditions. In FEM, the trial functions are defined only within the elements, not for the whole domain. Hence the trial function in FEM only has to satisfy certain continuity conditions within elements. The elements generated in FEM are simple in shape but collectively can represent very complex geometries. Hence FEM is far more versatile than the Ritz method. In other words, FEM becomes a special case of the Ritz method when the piecewise continuous trial functions obey certain continuity and completeness conditions.

5.4.2.1 Domain discretization

The principle of domain discretization is the same as that described in Sect. 4.4. For 2-D problem, the simplest shape of the element is a triangle (3-node) or a quadrangle (4-node).

For first order triangular elements (3-node), the trial function is assumed to be a linear function of x and y

$$u(x, y) = \alpha_1 + \alpha_2 x + \alpha_3 y = \sum_{k=1}^{3} N_k^e u_k$$

$$= \frac{1}{2S}\left[(a_i + b_i x + c_i y)u_i + (a_j + b_j x + c_j y)u_j\right.$$

$$\left. + (a_m + b_m x + c_m y)u_m\right] \quad (5.4.8)$$

where the subscripts i, j, m are the vertices of the triangle. This equation indicates that the function u within the element depends on the nodal values, u_k, and the nodal coordinates. The constants in Eq. (5.4.8) are the same as those derived in Chap. 4, i.e.

$$\begin{cases} a_i = x_j y_m - x_m y_j & b_i = y_j - y_m & c_i = x_m - x_j \\ a_j = x_m y_i - x_i y_m & b_j = y_m - y_i & c_j = x_i - x_m \\ a_m = x_i y_j - x_j y_i & b_m = y_i - y_j & c_m = x_j - x_i \end{cases} \quad (5.4.9)$$

The shape functions N_k^e contained in Eq. (5.4.8) are:

$$\begin{cases} N_i^e(x, y) = \dfrac{1}{2S}(a_i + b_i x + c_i y) \\ N_j^e(x, y) = \dfrac{1}{2S}(a_j + b_j x + c_j y) \\ N_m^e(x, y) = \dfrac{1}{2S}(a_m + b_m x + c_m y) . \end{cases} \qquad (5.4.10)$$

In axisymmetric coordinates, the trial functions and the shape functions are:

$$u(x, y) = \alpha_1 + \alpha_2 r + \alpha_3 z = \sum_{k=1}^{3} N_k^e u_k \qquad (5.4.11)$$

$$\begin{cases} N_i^e(r, z) = \dfrac{1}{2S}(a_i + b_i r + c_i z) \\ N_j^e(r, z) = \dfrac{1}{2S}(a_i + b_i r + c_i z) \\ N_m^e(r, z) = \dfrac{1}{2S}(a_i + b_i r + c_i z) . \end{cases} \qquad (5.4.12)$$

The next step of FEM is to substitute the trial function into the equivalent functional and to determine the constant values u_k by the principle of variations.

5.4.2.2 Finite element equation of a Laplacian problem

The equivalent functional of Laplace's equation in a 2-D case is:

$$I(\varphi) = \int_s \frac{\varepsilon}{2}\left[\left(\frac{\partial \varphi}{\partial x}\right)^2 + \left(\frac{\partial \varphi}{\partial y}\right)^2\right] dxdy = \Sigma_e I_e(\varphi) \quad \text{in } \Omega \qquad (5.4.13)$$

$$I_e(\varphi) = \int_{S_e} \tfrac{1}{2}\varepsilon |\nabla \varphi|^2 dxdy \quad \text{in } \Omega_e \qquad (5.4.14)$$

where $I_e(\varphi)$ is the functional within each element. Assume

$$\varphi(x, y) = \alpha_1 + \alpha_2 x + \alpha_3 y = \sum_{k=1}^{3} N_k \varphi_k \qquad (5.4.15)$$

5.4 Variational finite element method

and consider that:

$$\begin{cases} \dfrac{\partial \varphi}{\partial x} = \left[\dfrac{\partial N}{\partial x}\right]_e \{\varphi\} = \left[\dfrac{\partial N_i}{\partial x} \dfrac{\partial N_j}{\partial x} \dfrac{\partial N_m}{\partial x}\right] \begin{Bmatrix} \varphi_i \\ \varphi_j \\ \varphi_m \end{Bmatrix} = \dfrac{1}{2S}[b_i \; b_j \; b_m] \begin{Bmatrix} \varphi_i \\ \varphi_j \\ \varphi_m \end{Bmatrix} \\ \\ \dfrac{\partial \varphi}{\partial y} = \left[\dfrac{\partial N}{\partial y}\right]_e \{\varphi\} = \left[\dfrac{\partial N_i}{\partial y} \dfrac{\partial N_j}{\partial y} \dfrac{\partial N_m}{\partial y}\right] \begin{Bmatrix} \varphi_i \\ \varphi_j \\ \varphi_m \end{Bmatrix} = \dfrac{1}{2S}[c_i \; c_j \; c_m] \begin{Bmatrix} \varphi_i \\ \varphi_j \\ \varphi_m \end{Bmatrix} \end{cases}$$

(5.4.16)

The compact form of Eq. (5.4.16) is:

$$(\nabla \varphi) = \begin{Bmatrix} \dfrac{\partial \varphi}{\partial x} \\ \dfrac{\partial \varphi}{\partial y} \end{Bmatrix} = \dfrac{1}{2S}\begin{bmatrix} b_i & b_j & b_m \\ c_i & c_j & c_m \end{bmatrix} \begin{Bmatrix} \varphi_i \\ \varphi_j \\ \varphi_m \end{Bmatrix} = \mathbf{B}_e \{\varphi\} \qquad (5.4.17)$$

where

$$\mathbf{B}_e = \dfrac{1}{2S}\begin{bmatrix} b_i & b_j & b_m \\ c_i & c_j & c_m \end{bmatrix} \qquad (5.4.18)$$

and S is the area of Δ_{ijm}. Substitution of Eq. (5.4.17) into Eq. (5.4.14) leads to:

$$I_e(\varphi) = \int_{S_e} \tfrac{1}{2}\varepsilon \{\nabla \varphi\}^T \{\nabla \varphi\} \, dxdy = \tfrac{1}{2}\varepsilon \int_{S_e} [\mathbf{B}_e\{\varphi\}]^T [\mathbf{B}_e\{\varphi\}] \, dxdy$$

$$= \tfrac{1}{2}\varepsilon \{\varphi\}^T \int_S \mathbf{B}_e^T \mathbf{B}_e \, dxdy \{\varphi\} = \tfrac{1}{2}\{\varphi\}^T \mathbf{K}_e \{\varphi\} \qquad (5.4.19)$$

where

$$\{\nabla \varphi\}^T = \begin{bmatrix} \dfrac{\partial \varphi}{\partial x} & \dfrac{\partial \varphi}{\partial y} \end{bmatrix}. \qquad (5.4.20)$$

Since $\mathbf{B}_e$ is a constant matrix, the integration of Eq. (5.4.19) is easy to evaluate, i.e.

$$\mathbf{K}_e = \int_{S_e} \varepsilon \mathbf{B}_e^T \mathbf{B}_e \, dxdy = \dfrac{\varepsilon}{4S} \begin{bmatrix} b_i & c_i \\ b_j & c_j \\ b_m & c_m \end{bmatrix} \begin{bmatrix} b_i & b_j & b_m \\ c_i & c_j & c_m \end{bmatrix} \qquad (5.4.21)$$

where $\mathbf{K}_e$ is the symmetric 3×3 element coefficient matrix. The functionals in

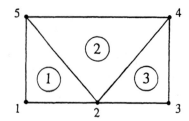

Fig. 5.4.1. Discretization of a 2-D problem

elements 1, 2 and 3 of Fig. 5.4.1 are:

$$\begin{cases} I_{e_1}(\varphi) = \frac{1}{2}\{\varphi_1 \ \varphi_2 \ \varphi_5\} \mathbf{K}_{e_1} \begin{Bmatrix} \varphi_1 \\ \varphi_2 \\ \varphi_5 \end{Bmatrix} \\[2mm] I_{e_2}(\varphi) = \frac{1}{2}\{\varphi_5 \ \varphi_2 \ \varphi_4\} \mathbf{K}_{e_2} \begin{Bmatrix} \varphi_5 \\ \varphi_2 \\ \varphi_4 \end{Bmatrix} \\[2mm] I_{e_3}(\varphi) = \frac{1}{2}\{\varphi_2 \ \varphi_3 \ \varphi_4\} \mathbf{K}_{e_3} \begin{Bmatrix} \varphi_2 \\ \varphi_3 \\ \varphi_4 \end{Bmatrix} \end{cases} \quad (5.4.22)$$

where

$$\mathbf{K}_{e_1} = \begin{bmatrix} k_{11}^{(1)} & k_{12}^{(1)} & k_{15}^{(1)} \\ k_{21}^{(1)} & k_{22}^{(1)} & k_{25}^{(1)} \\ k_{51}^{(1)} & k_{52}^{(1)} & k_{55}^{(1)} \end{bmatrix}$$

$$\mathbf{K}_{e_2} = \begin{bmatrix} k_{22}^{(2)} & k_{24}^{(2)} & k_{25}^{(2)} \\ k_{42}^{(2)} & k_{44}^{(2)} & k_{45}^{(2)} \\ k_{52}^{(2)} & k_{54}^{(2)} & k_{55}^{(2)} \end{bmatrix} \quad (5.4.23)$$

$$\mathbf{K}_{e_3} = \begin{bmatrix} k_{22}^{(3)} & k_{23}^{(3)} & k_{24}^{(3)} \\ k_{32}^{(3)} & k_{33}^{(3)} & k_{34}^{(3)} \\ k_{42}^{(3)} & k_{43}^{(3)} & k_{44}^{(3)} \end{bmatrix}.$$

Substituting Eq. (5.4.22) into Eq. (5.4.13) gives the total functional:

$$I(\varphi) = \sum_{e=1}^{3} I_e(\varphi) = \frac{1}{2} \sum_{i=1}^{5} \sum_{j=1}^{5} K_{ij} \varphi_i \varphi_j$$

$$= \frac{1}{2} [\varphi_1 \ \varphi_2 \ \varphi_3 \ \varphi_4 \ \varphi_5]^T \mathbf{K} [\varphi_1 \ \varphi_2 \ \varphi_3 \ \varphi_4 \ \varphi_5] \quad (5.4.24)$$

5.4 Variational finite element method

where

$$\mathbf{K} = \begin{bmatrix} k_{11} & k_{12} & 0 & 0 & k_{15} \\ k_{21} & k_{22} & k_{23} & k_{24} & k_{25} \\ 0 & k_{32} & k_{33} & k_{34} & 0 \\ 0 & k_{42} & k_{43} & k_{44} & k_{45} \\ k_{51} & k_{52} & 0 & k_{54} & k_{55} \end{bmatrix}. \tag{5.4.25}$$

The elements of the system matrix are the sum of the relevant coefficients of the element matrix, e.g. $k_{22} = k_{22}^{(1)} + k_{22}^{(2)} + k_{22}^{(3)}$, $k_{24} = k_{24}^{(2)} + k_{24}^{(3)}$, and so on.

Using the extremum condition of the functional:

$$\frac{\partial I(\varphi)}{\partial \varphi_i} = 0 \tag{5.4.26}$$

the system matrix equation is:

$$\mathbf{K}\{\varphi\} = 0. \tag{5.4.27}$$

Equation (5.4.27) is the finite element equation of a Laplacian problem. $\mathbf{K}$ is a sparse, symmetric and banded matrix of a size $N \times N$. (N is the total number of nodes.) The diagonal elements of the matrix are pivotal compared to the others and all of the values of the sub-determinates of the matrix are positive. Hence $\mathbf{K}$ is a positive definite matrix. This result can also be obtained by the property of the 'operator'. If the operator is completely continuous, the resulting matrix is positive definite. This property is sufficient to guarantee the convergence of the solution of the matrix equation.

For a 2-D Poisson's equation, the equivalent functional is:

$$I(\varphi) = \int_s \frac{1}{2} \left\{ \varepsilon \left[\left(\frac{\partial \varphi}{\partial x}\right)^2 + \left(\frac{\partial \varphi}{\partial y}\right)^2 \right] - 2\rho\varphi \right\} \mathrm{d}x\mathrm{d}y$$

$$= I_1(\varphi) - I_2(\varphi) = \sum_{e=1}^{E_0} [I_{e_1}(\varphi) - I_{e_2}(\varphi)]. \tag{5.4.28}$$

The first term in the square bracket of Eq. (5.4.28) is the same as the RHS of Eq. (5.4.19). The second term in the brackets of this equation is expanded as follows:

$$I_{e_2}(\varphi) = \int_{S_e} \rho[N_i\varphi_i + N_j\varphi_j + N_m\varphi_m]\mathrm{d}x\mathrm{d}y. \tag{5.4.29}$$

According to the extremum principle of a functional:

$$\frac{\partial I_2(\varphi)}{\partial \varphi_i} = \sum_{e=1}^{E_0} \frac{\partial I_{e_2}(\varphi)}{\partial \varphi_i} = \sum_{e=1}^{E_0} \int_{S_e} \rho'(N_i + N_j + N_m)\mathrm{d}x\mathrm{d}y. \tag{5.4.30}$$

If the charge density ρ' within the elements is assumed uniform, then the integral of Eq. (5.4.30) is:

$$\{p\}_e = \int_{S_e} \rho'[N^e]^T dxdy = \frac{S}{3}\rho'\begin{Bmatrix}1\\1\\1\end{Bmatrix} \tag{5.4.31}$$

where S is the area of the triangle. The formulation of the integration can refer to Eqs. (4.2.28) and (4.2.31). Then the finite element equation is:

$$\mathbf{K}\{\phi\} = \{P\} \tag{5.4.32}$$

where

$$\{P\} = \sum_{e=1}^{E_0} \{P\}_e. \tag{5.4.33}$$

For an axisymmetric problem

$$I(\varphi) = \int_s \frac{1}{2}\varepsilon\left[\left(\frac{\partial\varphi}{\partial r}\right)^2 + \left(\frac{\partial\varphi}{\partial z}\right)^2\right] \cdot 2\pi r dr\, dz \tag{5.4.34}$$

since

$$\int_{S_e} r\, drdz = (r_i + r_j + r_m)S/3 = r_0 S \tag{5.4.35}$$

$$r_0 = \tfrac{1}{3}(r_i + r_j + r_m)$$

then

$$\mathbf{K}_e = 2\pi \int_{S_e} \mathbf{B}_e^T \mathbf{B}_e r\, drdz$$

$$= \frac{2\pi\varepsilon}{4S}(r_0)\begin{bmatrix} b_i^2 + c_i^2 & b_i b_j + c_i c_j & b_i b_m + c_i c_m \\ b_j b_i + c_j c_i & b_j^2 + c_j^2 & b_j b_m + c_j c_m \\ b_m b_i + c_m c_i & b_m b_j + c_m c_j & b_m^2 + c_m^2 \end{bmatrix}. \tag{5.4.36}$$

The parameters $a_k, b_k, c_k (k = i, j, m)$ are similar to Eq. (5.4.9), i.e.

$$\begin{cases} a_i = r_j z_m - r_m z_j & a_j = r_m z_i - r_i z_m & a_m = r_i z_j - r_j z_i \\ b_i = z_j - z_m & b_j = z_m - z_i & b_m = z_i - z_j \\ c_i = r_m - r_j & c_j = r_i - r_m & c_m = r_j - r_i \end{cases}. \tag{5.4.37}$$

The element of the column matrix $\{P\}$ is:

$$\{P\}_e = 2\pi\int_{S_e}\rho'[N]_e^T r\, drdz = \frac{2\pi}{9}S\rho'(r_i + r_j + r_m)\begin{Bmatrix}1\\1\\1\end{Bmatrix}$$

$$= \frac{2\pi}{3}S\rho' r_0 \begin{Bmatrix}1\\1\\1\end{Bmatrix}. \tag{5.4.38}$$

5.4 Variational finite element method

5.4.2.3 Finite element equation for 2-D magnetic fields

The problem of 2-D static magnetic fields are described by:

$$\begin{cases} I(A) = \int_\Omega \frac{1}{2}\left[\frac{1}{\mu}\left[\left(\frac{\partial A}{\partial x}\right)^2 + \left(\frac{\partial A}{\partial y}\right)^2\right] - 2JA\right]dxdy \\ \quad + \oint_{\Gamma_2} \frac{1}{\mu}\left(\frac{1}{2}f_1 A^2 - f_2 A\right)dl = \min \\ A|_{\Gamma_1} = g(L) \end{cases} \quad (5.4.39)$$

where $A = A_z$, Γ_1, Γ_2 are the contours of the boundary with the boundary conditions of the first and second kind, respectively. Compare the first integral of Eq. (5.4.39) with Eq. (5.4.28); the difference is $A \to \varphi$, $1/\mu \to \varepsilon$, $J \to \rho$. The boundary integral of Eq. (5.4.39) has been discussed in Sect. 4.2.2.1.

Therefore rewrite the first equation of (5.4.39) as:

$$I(A) = \sum_e [I_{e_1}(A) + I_{e_2}(A) + I_{e_3}(A)] \quad (5.4.40)$$

where

$$I_{e_1}(A) = \tfrac{1}{2}\{A\}_e^T \mathbf{K}\{A\}_e . \quad (5.4.41)$$

Consider in cylindrical coordinates, the elements of matrix $\mathbf{K}_e$ as:

$$K_{rs}^e = K_{sr}^e = \frac{2\pi}{\mu} \cdot \frac{r_0}{4S}(b_r b_s + c_r c_s) \quad (r, s = i, j, m) . \quad (5.4.42)$$

The second part of Eq. (5.4.40) is:

$$I_{e_2}(A) = 2\pi J' A_0 r_0 S = \frac{2\pi}{3} J' r_0 S(A_i + A_j + A_m)$$

$$= \{A_i \ A_j \ A_m\} \begin{bmatrix} \frac{2\pi}{3} J' r_0 S \\ \frac{2\pi}{3} J' r_0 S \\ \frac{2\pi}{3} J' r_0 S \end{bmatrix} = \{A\}_e^T \{P\}_e \quad (5.4.43)$$

For inhomogeneous boundary conditions of the third kind, the contributions of the boundary value are only given by the boundary edges of the element which lies on the boundary such as the edge jm in Fig. 4.2.2. Suppose the vector potential along the edge of each element varies linearly, i.e.:

$$A = A_j + (A_m - A_j)\frac{l}{l_0} = \left(1 - \frac{l}{l_0}\right)A_j + \frac{l}{l_0}A_m$$

$$= (1 - t)A_j + tA_m \quad (5.4.44)$$

Then the third part of (5.4.40) is:

$$I_{e_9}(A) = 2\pi \int_{jm} \frac{1}{\mu} \left(\frac{1}{2} f_1 A^2 - f_2 A \right) r \, dl$$

$$= 2\pi \int \frac{1}{\mu} \frac{1}{2} f_1 \left[(1-t) A_j + t A_m \right]^2 - f_2 \left[(1-t) A_j + t A_m \right]$$

$$\times \left[(1-t) r_j + r_m \right] l \, dt \qquad (5.4.45)$$

Integration of Eq. (5.4.45), and substitution of the boundary conditions f_1 and f_2 by using the average value f_{1a}, f_{2a} along the edge $\overline{jm}$, thus the total matrix corresponding to the $I_{e_2}(A)$ and $I_{e_3}(A)$ is:

$$\{P_t\}_e = \begin{Bmatrix} P_i^e \\ p_j^e + \dfrac{2\pi l_0}{3\mu} f_{2a} \left(r_j + \dfrac{1}{2} r_m \right) \\ p_m^e + \dfrac{2\pi l_0}{3\mu} f_{2a} \left(r_m + \dfrac{1}{2} r_j \right) \end{Bmatrix} \qquad (5.4.46)$$

The element matrix coefficients considering of the element along the inhomogeneous boundary is:

$$\mathbf{K}_e = \begin{bmatrix} k_{ii}^e & k_{ij}^e & & k_{im}^e \\ k_{ji}^e & k_{jj}^e + \dfrac{\pi l_0 f_{1a}}{2\mu} \left(r_j + \dfrac{r_m}{3} \right) & k_{jm}^e + \dfrac{\pi l_0 f_{1a}}{6\mu} (r_j + r_m) \\ k_{mi}^e & k_{mj}^e + \dfrac{1}{6\mu} \pi l_0 f_{1a} (r_j + r_m) & k_{mm}^e + \dfrac{1}{2\mu} \pi l_0 f_{1a} \left(r_j + \dfrac{r_m}{3} \right) \end{bmatrix}$$

(5.4.47)

where the l_0 is the length of the edge, i.e.

$$l_0^2 = \left[(x_j - x_m)^2 + (y_j - y_j)^2 \right]. \qquad (5.4.48)$$

Then the final equation is obtained:

$$\mathbf{K}\{A\} = \{P\}. \qquad (5.4.49)$$

If the axisymmetric field is calculated and the flux density B is of interest, the relationship between B and A is noted.

$$B_r = -\frac{\partial A}{\partial z} \qquad B_z = \frac{A}{r} + \frac{\partial A}{\partial r} \qquad (5.4.50)$$

5.4 Variational finite element method

where $A = A_r$. The first equation of Eq. (5.4.39) becomes:

$$I(A) = \int_\Omega \frac{1}{2\mu} B^2 r\,dr\,dz - 2\pi \int_\Omega JAr\,dr\,dz$$

$$+ 2\pi \int_{\Gamma_2} \frac{1}{\mu}\left(\frac{1}{2}f_1 A^2 - f_2 A\right) r\,dl = \min. \qquad (5.4.51)$$

The first integral term of Eq. (5.4.51) is the magnetic energy stored in the region with unit length along the z-axis. If the potential A is assumed by a linear function, i.e.

$$A = \alpha_1 + \alpha_2 r + \alpha_3 z$$

then

$$\begin{cases} B_r = -\dfrac{\partial A_r}{\partial z} = -\alpha_3 \\ B_z = \dfrac{A}{r} + \dfrac{\partial A}{\partial r} = \dfrac{A}{r} + \alpha_2 \end{cases} \qquad (5.4.52)$$

$$B^2 = B_r^2 + B_z^2 = \alpha_3^2 + \alpha_2^2 + 2\alpha_2 \frac{A_0}{r_0} + \left(\frac{A_0}{r_0}\right)^2 \qquad (5.4.53)$$

where

$$A_0 = \tfrac{1}{3}(A_i + A_j + A_m) \qquad r_0 = \tfrac{1}{3}(r_i + r_j + r_m) \qquad (5.4.54)$$

$\alpha_1, \alpha_2, \alpha_3$ are similar to Eq. (4.2.8).

5.4.2.4 Finite element equation for non-linear magnetic fields

The equivalent variational problem corresponding to the magnetic field with non-linear permeability is:

$$\begin{cases} I(A) = \int_\Omega \left\{\left[\dfrac{\partial}{\partial r}\left(\dfrac{1}{\mu}\dfrac{\partial A}{\partial r}\right) + \dfrac{\partial}{\partial z}\left(\dfrac{1}{\mu}\dfrac{\partial A}{\partial z}\right)\right] - 2JA\right\} 2\pi r\,dr\,dz \\ \qquad + \oint_{\Gamma_2} \dfrac{1}{\mu}\left(\dfrac{1}{2}f_1 A^2 - f_2 A\right) dl = \min \\ A|_{\Gamma_1} = \text{const.} \end{cases} \qquad (5.4.55)$$

To avoid the repetition of tedious formulae, the assumption is made that the problem statisfies the homogeneous boundary condition of the second kind, and in each element the permeability is a constant. Thus

$$\frac{\partial I}{\partial A_k} = \sum_{e=1}^{E_0} \frac{\partial I_e}{\partial A_t} = \sum_{e=1}^{E_0} \int_{\Omega_e} \left[\frac{\partial}{\partial r} \left(\frac{1}{\mu} \frac{\partial}{\partial r} \sum_{k=1}^{3} N_k^e A \right) \right.$$
$$\left. + \frac{\partial}{\partial z} \left(\frac{1}{\mu} \frac{\partial}{\partial z} \sum_{k=1}^{3} N_k^e A \right) \right] 2\pi r \, dr dz \,. \qquad (5.4.56)$$

If $k = i$,

$$\left. \frac{\partial I}{\partial A_k} \right|_{k=i} = \frac{1}{\mu_e} (k_{ii}^e A_i + k_{ij}^e A_j + k_{im}^e A_m) \,. \qquad (5.4.57)$$

The formulations of $k_{rs}(r, s = i, j, m)$ satisfy the recurrence form and are expressed as follows [14]:

$$k_{rs}^e = k_{sr}^e = 2\pi \left[\frac{r_0}{4S} (b_r b_s + c_r c_s) + \frac{1}{6} (b_r + b_s) + \frac{S}{9 r_0} \right]$$

$$r, s = i, j, m \,. \qquad (5.4.58)$$

The coefficients of the global matrix are the summation of the element matrices, i.e.

$$K_{ij} = \sum_{e=1}^{E_0} \frac{1}{\mu_e} k_{ij} \,. \qquad (5.4.59)$$

The total matrix equation is:

$$\mathbf{K}\{A\} - \{P\} = 0 \qquad (5.4.60)$$

where

$$P_i = \sum_{e=1}^{E_0} P_i^e \quad i = 1, 2, \ldots, N_0 \qquad P_i^e = \frac{2\pi S}{3} J_e r_0 \,. \qquad (5.4.61)$$

Equation (5.4.60) is a non-linear matrix equation. It can be solved by function minimization methods discussed in Chap. 11 and the method introduced in reference [15]. The non-linear expression of B(H) can refer to reference [16].

5.4.2.5 Finite element equation for Helmholtz's equation (2-D case)

The equivalent functional of a Helmholtz equation in 2-D with homogeneous boundary conditions is:

$$I(A) = \sum_{e=1}^{E_0} I_e(A) = \sum \frac{1}{2} \int_{\Omega_e} \left[\left(\frac{\partial A}{\partial x} \right)^2 + \left(\frac{\partial A}{\partial y} \right)^2 \right.$$

5.4 Variational finite element method

$$+ j\omega\mu\gamma A^2 \bigg] dxdy - \int_{\Omega_e} \mu J A dxdy$$

$$= \sum_{e=1}^{E_0} [I_{e_1}(A) + I_{e_2}(A)] . \tag{5.4.62}$$

Compared with Eq. (5.4.38), only the term of $\int_{S_e} j\omega\mu\gamma A^2 dxdy$ is added, thus

$$I_{e_1}(A) = \frac{1}{2} \int_{\Omega_e} \left[\left(\frac{\partial A}{\partial x}\right)^2 + \left(\frac{\partial A}{\partial y}\right)^2 - \beta^2 A^2 \right] dxdy$$

$$= \frac{1}{2} \int_{\Omega_e} [\mathbf{B}_e\{A\}_e]^T [\mathbf{B}_e\{A\}_e] dxdy$$

$$- \frac{\beta^2}{2} \int_{S_e} ([N]_e\{A\}_e)^T ([N]_e\{A\}_e) dxdy$$

$$= \frac{1}{2} \{A\}_e^T \mathbf{K}_e \{A\}_e - \frac{\beta^2}{2} \{A\}_e^T \mathbf{H}_e \{A\}_e . \tag{5.4.63}$$

In Eq. (5.4.63), $\mathbf{K}_e$ is the same as in Eq. (5.4.21). Let

$$\mathbf{H}_e = \int_{\Omega_e} [N]_e^T [N]_e dxdy = \begin{bmatrix} h_{ii} & h_{ij} & h_{im} \\ h_{ji} & h_{jj} & h_{jm} \\ h_{mi} & h_{mj} & h_{mm} \end{bmatrix} \tag{5.4.64}$$

where $[N]_e$ is the shape function of a first order triangular element. Then

$$h_{rs} = \int_{S_e} N_r N_s dxdy = S(1 + \delta_{rs})/12 \quad (r, s = i, j, m) \tag{5.4.65}$$

$$\delta_{rs} = \begin{cases} 1 & r = s \\ 0 & r \neq s \end{cases} . \tag{5.4.66}$$

The process of integration is given in reference [12]. The integration in Eq. (5.4.65) is carried out using area coordinates.

The last term in Eq. (5.4.62) is the same as the matrix $\{P\}_e$ in Eq. (5.4.33) with ρ replaced by the current density J. After extremization, the finite element equation is:

$$\mathbf{K}\{A\} - \beta^2 \mathbf{H}\{A\} = \{P\} . \tag{5.4.67}$$

This is an eigenvalue equation. The solution method of this equation is discussed in references [12, 17, 18].

5.5 Special problems using the finite element method

5.5.1 Approaching floating electrodes by the variational finite element method

As previously mentioned, there are three kinds of the boundary conditions. Suppose an uncharged conductor or a charged conductor with total charge of Q_0 is immersed in the electric field, as shown in Fig. 5.5.1.

The potential of these conductors are unknown constants denoted by φ_F (called floating potentials). The value of φ_F depends on their positions and the imposed field strength. At the boundary of these electrodes, the following condition is satisfied

$$\int_\Gamma \varepsilon \frac{\partial \varphi}{\partial n} = Q \tag{5.5.1}$$

where Q is the total charge on the electrode. It is proved [19] that the equivalent functional of the problem is:

$$\begin{cases} I(\varphi) = \int_\Omega \frac{\varepsilon}{2}\left[\left(\frac{\partial \varphi}{\partial x}\right)^2 + \left(\frac{\partial \varphi}{\partial y}\right)^2\right] dxdy - \int_\Omega \rho\varphi \, d\Omega - Q\varphi_F = \min \\ \varphi|_{\Gamma_0} = \varphi_0 \\ \varphi|_{\Gamma_F} = \varphi_F \end{cases} \tag{5.5.2}$$

where Γ_0 satisfies the Dirichlet boundary condition, and Γ_F is the boundary of the floating conductor. If $Q = 0$, the discretization of Eq. (5.5.2) is:

$$I(\varphi) = \tfrac{1}{2}\{\Phi\}^T K\{\Phi\} - \{\Phi\}^T [P] \tag{5.5.3}$$

$$\{\Phi\} = \{\varphi_1 \ \varphi_2 \ldots \varphi_n \ \varphi_{l_1} \ldots \varphi_{l_m} \ldots \varphi_{N-1} \ \varphi_N\}^T. \tag{5.5.4}$$

Assume the nodal numbers of the floating electrodes are inserted between the others. n is the nodal number counted before the floating electrodes. Where the subscripts l_1, l_m are the first and last nodal number of the floating electrode, respectively, N is the total number of the nodes of the problem. By using the

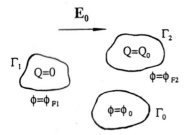

Fig. 5.5.1. Charged or uncharged conductor emerged in an exterior field

5.5 Special problems using the finite element method

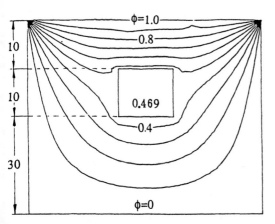

Fig. 5.5.2. Field distortion due to an uncharged conductor

condition $\partial I/\partial \varphi_i = 0$, $i = 1, \ldots, n, n + l_m + 1, \ldots, N$, one obtains the following equations:

$$\sum_{j=1}^{n} k_{ij}\varphi_j + \sum_{j=n+1}^{n+l_m} k_{ij}\varphi_j + \sum_{j=n+l_m+1}^{N} k_{ij}\varphi_j = 0. \tag{5.5.5}$$

For the nodes i, $i = l_1, \ldots l_m$, only one independent equation is obtained by $\partial I/\partial \varphi_i = 0$, i.e.

$$\sum_{j=1}^{n}\sum_{i=n+1}^{n+l_m} k_{ij}\varphi_j + \sum_{j=n+1}^{n+l_m}\sum_{i=n+1}^{n+l_m} k_{ij}\varphi_j$$
$$+ \sum_{j=n+l_m+1}^{N}\sum_{i=n+1}^{n+l_m} k_{ij}\varphi_j = 0. \tag{5.5.6}$$

Solve Eq. (5.5.5) and (5.5.6) simultaneously; the solution is then obtained.

Example 5.5.1. A long square conductor is enclosed in a grounded slot, as shown in Fig. 5.5.2. Using the above formulations, the field distortion by the square conductor is shown in the same figure.

5.5.2 Open boundary problems

5.5.2.1 Introduction

The variational method is not suitable for an infinite domain. The reason is that the variational method requires the operator to be positive definite. How to solve open boundary problems by using FEM has already been discussed in many papers. A good summary for solving the open boundary problems of different field problems is given in reference [20]. The simplest method is that of truncation. Other methods are ballooning [21], infinite element [22, 23], both mapping and infinite elements [24], mixed FEM and analytical methods [25], mixed FEM and BEM [26].

The method of truncation employed uses an infinite domain for a large but within a finite range. This is then considered as an approximation of the infinite domain. This method is simple but is uneconomical and computationally inefficient and, could even be inaccurate. The accuracy depends on the area of truncation. This method cannot be used in dynamic problems. The introduction of the terminating boundary leads to a reflection of waves by these boundaries. Consequently, the solution obtained from any such model is no longer be the original progressive wave problem.

In mapping, the domain z of the physical problem is mapped onto an image domain W. Then an infinite domain is mapped on to a small bounded domain. The standard FEM is then used in a finite domain W.

The mixed method combines both the advantages but avoids the disadvantages of the two methods.

In this section the ballooning method is chosen as an example to demonstrate the solution of an open boundary problem. The others can be found in the references.

5.5.2.2 Ballooning method

In the ballooning method, the whole space of the problem is divided into two parts. The inner part is closed by the boundary, Γ_0, and the outer part is extended to infinity, as shown in Fig. 5.5.3.

The inner part denoted by Ω_i in Fig. 5.5.3, contains the complex geometry, different materials and the area where the field distribution is of interest. In the outer part, the region is separated into several layers, $D_1, D_2, \ldots, D_n$, the outer boundary of these layers are $\Gamma_1, \Gamma_2, \ldots, \Gamma_n$. In these layers all the elements are

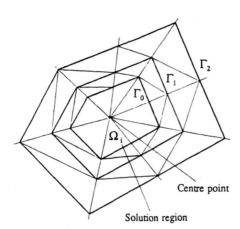

Fig. 5.5.3. Discretization scheme for ballooning method

5.5 Special problems using the finite element method

subdivided by the radial lines that start from one centre point. This ensures that the elements in each annular are similar to the next and the layers are able to be blown up rapidly. Thus the exterior boundary of the annular regions is extended to infinity by the speed of η^{2n}, where η is a constant. In the inner part, the conventional finite element mesh is used and a matrix $\mathbf{K}^i$ is established as usual.

In the annular region D_1, the matrix equation is:

$$\begin{bmatrix} \mathbf{K}^{(1)}_{11} & \mathbf{K}^{(1)}_{12} \\ \mathbf{K}^{(1)}_{21} & \mathbf{K}^{(1)}_{22} \end{bmatrix} \begin{bmatrix} \varphi_{\Gamma_0} \\ \varphi_{\Gamma_1} \end{bmatrix} = \mathbf{K}^{(1)} \begin{bmatrix} \varphi_{\Gamma_0} \\ \varphi_{\Gamma_1} \end{bmatrix} = 0 \tag{5.5.7}$$

where φ_{Γ_0}, φ_{Γ_1} are nodal potentials of the boundary Γ_0 and Γ_1. In the annular region D_2, the matrix equation is:

$$\begin{bmatrix} \mathbf{K}'_{11} & \mathbf{K}'_{12} \\ \mathbf{K}'_{21} & \mathbf{K}'_{22} \end{bmatrix} \begin{bmatrix} \varphi_{\Gamma_1} \\ \varphi_{\Gamma_2} \end{bmatrix} = \mathbf{K}' \begin{bmatrix} \varphi_{\Gamma_1} \\ \varphi_{\Gamma_2} \end{bmatrix} = 0 . \tag{5.5.8}$$

In a 2-D case, $\mathbf{K}'_{11} = \mathbf{K}^{(1)}_{11}$. Combination of Eqs. (5.5.7) and (5.5.8) leads to:

$$\begin{bmatrix} \mathbf{K}^{(2)}_{11} & \mathbf{K}^{(2)}_{12} \\ \mathbf{K}^{(2)}_{21} & \mathbf{K}^{(2)}_{22} \end{bmatrix} \begin{bmatrix} \varphi_{\Gamma_0} \\ \varphi_{\Gamma_2} \end{bmatrix} = \mathbf{K}^{(2)} \begin{bmatrix} \varphi_{\Gamma_0} \\ \varphi_{\Gamma_2} \end{bmatrix} = 0 . \tag{5.5.9}$$

The above equation shows that the boundary Γ_1 is eliminated. After repeating these recursive procedures, all of these interface boundaries are eliminated and the exterior region is extended to infinity. The recursive matrix is summarized as:

$$\mathbf{K}^{(n+1)} = \begin{bmatrix} \mathbf{K}^{(n)}_{11} & 0 \\ 0 & \mathbf{K}^{(n)}_{22} \end{bmatrix} - \begin{bmatrix} \mathbf{K}^{(n)}_{12} \mathbf{A}^{(n)} \mathbf{K}^{(n)}_{21} & \mathbf{K}^{(n)}_{12} \mathbf{A}^{(n)} \mathbf{K}^{(n)}_{12} \\ \mathbf{K}^{(n)}_{21} \mathbf{A}^{(n)} \mathbf{K}^{(n)}_{21} & \mathbf{K}^{(n)}_{21} \mathbf{A}^{(n)} \mathbf{K}^{(n)}_{12} \end{bmatrix} \tag{5.5.10}$$

where

$$\mathbf{A}^{(n)} = (\mathbf{K}^{(n)}_{22} + \mathbf{K}^{(n)}_{11})^{-1} . \tag{5.5.11}$$

Finally, combining the interior matrix $\mathbf{K}^i$ and the recursive matrix $\mathbf{K}^{(n+1)}$, results in:

$$\begin{bmatrix} \mathbf{K}^i_{11} & \mathbf{K}^i_{12} \\ \mathbf{K}^i_{21} & \mathbf{K}^i_{22} + \mathbf{K}^{(n+1)}_{11} \end{bmatrix} \begin{bmatrix} \varphi_i \\ \varphi_{\Gamma_0} \end{bmatrix} = 0 \tag{5.5.12}$$

For an axisymmetric field, the matrix $\mathbf{K}'_{11} = \eta \mathbf{K}^{(1)}_{11}$. Therefore,

$$\mathbf{K}^{n+1} = \begin{bmatrix} \mathbf{K}^{(n)}_{11} & 0 \\ 0 & \eta^{2n-1} \mathbf{K}^{(n)}_{22} \end{bmatrix} - \begin{bmatrix} \mathbf{K}^{(n)}_{12} \mathbf{A}^{(n)} \mathbf{K}^{(n)}_{21} & \eta^{2n-1} \mathbf{A}^{(n)} \mathbf{K}^{(n)}_{12} \\ \eta^{2n-1} \mathbf{A}^{(n)} \mathbf{K}^{(n)}_{21} & \eta^{2n-1} \mathbf{A}^{(n)} \mathbf{K}^{(n)}_{12} \end{bmatrix} . \tag{5.5.13}$$

In this case

$$\mathbf{A}^{(n)} = (\mathbf{K}^{(n)} + \eta^{2n-1} \mathbf{K}^{(n)})^{-1} . \tag{5.5.14}$$

By using these formulations, the field distribution of the 220 kV zinc-oxide arrester was obtained (Fig. 5.5.4) [27].

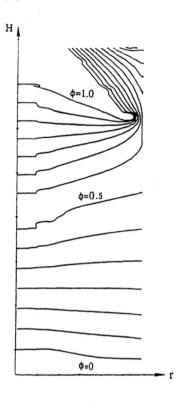

Fig. 5.5.4. Potential distribution of a 220 kV zinc-oxide arrester

5.6 Summary

The variational finite element method has been discussed in this chapter. It was shown that the extremum function of the functional is the solution of the corresponding Euler's equation of the functional. In variational FEM, the solution of the partial differential equation is obtained by solving the variation of the equivalent functional of the physical problem described by a partial differential equation or an integral equation.

The discussion started by deriving the equivalent functional of various electromagnetic field problems. The functional of the differential equation was derived by energy minimization or by using a mathematical approach. In doing this, the related concepts of the functional and variations were reviewed first.

It has been shown that the boundary conditions of the third kind (including the boundary conditions of the second kind) are contained in the equivalent functional for problems where the divergence theorem can be used. Therefore, the homogeneous boundary conditions of the second kind and the interface boundary conditions of different materials are automatically satisfied in the mean or in a least square sense in the process of finding the extremum of the functional. These boundary conditions need not be dealt with in the discretization process of the matrix equation; only the Dirichlet boundary condition is considered as a constraint on the functional. This is one of the most important advantages of FEM.

The equivalent functionals of some electromagnetic problems are derived in Sect. 5.3.

The method for dealing with some advanced problems,. such as the open boundary problem and the problem containing floating conductors, are discussed in the last section. These methods complete the applications of FEM.

References

1. Richard P. Feynman, Lectures on Physics, Vol. 2, Addison-Wesley, 1964
2. J.A. Stratton, Electromagnetic Theory, McGraw-Hill, 1941
3. P. Hammond, J. Penman, Calculation of Inductance and Capacitance by Means of Dual Energy Principles, Proc. IEE(A), 123(6), 554–559, 1976
4. P. Hammond, Energy Methods in Electromagnetism, Clarendon Press, 1981
5. I. Stakgold, Boundary Value Problems of Mathematical Physics, Ch. 8, Macmillan, New York, 1967
6. R. Courant, Methods of Mathematical Physics, Interscience, New York, 1953
7. S.G. Mikhlin, Variational Methods in Mathematical Physics, Macmillan, 1964
8. Jon Mathews, R.L. Walker. Mathematical Methods of Physics. W.A. Benjamin Inc. 1965
9. B.A. Finlayson, The Method of Weighted Residuals and Variational Principles, Academic Press, 1972
10. D.H. Norrie, The Finite Element Method, Academic Press Inc. 1973
11. Zhang Wenxun, Functional Methods in Electromagnetic Engineering (in Chinese), Science Press, Beijing, 1986
12. P.P. Silvester, R.L. Ferrari, Finite Elements for Electrical Engineers (2nd edn), Cambridge University Press, 1990
13. Kenneth H. Huebner, Earl A. Thornton, The Finite Element Method for Engineers (2nd edn), Wiley, 1982
14. Sheng Jian-ni (ed.), Numerical Analysis of Electromagnetic Fields, Science Press (in Chinese), Beijing, 1984
15. P. Rabinowitz, Numerical Methods for Nonlinear Algebraic Equations, Gordon and Breach, 1970
16. S. Ratnajeevan, H. Hoole, Computer-Aided Analysis and Design of Electromagnetic Devices, 1989
17. Gilbert Strang, Linear Algebra and Its Applications, Academic Press. 1976
18. H.R. Schwarz, Finite Element Methods, Academic Press, 1988
19. Wang Guo-hong, Numerical Calculation of Zinc-Oxide Surge Arresters, M.Sc. Thesis, Xi'an Jiaotong University, 1988
20. P. Bettess, Finite Element Modelling of Exterior Electromagnetic Problems, IEEE Trans. Mag-24(1), 238–243, 1988
21. P. Silvester, D.A. Lowther, C.T. Capenter, Exterior Finite Elements for 2-Dimensional Field Problems with Open Boundaries, Proc. IEE, 124(12), 1267–1270, 1977
22. P. Bettess, Infinite Elements. Int. J. Num. Meth. Engng. (11), 53–64, 1977
23. Nathan Ida, Efficient Treatment of Infinite Boundaries in Electromagnetic Field Problems, COMPEL, 6(3), 137–149, 1987
24. B. Nath, The W-Plane Finite Element Method for the Solution of Scalar Field Problems in Two Dimensions, Int. J. Num. Meth. Engng. (15) 361–379, 1980
25. M.V.K. Chari, Electromagnetic Field Computation of Open Boundary Problems by a Semi-analytic Approach, IEEE, Trans Mag-23(5) 3566–3568, 1987
26. S.J. Salon, J. D'Angelo, Application of the Hybrid Finite Element – Boundary Element Method in Electromagnetics, IEEE Trans. Mag-24(1), 80–85, 1988
27. Zhou Peibai, Wang Guohong, Analysis of the Potential Distribution of Gapless Surge Arrester, Proc. of 6th Int. Sym. on HVE, No. 26.09. 1989

Chapter 6

Elements and Shape Functions

6.1 Introduction

Domain discretization is one of the most important steps in many numerical methods to solve boundary value problems. In finite element method (FEM), the whole domain is discretized by elements. In boundary element method (BEM) the boundary of the domain is discretized by elements. The choice of the geometry of the element and the form of the approximating function to represent the behaviour of field variables (such as the potential φ, field strength E H and so on) within each element are both extremely important. They strongly influence the accuracy of the results, the computing time and the software engineering of computer programs. Hence the problem of element discretization is a generic problem and is common to element approximate methods as well as FEM. Hence element discretization techniques are discussed in one chapter.

It has been shown in former chapters (4 and 5) that a matrix equation is used for the solution of partial differential equations. The coefficients of the matrix are calculated via the partial derivatives of the shape functions. Hence the expressions of shape functions of various elements are the first to be formulated in the use of FEM. The shape functions of a 3-node triangular element, corresponding to linear approximation, has been introduced in Chaps. 4 and 5 to convey the principles underlying FEM. To obtain more accurate results and to solve different problems, many different kinds of elements and approximating functions can be used. For instance, both theory and experience indicate that, for many two dimensional problems, it is best to subdivide the problem region into the smallest possible number of large triangles, and to achieve the desired accurate solution by using high-order polynomial approximations on this very coarse mesh.

This chapter intends to provide the principles required to construct the various shape functions belonging to Lagrange and Hermite polynomials. For easy to evaluate element coefficients, the shape functions of different elements are formulated in local coordinates. The most effective elements belonging to the isoparametric family with linear and high order interpolations are the main topic in this chapter.

6.2 Types and requirements of the approximating functions

The functions used to represent the behaviour of field variables specified by the governing equations are called interpolation functions or approximating functions. In element discretization methods, the approximating function $u(x, y)$ is defined for each element. It is expressed by the following equation:

$$u(x, y) \cong \tilde{u}(x, y) = \sum_{k=1}^{m} \alpha_k \psi_k(x, y) = \sum_{k=1}^{m} \alpha_k N_k(x, y) \qquad (6.2.1)$$

In Eq. (6.2.1), α_k are undetermined parameters. They may be the value of potentials or other physical quantities defined in a governing equation, hence they are called generalized coordinates. As shown in Chaps. 4 and 5, α_k are nodal values of a potential. ψ_k is a set of independent basis functions. Thus, the approximating function is a linear combination of a set of discretized values and the basis functions.

In the method using weighted residuals, the interpolation function needs to ensure that

$$\int_\Omega RW \, d\Omega = 0 \qquad (6.2.2)$$

where R is the residual due to the approximation and W is a weighting function. This equation guarantees that the integral of the weighted residuals over the whole domain approaches zero.

In the variational principle, the generalized coordinates are determined by the extremum principle:

$$\frac{\partial I(u)}{\partial \alpha_k} = 0 \qquad (6.2.3)$$

where $I(u)$ is the equivalent functional of the problem to be solved. In other words, these two methods have to ensure that the approximate solution must be convergent to the exact solution. Hence the interpolation function needs to satisfy some requirements discussed in the following subsection.

6.2.1 Lagrange and Hermite shape functions

In FEM, there are two categories of elements known as Lagrange and Hermite. The former takes the values of the field variables at nodes of the element as being the unknowns. The latter takes both the unknown function and its partial derivative as unknowns. Thus, a Hermite element contains more than one unknown at each node. Usually the symbol N_f is used to represent the degrees of freedom (in structural machanics, the number of unknowns is called the degrees of freedom) at each node. $N_f = 1$ represents the Lagrange element, $N_f > 1$ indicates the

Hermite element. For example, for a 1-D problem, if $N_f = 2$, it means the values of the function and its first order derivative are the unknowns at each node. Applying this rule to electrostatic fields, if the potential φ and its first order partial derivatives $\partial \varphi/\partial x$, $\partial \varphi/\partial y$ are to be solved simultaneously then a Hermite element should be used.

6.2.2 Requirements of the approximating functions

Both in the method of weighted residuals and in the variational formulations, the derivatives of the unknowns contained in the integrand of the functional and the weighted residual formulations (Eq. (4.2.22)) is one order less than that which appears in differential equations. These formulations are called weak formulations. For example, the equivalent functional of Poisson's equation, $\nabla^2 u = f$, is

$$I(u) = \int_\Omega \left[\left(\frac{\partial u}{\partial x}\right)^2 + \left(\frac{\partial u}{\partial y}\right)^2 \right] dxdy + \int_\Omega uf\,dxdy . \quad (6.2.4)$$

The problem of first order derivatives of the unknowns contained in a weak formulation is called C^0-continuous. To solve this type of problem only the approximating function itself needs to be continuous within the element and along the side of adjacent elements. Hence a first order interpolation function as used in Chaps. 4 and 5 can be chosen as the approximating function. In a similar manner, a C^1-continuous problem is the one whose weak formulation contains at most second-order derivatives. In general, if there are derivatives of the order $(n + 1)$ contained in the integrand of the element equation, the problem is C^n-continuous. (Most of the potential problems in electromagnetic fields belong to C^0-continuous problems.)

In order to obtain convergent results as the element size is reduced to zero, the approximating function has to satisfy both *compatibility* and *completeness*, [1–4].

(a) Compatibility requirement

Compatibility means that the field variable and its partial derivative up to one order less than the highest-order derivative appearing in the weak formulation must be continuous along the element interfaces. In Eq. (6.2.1), $\psi_k(x)$ is constructed such that the generalized coordinates at any node of the common side of adjacent elements are the same regardless from which element the node is approached. In this case the generalized coordinates α_k and the interpolation function $\psi_k(x)$ are compatible. The elements exhibiting these characteristics are called conforming elements. Those that do not satisfy this requirement are said to be nonconforming elements. If a linear function $u(x, y) = \alpha_1 + \alpha_2 x + \alpha_3 y$ is used as the

6.2 Types and requirements of the approximating functions

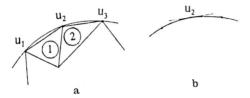

Fig. 6.2.1a, b. A confirming element

approximate function to solve a 2-D problem, the continuity of the zero order derivatives along the element interfaces is satisfied, but the first order derivatives along the element interfaces are discontinuous.

The above requirement is illustrated in Fig. 6.2.1(a) by considering the surface of the conductor. There are two linear elements. The value u_2 is the same whether it is obtained by element 1 or 2, but the derivative at node 2 is discontinuous. Figure 6.2.1(b) uses different kinds of shape function which ensures the continuity of the derivative as well.

(b) Completeness requirement

Completeness requires that within the element the partial derivatives of the interpolation function must be continuous in the same order as contained in the weak formulations of the problem. During the process of mesh refinement the interpolation function must remain unchanged. In other words for a C^0-continuous problem the approximating function must be capable of representing both a constant value of unknown function and a constant partial derivative of the highest order appearing in the weak formulation as the element size reduces to zero in a limited case. These properties ensure the approximate solution is convergent to the exact solution if enough numbers of elements are used in the whole domain.

The completeness requirement can be explained physically. For instance, the uniform value of the field variable is the most elementary type of variation. Thus the interpolation function must be able to give a constant value of the function and its partial derivative appearing in the functional as the element is reduced to zero. For example, in electrostatic fields, if the potential is considered as unknown, then only $\partial\varphi/\partial x$, $\partial\varphi/\partial y$ are contained in the functional, hence it is a C^0-continuous problem. If the linear function $\varphi = a + bx + cy$ is chosen as the interpolation function, when the element is reduced to a point, then $\varphi = a$, which is a constant and in addition $\partial\varphi/\partial x$, $\partial\varphi/\partial y$ are also constants as required.

To construct elements and interpolation functions to achieve C^0 continuity is not especially difficult but the difficulty increases rapidly when high order continuity is desired. Fortunately, analysts have developed a variety of elements applicable to many different types of problems. Each type of element in the catalogue is characterized by several features. These are the shape of the element, the number of nodes, the type of nodal variables and the type of interpolation function. If any one of these characterizing features is lacking, the description of an element is incomplete.

The complete polynomials with a specific order such as the following expressions satisfy the compatibility and completeness requirements

$$u(x, y) = \alpha_1 + \alpha_2 x + \alpha_3 y \tag{6.2.5}$$

$$u(x, y) = \alpha_1 + \alpha_2 x + \alpha_3 y + \alpha_4 x^2 + \alpha_5 xy + \alpha_6 y^2 . \tag{6.2.6}$$

The bilinear approximating (four node quadrilateral) and the incomplete third-order polynomial (9 node triangle) are appropriate for 2-D problems, because they are always complete polynomials of one variable. That polynomial functions have been widely used as the approximating function is due to the following reasons:

a) The polynomial functions are inherently continuous.

b) It is possible to improve the accuracy of the results by increasing the order of the polynomial. Theoretically, a polynomial of infinite order corresponds to the exact solution.

c) The polynomial is easy to differentiate and integrate.

6.3 Global, natural and local coordinates

In Chaps. 4 and 5, the shape functions of the first order triangular element are derived in Cartisian coordinates. The real positions of every node in the chosen coordinates are called global coordinates. If a high order polynomial is chosen as the approximating function, the derivation of the shape function will be more complicated and consequently the derivation of the element matrix equation is also cumbersome. The use of local coordinates will simplify any calculation of the element matrix greatly for any kind of elements.

6.3.1 Natural coordinates

In natural coordinates [3], *the position of any point of the element is defined by a dimensionless quantity which varies between 0 and 1. At the end nodes of the element, the natural coordinate is either 1 or 0, within the element, it varies. These values are independent of the geometry of the element only on the location of point.* The use of natural coordinates is illustrated in the following subsections.

(a) Linear elements in a one-dimensional case

Consider a one-dimensional problem. The whole region $[a, b]$ is subdivided into N straight line filaments with different lengths. Assume the end nodes of any element are denoted by x_i and x_j as shown in Fig. 6.3.1(a).

6.3 Global, natural and local coordinates

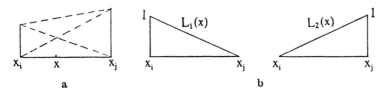

Fig. 6.3.1a, b. Natural coordinates for a 1-D case

The position of any point P within the element is expressed by:

$$x = \frac{(x_j - x)x_i}{x_j - x_i} + \frac{(x - x_i)x_j}{x_j - x_i} = L_1 x_i + L_2 x_j \qquad (6.3.1)$$

where the two dimensionless coefficients L_1 and L_2, are defined as natural coordinates, i.e.

$$L_1(x) = \frac{x_j - x}{x_j - x_i} \quad \text{and} \quad L_2(x) = \frac{x - x_i}{x_j - x_i} \qquad (6.3.2)$$

where L_1 and L_2 are ratios of lengths. At node i, $L_1 = 1$, $L_2 = 0$ and at node j, $L_1 = 0$, $L_2 = 1$. They automatically vary between 0 and 1. The natural coordinates L_1 and L_2 may be interpreted as weighting functions relating the coordinates of the end nodes to the coordinates of any interior point. From the definition of Eq. (6.3.2) L_1 and L_2 must satisfy the following constraint.

$$L_1 + L_2 = 1. \qquad (6.3.3)$$

Hence only one of the natural coordinates is independent.

Equation (6.3.1) expresses that the position of any interior point is the linear combination of the natural coordinates and nodal values. L_1 and L_2 are called the shape functions of a 1-D linear element. For a 1-D linear element there are two shape functions but only one of them is independent. At the end nodes of the element, the shape functions are equal to those of the natural coordinates.

Furthermore, if the unknown function varies linearly along the element as shown in Fig. 6.3.2, then the function u is expressed by

$$u(x) = \alpha_1 + \alpha_2 x = \begin{bmatrix} 1 & x \end{bmatrix} \begin{Bmatrix} \alpha_1 \\ \alpha_2 \end{Bmatrix} \qquad (6.3.4)$$

At nodes i and j, $u(x)$ becomes

$$\begin{Bmatrix} u_i \\ u_j \end{Bmatrix} = \begin{bmatrix} 1 & x_i \\ 1 & x_j \end{bmatrix} \begin{Bmatrix} \alpha_1 \\ \alpha_2 \end{Bmatrix}. \qquad (6.3.5)$$

Inverse of the above equation yields

$$\begin{Bmatrix} \alpha_1 \\ \alpha_2 \end{Bmatrix} = \begin{bmatrix} 1 & x_i \\ 1 & x_j \end{bmatrix}^{-1} \begin{Bmatrix} u_i \\ u_j \end{Bmatrix}. \qquad (6.3.6)$$

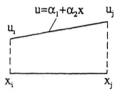

Fig. 6.3.2. Linear interpolation for a 1-D case

Substitution of this equation into Eq. (6.3.4) gives

$$u = [1 \ x] \begin{bmatrix} 1 & x_i \\ 1 & x_j \end{bmatrix}^{-1} \begin{Bmatrix} u_i \\ u_j \end{Bmatrix} = \frac{x_j - x}{x_j - x_i} u_i + \frac{x - x_i}{x_j - x_i} u_j \quad (6.3.7)$$

i.e.

$$u = L_1 u_i + L_2 u_j \quad (6.3.8)$$

where u_i and u_j are values of the unknown function at the end nodes. Consequently, the value of the function along the element can be expressed in terms of the nodal values and the natural coordinates as given in Eq. (6.3.8). Examination of Eqs. (6.3.1) and (6.3.7) shows that both the coordinates x and y and the function u have the same formulations expressed by the natural coordinates L_1 and L_2. This kind of element is called an *isoparametric element*.

(b) **Linear elements in a two-dimensional case**

In a 2-D triangular element, any point within the element is determined by three natural coordinates L_1, L_2 and L_3 [5]

$$\begin{cases} L_1 = \Delta_1/\Delta \\ L_2 = \Delta_2/\Delta \\ L_3 = \Delta_3/\Delta \end{cases} \quad (6.3.9)$$

where Δ is the area of triangle ijm and Δ_1, Δ_2 and Δ_3 are the areas of triangles Pjm and Pmi, Pij, respectively, as shown in Fig. 6.3.3(a). L_1, L_2 and L_3 are called natural or area coordinates. They represent the ratios of areas. Each of these have values varying from 0 to 1 and satisfy the following constraint.

$$L_1 + L_2 + L_3 = 1 \ . \quad (6.3.10)$$

Only two of the natural coordinates are independent. All points located on a line parallel to any edge of the triangle have the same natural coordinates as shown in Fig. 6.3.3(b). The natural coordinates of the nodal points and the middle points of each edge are shown in Fig. 6.3.3(c)

$$\begin{array}{llll} \text{at node } i & L_1 = 1 & L_2 = 0 & L_3 = 0 \\ \text{at node } j & L_1 = 0 & L_2 = 1 & L_3 = 0 \\ \text{at node } m & L_1 = 0 & L_2 = 0 & L_3 = 1 \ . \end{array} \quad (6.3.11)$$

6.3 Global, natural and local coordinates

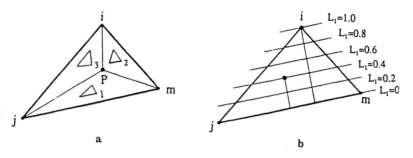

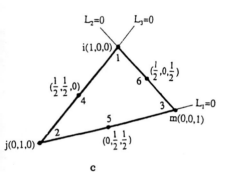

Fig. 6.3.3a–c. Natural coordinates for a 3-D case

In Fig. 6.3.3(a) if node P is located on line ij, then the area $\Delta_3 = 0$, thus $L_3 = \Delta_3/\Delta = 0$, hence line ij is the axis of $L_3 = 0$.

The relationship between the natural coordinates and the global coordinates of a 3-node triangular element are classified as below. The area of the triangle is determined by the coordinates of the three vertices:

$$S = \frac{1}{2} \begin{vmatrix} 1 & x_i & y_i \\ 1 & x_j & y_j \\ 1 & x_m & y_m \end{vmatrix}. \qquad (6.3.12)$$

The areas of the three subordinate triangles are given by

$$S_i = \frac{1}{2} \begin{bmatrix} 1 & x & y \\ 1 & x_j & y_j \\ 1 & x_m & y_m \end{bmatrix} = \frac{1}{2}[(x_j y_m - x_m y_j) + x(y_j - y_m) + y(x_m - x_j)]$$

$$= \tfrac{1}{2}(a_i + b_i x + c_i y). \qquad (6.3.13)$$

Substitution of Eqs. (6.3.12) and (6.3.13) into Eq. (6.3.9), gives the three natural coordinates

$$\begin{cases} L_i(x, y) = (a_i + b_i x + c_i y)/2S \\ L_j(x, y) = (a_j + b_j x + c_j y)/2S \\ L_m(x, y) = (a_m + b_m x + c_m y)/2S \end{cases} \quad (6.3.14)$$

where S is the area of triangle ijm. The formulae of a_k, b_k and c_k ($k = i, j, m$) are the same as those given in Eq. (4.2.10). *Comparison of Eq. (6.3.14) and Eq. (4.2.13) shows that the shape functions of a 3-node triangle are the same as those of the natural coordinates of the same triangle.* In other words, the shape functions of a 3-node triangle can be expressed by area coordinates as follows:

$$N_i = L_i \qquad N_j = L_j \qquad N_m = L_m . \quad (6.3.15)$$

The global coordinates of any point within the element is now linearly related to the natural coordinates as follows

$$\begin{cases} x = L_i x_i + L_j x_j + L_m x_m = \sum_k L_k x_k \quad (k = i, j, m) \\ y = L_i y_i + L_j y_j + L_m y_m = \sum_k L_k y_k \\ L_i + L_j + L_m = 1 . \end{cases} \quad (6.3.16)$$

These equations can be expressed in matrix form

$$x = [L_i \ L_j \ L_m] \begin{Bmatrix} x_i \\ x_j \\ x_m \end{Bmatrix} = [N_i \ N_j \ N_m] \begin{Bmatrix} x_i \\ x_j \\ x_m \end{Bmatrix}$$

$$y = [L_i \ L_j \ L_m] \begin{Bmatrix} y_i \\ y_j \\ y_m \end{Bmatrix} = [N_i \ N_j \ N_m] \begin{Bmatrix} y_i \\ y_j \\ y_m \end{Bmatrix} . \quad (6.3.17)$$

A similar conclusion has been shown in the linear element in a 1-D case.

If the function u within the element is assumed to be various and linearly dependent upon the coordinates x and y, then

$$u = \sum_{k=1}^{n} L_k u_k . \quad (6.3.18)$$

The derivation of $\partial u/\partial x$ and $\partial u/\partial y$ can be calculated in natural coordinates as follows

$$\begin{cases} \dfrac{\partial u}{\partial x} = \dfrac{\partial u}{\partial L_i} \cdot \dfrac{\partial L_i}{\partial x} + \dfrac{\partial u}{\partial L_j} \cdot \dfrac{\partial L_j}{\partial x} + \dfrac{\partial u}{\partial L_m} \cdot \dfrac{\partial L_m}{\partial x} \\ \dfrac{\partial u}{\partial y} = \dfrac{\partial u}{\partial L_i} \cdot \dfrac{\partial L_i}{\partial y} + \dfrac{\partial u}{\partial L_j} \cdot \dfrac{\partial L_j}{\partial y} + \dfrac{\partial u}{\partial L_m} \cdot \dfrac{\partial L_m}{\partial y} \end{cases} \quad (6.3.19)$$

6.3 Global, natural and local coordinates

where

$$\begin{cases} \dfrac{\partial L_k}{\partial x} = \dfrac{b_k}{2S} \\ \dfrac{\partial L_k}{\partial y} = \dfrac{c_k}{2S} \end{cases} (k = i, j, m).$$

Because the interpolation functions are assumed to be linear, hence the partial derivatives of natural coordinates to the variables of coordinates are constants.

(c) Linear element in a three-dimensional case

By extending the natural coordinates defined in a 2-D case, the natural coordinates in a 3-D case are defined as the ratios of volumes. Figure 6.3.4 shows a tetrahedral element. Any point P inside the tetrahedron is expressed by the natural coordinates L_1, L_2, L_3 and L_4. They are defined as:

$$L_i = \frac{V_i}{V} \quad i = 1, 2, 3, 4 \tag{6.3.20}$$

subject to

$$L_1 + L_2 + L_3 + L_4 = 1 \tag{6.3.21}$$

where V is the volume of tetrahedron 1234 and $V_1, \ldots, V_4$ are volumes of the sub-tetrahedra P234, P341, P421 and P123, respectively. Hence

$$\begin{array}{llll}
\text{at node 1:} & L_1 = 1, & L_2 = 0, & L_3 = 0, & L_4 = 0 \\
\text{at node 2:} & L_1 = 0, & L_2 = 1, & L_3 = 0, & L_4 = 0 \\
\text{at node 3:} & L_1 = 0, & L_2 = 0, & L_3 = 1, & L_4 = 0 \\
\text{at node 4:} & L_1 = 0, & L_2 = 0, & L_3 = 0, & L_4 = 1.
\end{array} \tag{6.3.22}$$

The global Cartesian coordinates and the natural coordinates are related by:

$$\begin{Bmatrix} x \\ y \\ z \\ 1 \end{Bmatrix} = \begin{bmatrix} x_1 & x_2 & x_3 & x_4 \\ y_1 & y_2 & y_3 & y_4 \\ z_1 & z_2 & z_3 & z_4 \\ 1 & 1 & 1 & 1 \end{bmatrix} \begin{Bmatrix} L_1 \\ L_2 \\ L_3 \\ L_4 \end{Bmatrix} \tag{6.3.23}$$

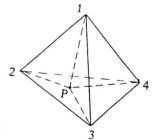

Fig. 6.3.4. Natural coordinates for a 3-D case

The solution of Eq. (6.3.23) is shown in the results of L_1, L_2, L_3, and L_4. For example

$$L_1 = \frac{1}{6V}(a_1 + b_1 x + c_1 y + d_1 z)$$

$$= \frac{1}{6V}\left[\begin{vmatrix} x_2 & y_2 & z_2 \\ x_3 & y_3 & z_3 \\ x_4 & y_4 & z_4 \end{vmatrix} - \begin{vmatrix} 1 & y_2 & z_2 \\ 1 & y_3 & z_3 \\ 1 & y_4 & z_4 \end{vmatrix} x \right.$$

$$\left. + \begin{vmatrix} 1 & x_2 & z_2 \\ 1 & x_3 & z_3 \\ 1 & x_4 & z_4 \end{vmatrix} y - \begin{vmatrix} 1 & x_2 & y_2 \\ 1 & x_3 & y_3 \\ 1 & x_4 & y_4 \end{vmatrix} z \right] \quad (6.3.24)$$

where the volume V is determined by the position of nodes 1, 2, 3 and 4, i.e.

$$V = \frac{1}{6}\begin{vmatrix} 1 & x_1 & y_1 & z_1 \\ 1 & x_2 & y_2 & z_2 \\ 1 & x_3 & y_3 & z_3 \\ 1 & x_4 & y_4 & z_4 \end{vmatrix}. \quad (6.3.25)$$

To simplify the expression, the natural coordinates for a tetrahedron are

$$L_i = \frac{1}{6V}(a_i + b_i x + c_i y + d_i z), \quad i = 1, 2, 3 \text{ and } 4. \quad (6.3.26)$$

The constants $a_2, \ldots, d_4$ are obtained via a cyclic permutation of subscripts 1, 2, 3 and 4 as in Eq. (6.3.23). Since the constants in Eq. (6.3.26) are cofactors of the determinant of Eq. (6.3.25), only the appropriate sign +ve or −ve should be given before each term. Equation (6.3.26) is valid only when the nodes are numbered counterclockwise from node 1.

In summary of the previous section, *the natural coordinates are defined to be the ratios of lengths, areas or volumes corresponding to the one, two and three dimensional cases, respectively.*

6.3.2 Local coordinates

As shown in the previous sections, for linear elements in 1-D, 2-D and 3-D cases, there are 1, 2 and 3 independent natural coordinates. In order to simplify the evaluation of the derivatives of the shape function, it will be convenient to define a set of local coordinates. In Fig. 6.3.5(a), let the line of $L_1 = 0$ denote the η' axis and the line of $L_2 = 0$ denote the ζ' axis. Then the relationship between the

6.4 Lagrange shape function

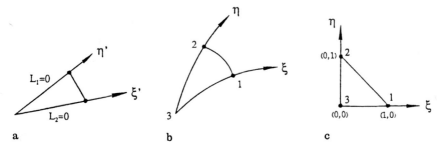

Fig. 6.3.5a–c. Local coordinates for a 2-D case

coordinates of ξ', η' and the natural coordinates are

$$\eta' = L_1$$
$$\xi' = L_2 \qquad (6.3.27)$$
$$1 - \xi' - \eta' = L_3$$

Furthermore if the coordinate transformation is defined as:

$$\begin{cases} \xi = \xi' \\ \eta = \eta' \end{cases} \qquad (6.3.28)$$

then

$$\begin{cases} L_1 = \eta \\ L_2 = \xi \\ L_3 = 1 - \xi - \eta \end{cases} \qquad (6.3.29)$$

The coordinates $\xi - \eta$ are called local coordinates. This transformation allows an arbitrary shaped triangle and a curvilinear triangle (Fig. 6.3.5(b)) into a right angled triangle as shown in Fig. 6.3.5(c).

Here the discussion is limited to a 2-D case. The extension of local coordinates to one- and three-dimensional cases is straightforward. The construction of the shape functions of various elements in local coordinates will be given in the next section.

6.4 Lagrange shape function

It is indicated in Sect. 6.2, the degrees of freedom at each node could be 1, 2 or even more. When $N_f = 1$, the shape function is called a Lagrange polynomial or a Lagrange shape function. It is used to analyse problems where only the nodal values of the function u need to be determined. In this case, the problem requires the continuity of u at the element interfaces. The normal derivative of u at these

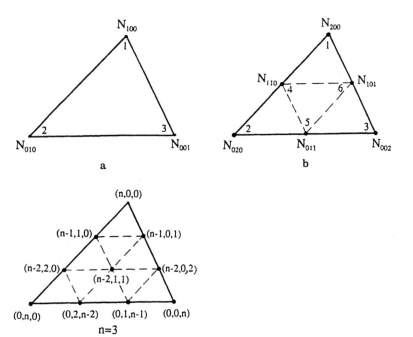

Fig. 6.4.1a–c. Shape function in local coordinates

boundaries is discontinuous. This polynomial has been defined as C^0 continuous. In the following subsections, the Lagrange shape functions of triangular and quadrilateral elements with different orders will be derived in local coordinates.

6.4.1 Triangular elements

As shown in Sect. 6.3, the number of shape functions is coincident with the number of the nodes. For example, there are three shape functions for a 3-node triangular element. For a 6-node triangle, there are 6 shape functions.

Silvester [6] pointed out that the shape functions of a nth-order triangular element can be expressed by the following systematic formulae

$$N_{\alpha\beta\gamma}(L_1, L_2, L_3) = N_\alpha(L_1) \cdot N_\beta(L_2) \cdot N_\gamma(L_3) \tag{6.4.1}$$

$$\begin{cases} N_\alpha(L_1) = \prod_{i=1}^{\alpha} \left(\dfrac{nL_1 - i + 1}{i} \right) & \alpha \geq 1 \\ N_\alpha(L_1) = 1 & \alpha = 0 \end{cases} \tag{6.4.2}$$

The subscripts α, β and γ are integers satisfying the constraint of

$$\alpha + \beta + \gamma = n \tag{6.4.3}$$

6.4 Lagrange shape function

where n is the order of the interpolation polynomial of the approximating function. For $N_\beta(L_2)$ and $N_\gamma(L_3)$, the formulae have the same form as those in Eq. (6.4.2).

A 3-node element is now used to explain the use of Eq. (6.4.2). It should be recalled that for a 3-node element, the order of the approximating function is $n = 1$. Hence $\alpha + \beta + \gamma = 1$. Considering node 1 in Fig. 6.4.1(a), the shape function is designated as $N_{100}(\alpha = 1, \beta = 0, \gamma = 0)$. From Eq. (6.4.1) and Eq. (6.4.2),

$$N_{100}(L_1, L_2, L_3) = N_1(L_1) \cdot N_0(L_2) \cdot N_0(L_3) = L_1 . \qquad (6.4.4)$$

Using the same treatment at node 2 and 3, similar results are obtained:

$$N_{010}(L_1, L_2, L_3) = N_0(L_1) \cdot N_1(L_2) \cdot N_0(L_3) = L_2 \qquad (6.4.5)$$

$$N_{001}(L_1, L_2, L_3) = N_0(L_1) \cdot N_0(L_2) \cdot N_1(L_3) = L_3 . \qquad (6.4.6)$$

These results are identical to those given in Eq. (6.3.15).

With the above method, the shape functions of a 6-node triangular element ($n = 2$) shown in Fig. 6.4.1(b) may be obtained more easily by the following manipulations. Designate node 1 as $N_{\alpha\beta\gamma} = N_{200}$, then

$$\begin{cases} N_{200} = N_2(L_1) \cdot N_0(L_2) \cdot N_0(L_3) \\ N_\alpha(L_1) = \prod_{i=1}^{2} \left(\dfrac{2L_1 - i + 1}{i} \right) = \left(\dfrac{2L_1 - 1 + 1}{1} \right) \\ \qquad \cdot \left(\dfrac{2L_1 - 2 + 1}{2} \right) = L_1(2L_1 - 1) \\ N_\beta(L_2) = 1 \\ N_\gamma(L_3) = 1 \end{cases} \qquad (6.4.7)$$

Hence $N_{200} = L_1(2L_1 - 1)$. Similarly, for node 2, 3, 4, 5 and 6, the shape functions are:

$$\begin{cases} N_{020} = N_0(L_1) \cdot N_2(L_2) \cdot N_0(L_3) \\ \qquad = (2L_2 - 1 + 1) \cdot 1/2(2L_2 - 2 + 1) = L_2(2L_2 - 1) \\ N_{002} = L_3(2L_3 - 1) \\ N_{110} = N_1(L_1) \cdot N_1(L_2) \cdot N_0(L_3) = 2L_1(2L_2) = 4L_1 L_2 \\ N_{011} = 4L_2 L_3 \\ N_{101} = 4L_1 L_3 . \end{cases} \qquad (6.4.8)$$

Equation (6.4.8) can be summarized as

$$\begin{cases} N_i = \zeta_i(2\zeta_i - 1) \quad i = 1, 2, 3 \\ N_4 = 4\zeta_1 \zeta_2 \\ N_5 = 4\zeta_2 \zeta_3 \qquad \zeta_3 = 1 - \zeta_1 - \zeta_2 \\ N_6 = 4\zeta_3 \zeta_1 \end{cases} \qquad (6.4.9)$$

These are shape functions of a 6-node triangular element. It should be noted, that the area coordinates L_1, L_2 and L_3 are determined by the position of the nodes. The shape functions of N_5, N_6 are the permutations of N_4 as shown in Eq. (6.4.9). It can be seen clearly that the area coordinates, often referred to as the natural coordinates are very useful for constructing the shape functions. The shape functions of a 9-node triangular element can be obtained by the same formulae with the help of Fig. 6.4.1(c).

6.4.2 Quadrilateral elements

Figure 6.4.2(a) shows a rectangular element with lengths 2a and 2b. It can be mapped to local coordinates as a square element (Fig. 6.4.2(b)) by the transformation equation

$$\xi = \frac{x - x_0}{a} \qquad \eta = \frac{y - y_0}{b}. \qquad (6.4.10)$$

In local coordinates, the shape function of the quadrilateral element can be constructed according to the following formula

$$N_i^e = \left(\prod_{j=1}^{4} F_j \right)(\alpha_1 + \alpha_2 \xi + \alpha_3 \eta + \alpha_4 \xi^2 + \alpha_5 \xi \eta + \alpha_6 \eta^2 + \alpha_7 \xi^2 \eta \\ + \alpha_8 \xi \eta^2 + \alpha_9 \xi^3 + \alpha_{10} \eta^3) \qquad (6.4.11)$$

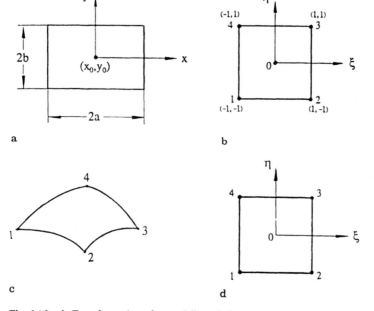

Fig. 6.4.2a–d. Transformation of a quadrilateral element

6.4 Lagrange shape function

where

$$F_j = \begin{cases} G_j & \text{node } i \text{ is not located on the edge } j \\ 1 & \text{node } i \text{ is located on the edge } j \end{cases} \quad (j = 1, 2, 3, 4).$$

(6.4.12)

Assume N is the total number of nodes of the element. The number of terms contained in the second parenthesis of Eq. (6.4.11) is $N - n_1$, n_1 is the nodal number on two edges which cross at node i. $G_1 - G_4$ are symbols to express the equations of the four edges, $\overline{12}, \overline{23}, \overline{34}, \overline{41}$, e.g.

$$G_1 = \eta + 1, \quad G_2 = -\xi + 1,$$
$$G_3 = -\eta + 1, \quad G_4 = \xi + 1. \tag{6.4.13}$$

Consider a 4-node quadrilateral element corresponding to the bilinear interpolation function

$$u(x, y) = \alpha_1 + \alpha_2 x + \alpha_3 y + \alpha_4 xy. \tag{6.4.14}$$

Then the function value and the coordinates within the element are expressed by shape functions and the nodal values as:

$$u(\xi, \eta) = N_1(\xi, \eta)u_1 + N_2(\xi, \eta)u_2 + N_3(\xi, \eta)u_3 + N_4(\xi, \eta)u_4 \tag{6.4.15}$$

$$x(\xi, \eta) = \sum_{i=1}^{4} N_i x_i$$

$$y(\xi, \eta) = \sum_{i=1}^{4} N_i y_i. \tag{6.4.16}$$

Using Eqs. (6.4.11) and (6.4.12), the four shape functions N_1 to N_4 are evaluated as follows

due to $N = 4$, $n_1 = 3$,

$$N_1(\xi, \eta) = G_2 G_3 \alpha_1 = (-\xi + 1)(-\eta + 1)\alpha_1 \tag{6.4.17}$$

$$N_2(\xi, \eta) = G_3 G_4 \alpha_1 = (\xi + 1)(-\eta + 1)\alpha_1 \tag{6.4.18}$$

$$N_3(\xi, \eta) = G_1 G_4 \alpha_1 = (\xi + 1)(\eta + 1)\alpha_1 \tag{6.4.19}$$

$$N_4(\xi, \eta) = G_1 G_2 \alpha_1 = (-\xi + 1)(\eta + 1)\alpha_1. \tag{6.4.20}$$

At node 1, $N_1(-1, -1) = 1$. Substitution of this condition into Eq. (6.4.17), yields $\alpha = \frac{1}{4}$. Consequently, the four shape functions of a 4-node quadrilateral element become

$$N_1(\xi, \eta) = (1 - \xi)(1 - \eta)/4$$
$$N_2(\xi, \eta) = (1 + \xi)(1 - \eta)/4$$
$$N_3(\xi, \eta) = (1 + \xi)(1 + \eta)/4$$
$$N_4(\xi, \eta) = (1 - \xi)(1 + \eta)/4. \tag{6.4.21}$$

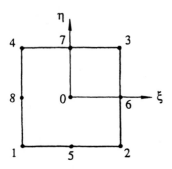

Fig. 6.4.3. 8-node quadrilateral element

At the four corners, the following equations hold

$$N_i(\xi_i, \eta_i) = 1 \quad (i = 1, 2, 3, 4)$$
$$N_i(\xi_j, \eta_j) = 0 \quad (i \neq j, j = 1, 2, 3, 4) \, . \tag{6.4.22}$$

Equation (6.4.21) can be simplified as

$$N_i^e = (1 + \xi_i\xi)(1 + \eta_i\eta)/4 \quad (i = 1, 2, 3, 4)$$
$$[\xi_i] = [-1 \quad 1 \quad 1 \quad -1]^T \tag{6.4.23}$$
$$[\eta_i] = [-1 \quad -1 \quad 1 \quad 1]^T \, .$$

By using the transformation of coordinates, the curved sided quadrilateral element (Fig. 6.4.2(c)) also can be mapped to the local coordinates as a square element (Fig. 6.4.2(d)). For a 8-node quadrilateral element as shown in Fig. 6.4.3, the 8 shape functions can be obtained by the same method. The formulations of these shape functions are given in the table of Appendix 6.1.

6.4.3 Tetrahedral and hexahedral elements

The simplest element in 3-dimensional cases is the four node tetrahedron (Fig. 6.4.4(a)). A linear approximating function for u within the element is

$$u(x, y, z) = \alpha_1 + \alpha_2 x + \alpha_3 y + \alpha_4 z \, . \tag{6.4.24}$$

A quadratic approximating function corresponding to Fig. 6.4.4(b) is

$$u(x, y, z) = \alpha_1 + \alpha_2 x + \alpha_3 y + \alpha_4 z + \alpha_5 x^2 + \alpha_6 y^2 + \alpha_7 z^2$$
$$+ \alpha_8 xy + \alpha_9 yz + \alpha_{10} xz \, . \tag{6.4.25}$$

The shape functions of these elements are obtained by the following equation similar to Eq. (6.4.1)

$$N_{\alpha\beta\gamma\delta}(L_1, L_2, L_3, L_4) = N_\alpha(L_1)N_\beta(L_2)N_\gamma(L_3)N_\delta(L_4) \tag{6.4.26}$$

$$\begin{cases} N_\alpha(L_1) = \prod_{i=1}^{\alpha} \left(\dfrac{nL_1 - i + 1}{i} \right) & \alpha \geq 1 \\ N_\alpha(L_1) = 1 & \alpha = 0 \, . \end{cases} \tag{6.4.27}$$

6.5 Parametric elements

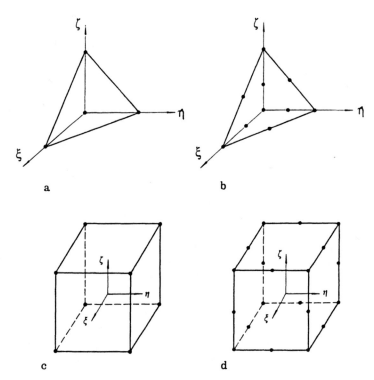

Fig. 6.4.4a–d. Elements in a 3-D case

Figures 6.4.4(c) and (d) show the hexahedral element by using linear and quadratic interpolation functions, the shape functions are determined by

$$N_k(\xi, \eta, \zeta) = L_k(\xi) L_k(\eta) L_k(\zeta) \ . \tag{6.4.28}$$

If a linear interpolation function is used, then

$$N_i = (1 + \xi \xi_i)(1 + \eta \eta_i)(1 + \zeta \zeta_i)/8 \ . \tag{6.4.29}$$

Detailed formulations are listed in Appendix 6.2.

6.5 Parametric elements

According to the type of the approximating functions of the field variables and the geometry, there are three types of elements: isoparametric element, sub-parametric element and super-parametric element.

The isoparametric element is defined as the unknown function and the coordinates of geometry are approximated by the shape functions with the same order. For example in a 3-node triangle, the function u and the coordinates are approximated in the same form, e.g.

$$u(x, y) = \alpha_1 + \alpha_2 x + \alpha_3 y = \sum_k N_k^e u_k \qquad (6.5.1)$$

$$\begin{aligned} x &= \sum_{k=1}^{3} N_k x_k \\ y &= \sum_{k=1}^{3} N_k y_k \end{aligned} \qquad k = i, j, m \qquad (6.5.2)$$

where

$$N_k^e = (a_k + b_k x + c_k y)/2s \quad k = i, j, m . \qquad (6.5.3)$$

The comparison of Eq. (6.5.1) and Eqs. (6.5.2) and (6.5.3) shows that the interpolation function for $u(x, y)$ has the same form and the same order as the expressions for the coordinates x and y, which both contain 3 unknown parameters, u_k or x_k, y_k. If the parameters are the nodal values of the function, the formula becomes the approximate expression of the function u. If the parameters are the values of the nodal coordinates, the formulae represent the actual coordinates. The element which exhibits these properties is called an isoparametric element. *In other words, for an isoparametric element, the interpolation function for the unknowns within any element has the same shape function N_k^e as those in the coordinates.*

It is not necessary to use an interpolation function with the same order when describing both the geometry and field variable of an element. *For a subparametric element, the order of the shape function of the coordinates is lower than that of the unknown function itself. On the other hand, for a superparameter element, the order of the interpolation function of the unknown is lower than that of the coordinates.* In some cases, the subparametric element is beneficial.

When a 3-node isoparametric triangle is chosen as a subdivision unit. It is assumed that the unknown function u varies linearly in each element. Hence the derivative of function u becomes a constant within the element. Therefore, the derivative of function u is discontinuous both at the nodes and along the edges of the element. In general, this leads to unsatisfactory results. In order to increase the accuracy of the results, a high order interpolation function is considered. Examples for these polynomials are

$$u(x, y) = \alpha_1 + \alpha_2 x + \alpha_3 y + \alpha_4 x^2 + \alpha_5 xy + \alpha_6 y^2 \qquad (6.5.4)$$

$$u(x, y) = \alpha_1 + \alpha_2 x + \alpha_3 y + \alpha_4 x^2 + \alpha_5 xy$$
$$+ \alpha_6 y^2 + \alpha_7 x^3 + \alpha_8 x^2 y + \alpha_9 xy^2 + \alpha_{10} y^3 . \qquad (6.5.5)$$

Equations (6.5.4) and (6.5.5) are complete polynomials of the second and third order, respectively. By increasing the order of the interpolation function, the

6.6 Element matrix equation

form of the polynomial can be expressed by the Pascal triangle as shown below:

zero order $\quad\quad\quad\quad\quad\quad\quad\quad \alpha_1$

first order $\quad\quad\quad\quad\quad\quad\quad \alpha_2 x \quad \alpha_3 y$

second order $\quad\quad\quad\quad\quad \alpha_4 x^2 \quad \alpha_5 xy \quad \alpha_6 y^2$

third order $\quad\quad\quad\quad \alpha_7 x^3 \quad \alpha_8 x^2 y \quad \alpha_9 xy^2 \quad \alpha_{10} y^3$

fourth order $\quad \alpha_{11} x^4 \quad \alpha_{12} x^3 y \quad \alpha_{13} x^2 y^2 \quad \alpha_{14} xy^3 \quad \alpha_{15} y^4$.

It has been shown in Sect. 4.2.1 that the parameters $\alpha_1, \ldots, \alpha_n$ are determined by the nodal values u_k and the nodal coordinates x_k and y_k. In the isoparametric finite element method, 6-node or 10-node elements are associated with the second or third order interpolation function. This allows evaluation of the parameters $\alpha_1, \ldots, \alpha_6$ or $\alpha_1, \ldots, \alpha_{10}$. In order to satisfy the interelement continuity condition, the number of nodes along each side of the element must be sufficient to determine the variation of function along that side uniquely. If a quadratic interpolation function is assumed to retain the quadratic behaviour along the sides of the element, three nodes are required at each side. For a cubic interpolation function, four nodes are necessary along each side of the element. Hence Eqs. (6.5.4) and (6.5.5) correspond to a 6-node triangle and 10-node triangle, respectively. Consequently, the order of the interpolation function and the number of nodes of the element are matched to each other. In 2-D cases, the relationship between the order of polynomial and the numbers of nodes is given by:

$$P_n(x, y) = \sum_{k=1}^{T_n^{(2)}} \alpha_k x^i y^j \quad i + j \leq n. \tag{6.5.6}$$

For a complete polynomial $P_n(x, y)$, the number of terms in the summation is

$$T_n^{(2)} = (n + 1)(n + 2)/2. \tag{6.5.7}$$

As an example,

for $n = 1$, $T_1^{(2)} = 3 \quad P_1(x, y) = \alpha_1 + \alpha_2 x + \alpha_3 y$ $\quad\quad$ (6.5.8)

for $n = 2$, $T_2^{(2)} = 6$ and so on.

The use of the isoparametric element results in the simplest derivation of finite element equations.

6.6 Element matrix equation

The solution of FEM is based on the element equation. After the equivalent functional is found and the form of the discretization element is chosen, the coefficients of the element matrix can be evaluated.

6.6.1 Coordinate transformations, Jacobian matrix

Section 6.3.3 shows that by using the coordinate transformation, a straight sided element in local coordinates corresponds to a curved sided element in global coordinates. Hence any curved boundary of the problems may be simulated by the curved element and all the computations carried out conveniently using local coordinates. In Sects. 6.4.1 to 6.4.3, the shape functions are constructed using local coordinates ξ, η and ζ. The relationships between local and global coordinates are

$$x = \sum_{i=1}^{n} N_i(\xi, \eta) x_i$$

$$y = \sum_{i=1}^{n} N_i(\xi, \eta) y_i \qquad (6.6.1)$$

$$u(\xi, \eta) = \sum_{i=1}^{n} N_i(\xi, \eta) u_i .$$

It has been shown in Sect. 4.2.2.1 that during the calculation of the coefficients of the element matrix, the partial derivatives of $\partial u/\partial x, \partial u/\partial y, \partial N_i/\partial x$ and $\partial N_i/\partial y$ must be evaluated. The relationship between $\partial N_i/\partial x, \partial N_i/\partial y$, and $\partial N_i/\partial \xi, \partial N_i/\partial \eta$ are related by a Jacobian matrix $\mathbf{J}$ which is derived below

$$\begin{Bmatrix} \dfrac{\partial N_i}{\partial \xi} \\ \dfrac{\partial N_i}{\partial \eta} \end{Bmatrix} = \begin{bmatrix} \dfrac{\partial x}{\partial \xi} & \dfrac{\partial y}{\partial \xi} \\ \dfrac{\partial x}{\partial \eta} & \dfrac{\partial y}{\partial \eta} \end{bmatrix} \begin{Bmatrix} \dfrac{\partial N_i}{\partial x} \\ \dfrac{\partial N_i}{\partial y} \end{Bmatrix} = \mathbf{J} \begin{Bmatrix} \dfrac{\partial N_i}{\partial x} \\ \dfrac{\partial N_i}{\partial y} \end{Bmatrix} \qquad (6.6.2)$$

where

$$\mathbf{J} = \begin{bmatrix} \dfrac{\partial x}{\partial \xi} & \dfrac{\partial y}{\partial \xi} \\ \dfrac{\partial x}{\partial \eta} & \dfrac{\partial y}{\partial \eta} \end{bmatrix} . \qquad (6.6.3)$$

This is called a Jacobian matrix. Hence the Jacobian is the transformation coefficient between the global and local coordinates. The inverse matrix of $[\mathbf{J}]$ is

$$\mathbf{J}^{-1} = \dfrac{1}{|\mathbf{J}|} \text{adj}[\mathbf{J}] = \dfrac{1}{|\mathbf{J}|} \begin{bmatrix} \dfrac{\partial y}{\partial \eta} & -\dfrac{\partial y}{\partial \xi} \\ -\dfrac{\partial x}{\partial \eta} & \dfrac{\partial x}{\partial \xi} \end{bmatrix} \qquad (6.6.4)$$

then

$$\begin{Bmatrix} \dfrac{\partial N_i}{\partial x} \\ \dfrac{\partial N_i}{\partial y} \end{Bmatrix} = \mathbf{J}^{-1} \begin{Bmatrix} \dfrac{\partial N_i}{\partial \xi} \\ \dfrac{\partial N_i}{\partial \eta} \end{Bmatrix} . \qquad (6.6.5)$$

6.6 Element matrix equation

Substitution of Eq. (6.6.1) into Eq. (6.6.3) and evaluation of the determinant of the Jacobian for 3-node triangular element gives

$$|J| = \begin{vmatrix} \dfrac{\partial x}{\partial \xi} & \dfrac{\partial y}{\partial \xi} \\ \dfrac{\partial x}{\partial \eta} & \dfrac{\partial y}{\partial \eta} \end{vmatrix} = \begin{vmatrix} x_j - x_i & y_j - y_i \\ x_m - x_i & y_m - y_i \end{vmatrix} = 2S. \tag{6.6.6}$$

This result shows that for a 3-node isoparametric element, the Jacobian is constant, and is identical both for the field variables and coordinates. Therefore the derivation of the element matrix may be easily obtained.

6.6.2 Evaluation of the Lagrangian element matrix

For 2-D Poissonian problems, the equivalent functional is

$$I(\varphi) = \Sigma I_e(\varphi) = \Sigma \int_\Delta \tfrac{1}{2}[\varepsilon |\nabla \varphi|^2 - 2\rho\varphi]\,dxdy$$

$$= \Sigma [I_{e1}(\varphi) - I_{e2}(\varphi)]. \tag{6.6.7}$$

Substitution of the approximating function of φ into Eq. (6.6.7) yields

$$I_{e1} = \int_\Delta \tfrac{1}{2}\varepsilon \left[\left(\sum_{k=1}^{n} \frac{\partial N_k}{\partial x}\varphi_k\right)^2 + \left(\sum_{k=1}^{n} \frac{\partial N_k}{\partial y}\varphi_k\right)^2 \right] dxdy \tag{6.6.8}$$

$$I_{e2} = \int_\Delta 2\rho \sum_{k=1}^{n} N_k \varphi_k \, dxdy. \tag{6.6.9}$$

According to the extremum principle

$$\frac{\partial I_{e1}}{\partial \varphi_k} = \int_\Delta \sum_{q=1}^{n} \varepsilon \left(\frac{\partial N_k}{\partial x}\frac{\partial N_q}{\partial x} + \frac{\partial N_k}{\partial y}\frac{\partial N_q}{\partial y} \right) \varphi_q \, dxdy \tag{6.6.10}$$

$$\frac{\partial I_{e2}}{\partial \varphi_k} = \int_\Delta 2 \sum_{k=1}^{n} \rho N_k \, dxdy \tag{6.6.11}$$

the coefficients of the element matrix are

$$k_{kq} = \int_\Delta \varepsilon \left(\frac{\partial N_k}{\partial x}\frac{\partial N_q}{\partial x} + \frac{\partial N_k}{\partial y}\frac{\partial N_q}{\partial y} \right) dxdy. \tag{6.6.12}$$

Since for a linear shape function $\partial N_k/\partial x$ and $\partial N_k/\partial y$ are constants, the evaluation of the integration of Eq. (6.6.12) is easy. If a high order interpolation function is used, the integrand in Eq. (6.6.12) is complicated. In order to simplify the calculation and to set up a general program in FEM, the coefficients of the element matrix for higher-order shape functions are calculated in local coordinates.

The relationship of the integration of an infinite small element between the global and local coordinates is

$$\int_\Delta dxdy = \int_\Delta |\mathbf{J}| \, d\xi \, d\eta . \tag{6.6.13}$$

This is because

$$ds = |\mathbf{a} \times \mathbf{b}| = |(a_x\mathbf{i} + a_y\mathbf{j}) \times (b_x\mathbf{i} + b_y\mathbf{j})| \tag{6.6.14}$$

where $\mathbf{a}$ and $\mathbf{b}$ are two infinitesimal vectors shown in Fig. 6.6.1. The components of the vectors are

$$\begin{cases} a_x = \lim_{\Delta\xi \to 0} [x(\xi + \Delta\xi, \eta) - x(\xi, \eta)] = \dfrac{\partial x}{\partial \xi} d\xi \\[6pt] a_y = \lim_{\Delta\xi \to 0} [y(\xi + \Delta\xi, \eta) - y(\xi, \eta)] = \dfrac{\partial y}{\partial \xi} d\xi \\[6pt] b_x = \dfrac{\partial x}{\partial \eta} d\eta \\[6pt] b_y = \dfrac{\partial y}{\partial \eta} d\eta . \end{cases} \tag{6.6.15}$$

Substitution of Eq. (6.6.15) into Eq. (6.6.14) yields

$$ds = |\mathbf{a} \times \mathbf{b}| = \frac{\partial x}{\partial \xi} d\xi \cdot \frac{\partial y}{\partial \eta} d\eta - \frac{\partial y}{\partial \xi} d\xi \cdot \frac{\partial x}{\partial \eta} d\eta = |\mathbf{J}| d\xi d\eta . \tag{6.6.16}$$

Substitution of Eq. (6.6.16) into Eq. (6.6.12), gives

$$k_{kq}^e = \int_0^1 \int_0^{1-\eta} \varepsilon \left(\frac{\partial N_k}{\partial x} \cdot \frac{\partial N_q}{\partial x} + \frac{\partial N_k}{\partial y} \cdot \frac{\partial N_q}{\partial y} \right) |\mathbf{J}| \, d\xi d\eta . \tag{6.6.17}$$

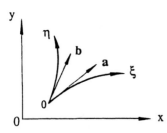

Fig. 6.6.1. Coordinate transformation of an infinitesimal area

6.6 Element matrix equation

For a quadrilateral element, the formula is

$$k_{kq}^e = \int_{-1}^{1}\int_{-1}^{1} \varepsilon \left(\frac{\partial N_k}{\partial x} \cdot \frac{\partial N_q}{\partial x} + \frac{\partial N_k}{\partial y} \cdot \frac{\partial N_q}{\partial y} \right) |J| d\xi d\eta \ . \tag{6.6.18}$$

In Eq. (6.6.17) and Eq. (6.6.18), the partial derivatives are given by

$$\begin{Bmatrix} \dfrac{\partial N_k}{\partial x} \\ \dfrac{\partial N_k}{\partial y} \end{Bmatrix} = \mathbf{J}^{-1} \begin{Bmatrix} \dfrac{\partial N_k}{\partial \xi} \\ \dfrac{\partial N_k}{\partial \eta} \end{Bmatrix} = \dfrac{1}{|\mathbf{J}|} \begin{bmatrix} \dfrac{\partial y}{\partial \eta} & -\dfrac{\partial y}{\partial \xi} \\ \dfrac{\partial x}{\partial \eta} & \dfrac{\partial x}{\partial \xi} \end{bmatrix} \begin{Bmatrix} \dfrac{\partial N_k}{\partial \xi} \\ \dfrac{\partial N_k}{\partial \eta} \end{Bmatrix} =$$

$$= \dfrac{1}{|\mathbf{J}|} \begin{bmatrix} \dfrac{\partial y}{\partial \eta} \cdot \dfrac{\partial N_k}{\partial \xi} & -\dfrac{\partial y}{\partial \xi} \cdot \dfrac{\partial N_k}{\partial \eta} \\ \dfrac{\partial x}{\partial \eta} \cdot \dfrac{\partial N_k}{\partial \xi} & +\dfrac{\partial x}{\partial \xi} \cdot \dfrac{\partial N_k}{\partial \eta} \end{bmatrix}. \tag{6.6.19}$$

To evaluate $\partial N_k/\partial x$, $\partial N_k/\partial y$ and to integrate Eq. (6.6.18) manually is a teious task. To ease computation, Silvester has provided an universal matrix [7–9] method which will be described in the following section.

6.6.3 Universal matrix

The evaluation of the element matrix is calculated by the following equations:

$$\frac{\partial N_k}{\partial x} = \sum_{i=1}^{3} \frac{\partial N_k}{\partial \zeta_i} \frac{\partial \zeta_i}{\partial x} = \sum_{i=1}^{3} \frac{b_i}{2S} \frac{\partial N_k}{\partial \zeta_i}$$

$$\frac{\partial N_k}{\partial y} = \sum_{i=1}^{3} \frac{\partial N_k \partial \zeta_i}{\partial \zeta_i \partial y} = \sum_{i=1}^{3} \frac{c_i}{2S} \frac{\partial N_k}{\partial \zeta_i} \tag{6.6.20}$$

where

$$b_i = y_{i+1}^\dagger - y_{i+2} \qquad c_i = x_{i+2} - x_{i+1} \ . \tag{6.6.21}$$

The subscript 'i' is changed cyclically around the three vertices of the triangle. The combination of Eqs. (6.6.12), (6.6.20) and (6.6.21) is rewritten as:

$$k_{kq} = \frac{1}{4S^2} \sum_{i=1}^{3} \sum_{j=1}^{3} (b_i b_j + c_i c_j) \int_\Delta \frac{\partial N_k}{\partial \zeta_i} \frac{\partial N_q}{\partial \zeta_j} dxdy. \tag{6.6.22}$$

The double summation of Eq. (6.6.22) can be reduced to a single summation if

---

† $i+1 = j$, $i+2 = m$

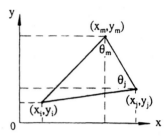

Fig. 6.6.2. Relationship between the variables for proof of Eq. (6.6.23)

the following relationships are introduced

$$b_i b_j + c_i c_j = -2S \cot \vartheta_m \qquad (6.6.23)$$

$$b_i^2 + c_i^2 = 2S(\cot \vartheta_j + \cot \vartheta_m) . \qquad (6.6.24)$$

These equations can be derived with the help of Fig. 6.6.2. A more detailed process is given in Appendix 1 of [9].

Then

$$k_{kq} = \frac{1}{2S} \sum_{i=1}^{3} \cot \vartheta_i \int \left[\left(\frac{\partial N_k}{\partial \zeta_j} - \frac{\partial N_k}{\partial \zeta_m} \right) \left(\frac{\partial N_q}{\partial \zeta_j} - \frac{\partial N_q}{\partial \zeta_m} \right) \right] d\zeta_j d\zeta_m . \qquad (6.6.25)$$

In Eq. (6.6.25), the value in the square brackets is dimensionless, they are independent of the geometry of the triangle and only depend on the approximation function. Hence, the stiffness element coefficient matrix $[k]_e$ may be expressed as:

$$[k]_e = \sum_{i=1}^{3} \cot \vartheta_i \mathbf{Q}_i . \qquad (6.6.26)$$

To calculate $\mathbf{Q}_i$, the integration in Eq. (6.6.25) is evaluated by using the following formula [9]

$$\int \zeta_1^m \zeta_2^n \, ds = 2S \int_0^1 \int_0^{1-\zeta_2} \zeta_1^m \zeta_2^n \, d\zeta_1 d\zeta_2 = 2S \frac{m! \, n!}{(m+n+2)!} . \qquad (6.6.27)$$

For a 3-node triangular element

$$[k]_e = \varepsilon \left\{ \frac{1}{2} \begin{bmatrix} 0 & 0 & 0 \\ 0 & 1 & -1 \\ 0 & -1 & 1 \end{bmatrix} \cot \vartheta_1 + \frac{1}{2} \begin{bmatrix} 1 & 0 & -1 \\ 0 & 0 & 0 \\ -1 & 0 & 1 \end{bmatrix} \cot \vartheta_2 \right.$$

$$\left. + \frac{1}{2} \begin{bmatrix} 1 & -1 & 0 \\ -1 & 1 & 0 \\ 0 & 0 & 0 \end{bmatrix} \cot \vartheta_3 \right\}$$

$$= \varepsilon \sum_{i=1}^{3} \mathbf{Q}_i \cot \vartheta_i . \qquad (6.6.28)$$

6.6 Element matrix equation

$$\begin{array}{c} \;1\;\;\;2\;\;\;3 \\ \begin{array}{c}1\\2\\3\end{array}\!\!\left[\begin{array}{ccc}0 & 0 & 0\\ 0 & 1 & -1\\ 0 & -1 & 1\end{array}\right]\end{array} \quad \begin{array}{c} \;3\;\;\;1\;\;\;2 \\ \begin{array}{c}3\\1\\2\end{array}\!\!\left[\begin{array}{ccc}1 & 0 & -1\\ 0 & 0 & 0\\ -1 & 0 & 1\end{array}\right]\end{array} \cdot \quad \begin{array}{c} \;2\;\;\;3\;\;\;1 \\ \begin{array}{c}2\\3\\1\end{array}\!\!\left[\begin{array}{ccc}1 & -1 & 0\\ -1 & 1 & 0\\ 0 & 0 & 0\end{array}\right]\end{array}$$

Fig. 6.6.3. The permutation of Q_i

Due to the property of natural coordinates, when Q_1 is calculated, Q_2 and Q_3 are the permutations of the first matrix Q_1. This is shown in Fig. 6.6.3. Using this rule, the computation time may be greatly reduced and no numerical integration will have to be repeated.

For a high order element, define a vector $N = [N_1 N_2 \ldots N_6]$ to express the shape function, then

$$\frac{\partial N}{\partial \zeta_1} = \frac{\partial}{\partial \zeta_1}[\zeta_1(2\zeta_1 - 1) \;\; 4\zeta_1\zeta_2 \;\; 4\zeta_1\zeta_3 \;\; \zeta_2(2\zeta_2 - 1) \;\; 4\zeta_2\zeta_3 \;\; \zeta_3(2\zeta_3 - 1)]$$

$$= [4\xi_1 - 1 \;\; 4\xi_2 \;\; 4\xi_3 \;\; 0 \;\; 0 \;\; 0]$$

$$= [\zeta_1 \;\; \zeta_2 \;\; \zeta_3] \begin{bmatrix} 3 & 0 & 0 & 0 & 0 & 0 \\ -1 & 4 & 0 & 0 & 0 & 0 \\ -1 & 0 & 4 & 0 & 0 & 0 \end{bmatrix} \quad (6.6.29)$$

$$\frac{\partial N}{\partial \zeta_2} = [\zeta_1 \;\; \zeta_2 \;\; \zeta_3] \begin{bmatrix} 0 & 4 & 0 & -1 & 0 & 0 \\ 0 & 0 & 0 & 3 & 0 & 0 \\ 0 & 0 & 0 & -1 & 4 & 0 \end{bmatrix} \quad (6.6.30)$$

$$\frac{\partial N}{\partial \zeta_3} = [\zeta_1 \;\; \zeta_2 \;\; \zeta_3] \begin{bmatrix} 0 & 0 & 4 & 0 & 0 & -1 \\ 0 & 0 & 0 & 0 & 4 & -1 \\ 0 & 0 & 0 & 0 & 0 & 3 \end{bmatrix}. \quad (6.6.31)$$

An important observation is that these matrices in Eq. (6.6.29) to Eq. (6.6.31) are independent of the shape of the triangle and so are called universal matrices. The elements of the matrix contained in Eq. (6.6.29) can be denoted by double notations as shown in Fig. 6.6.4(a). The rotating sequence of the nodes are shown in Fig. 6.6.5. Then the elements in the matrices of Eqs. (6.6.30) and (6.6.31) are obtained by using (b) and (c) shown in Fig. 6.6.4. After the computation, the matrix Q_1 of a 6-node triangular element is

$$\begin{array}{c}\begin{array}{cccccc}1&2&3&4&5&6\end{array}\\\begin{array}{c}1\\2\\3\end{array}\left[\begin{array}{cccccc}3&0&0&0&0&0\\-1&4&0&0&0&0\\-1&0&4&0&0&0\end{array}\right]\end{array}$$

a

$$\begin{array}{c}\begin{array}{cccccc}6&3&5&1&2&4\end{array}\\\begin{array}{c}3\\1\\2\end{array}\left[\begin{array}{cccccc}0&4&0&-1&0&0\\0&0&0&3&0&0\\0&0&0&-1&4&0\end{array}\right]\end{array}$$

b

$$\begin{array}{c}\begin{array}{cccccc}4&5&2&6&3&1\end{array}\\\begin{array}{c}2\\3\\1\end{array}\left[\begin{array}{cccccc}0&0&4&0&0&-1\\0&0&0&0&4&-1\\0&0&0&0&0&3\end{array}\right]\end{array}$$

c

Fig. 6.6.4a–c. The permutation of element matrices

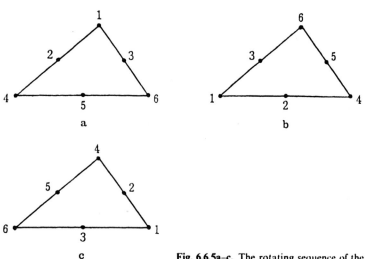

Fig. 6.6.5a–c. The rotating sequence of the nodes

$$\mathbf{Q}_1 = \frac{1}{6}\begin{bmatrix}0&&&&&\\0&8&\text{symmetry}&&&\\0&-8&8&&&\\0&0&0&3&&\\0&0&0&-4&8&\\0&0&0&1&-4&3\end{bmatrix} \qquad (6.6.32)$$

The matrices $\mathbf{Q}_2$ and $\mathbf{Q}_3$ will be obtained by rotation.

6.7 Hermite shape function

Finally, the computation of the element coefficient requires only the calculation of the coefficients b_i and c_i which are of the first order element. The universal matrices of axisymmetric field are given in Appendix A.6.3.

The construction of an element matrix of quadrilateral elements can also be obtained by the universal matrix method [3, 8]. For axisymmetric fields the universal matrix can also be derived [10].

6.7 Hermite shape function [2, 8]

As indicated in Sect. 6.4, the Lagrange polynomial only ensures the continuity of the unknown function at the nodes and along the edge of element. It will result in discontinuity of the derivatives of the function. In order to overcome this limitation, the derivatives of the unknowns are included in an approximating scheme. *The Hermite polynomial achieves this objective as it incorporates both the values and its derivations are specified as the unknowns directly at the nodes of element. It is C^1-continuous element. It ensures the continuity of unknowns and its normal derivatives along the side of element.* For example, in an one-dimensional linear element for a potential problem, both u and $\partial u/\partial n$ at two end nodes are assumed as unknowns and are continuous between neighbouring elements.

6.7.1 One-dimensional Hermite shape function

A Hermite shape function of order n in a one-dimensional Cartesian coordinate is denoted as $H^n(x)$. It is a polynomial of the order $2n+1$ of variable x. For the first order Hermite shape function, $n=1$, $H^1(x)$ is used to represent the first order approximation of the Hermite shape function. It represents a cubic approximate of x i.e.

$$u(x) = \alpha_1 + \alpha_2 x + \alpha_3 x^2 + \alpha_4 x^3 = \begin{bmatrix} 1 & x & x^2 & x^3 \end{bmatrix} \begin{Bmatrix} \alpha_1 \\ \alpha_2 \\ \alpha_3 \\ \alpha_4 \end{Bmatrix} \quad (6.7.1)$$

This is because at two terminal nodes of the linear element both u and $\partial u/\partial x$ are assumed as unknowns, they can be used to determine the four constants $\alpha_1 - \alpha_4$. If the interpolation function $u(x)$ is expressed by the shape functions

$$u(x) = H^1_{01} u_1 + H^1_{11} u_{x_1} + H^1_{02} u_2 + H^1_{12} u_{x_2} = [N]^e_H \{u\} \quad (6.7.2)$$

then the four constants $\alpha_1-\alpha_4$ or the four shape functions $H^1_{01} - H^1_{12}$ can be determined. In Eq. (6.7.2), the superscript 1 denotes the first order of the Hermite interpolation, the first subscripts 0 and 1 express the zero and the first order derivatives, the second subscripts 1 and 2 express the end nodes 1 and 2. The column matrix $\{u\}$ represents the four unknowns at two end nodes, i.e.

$$\{\mathbf{u}\} = \begin{Bmatrix} u_1 \\ u_{x_1} \\ u_2 \\ u_{x_2} \end{Bmatrix} \tag{6.7.3}$$

where

$$u_{x_1} = \left(\frac{\partial u}{\partial x}\right)_1 \quad \text{and} \quad u_{x_2} = \left(\frac{\partial u}{\partial x}\right)_2 .$$

In writing the first derivative of Eq. (6.7.1), one obtains:

$$\frac{\partial u}{\partial x} = \alpha_2 + 2\alpha_3 x + 3\alpha_4 x^2 . \tag{6.7.4}$$

Substitution of the four nodal values u_1, u_2 and u_{x_1}, u_{x_2} into Eq. (6.7.1) and Eq. (6.7.4) yields

$$\begin{Bmatrix} \alpha_1 \\ \alpha_2 \\ \alpha_3 \\ \alpha_4 \end{Bmatrix} = \begin{bmatrix} 1 & x_1 & x_1^2 & x_1^3 \\ 0 & 1 & 2x_1 & 3x_1^2 \\ 1 & x_2 & x_2^2 & x_2^3 \\ 0 & 1 & 2x_2 & 3x_2^2 \end{bmatrix}^{-1} \begin{Bmatrix} u_1 \\ u_{x1} \\ u_2 \\ u_{x2} \end{Bmatrix} . \tag{6.7.5}$$

By using the coordinate transformation $\xi = x - x_1/1$, $1 = x_2 - x_1$ and expressing the four shape functions by N_1^1, $\bar{N}_1^1$, N_2^1, $\bar{N}_2^1$, i.e.

$$[N]_H^e = [H_{01}^1 \quad H_{11}^1 \quad H_{02}^1 \quad H_{12}^1] = [N_1^1 \quad \bar{N}_1^1 \quad N_2^1 \quad \bar{N}_2^1] \tag{6.7.6}$$

then the four shape functions are formulated by

$$N_1^1 = 1 - 3\xi^2 + 2\xi^3 \qquad \bar{N}_1^1 = (\xi - 2\xi^2 + \xi^3)1$$
$$N_2^1 = 3\xi^2 - 2\xi^3 \qquad \bar{N}_2^1 = (-\xi^2 + \xi^3)1 . \tag{6.7.7}$$

They are the cubic functions of local coordinates.

The above procedures complete the analysis of a first order Hermite shape function of 1-D case. It can be expanded to derive the formulations for a 2-D triangular element.

6.7.2 Triangular Hermite shape functions

In 2-D cases, the 3-node triangle is the basic element for domain discretization. To ensure the continuity of the function u and its normal derivatives $\partial u/\partial n$ along the side of the element, the first order Hermite element is considered. According to the analysis in above section, the first order Hermite element represents the

6.7 Hermite shape function

cubic interpolation function which is used to approximate the unknown function. In a 2-D case, the complete cubic interpolation function is:

$$u(x, y) = \alpha_1 + \alpha_2 x + \alpha_3 y + \alpha_4 x^2 + \alpha_5 xy + \alpha_6 y^2 + \alpha_7 x^3 + \alpha_8 x^2 y + \alpha_9 xy^2 + \alpha_{10} y^3 \quad (6.7.8)$$

where there are ten constants α_1–α_{10} to be determined. However, for a 3-node triangle, there are 3 variables $(u, \partial u/\partial x, \partial u/\partial y)$ in each node. Thus the total degrees of freedom of a 3-node triangular element is 9. This is inefficient when determining the ten constants as constructing the C^1-continuious element is difficult to achieve. Instead the unknowns at each node $(u, \partial u/\partial x, \partial u/\partial y)_k$ $(k = i, j, m)$, can be used to determine the 9 shape functions. However according to Zienkicwicz, the constraint conditions of the ten constants are

$$2\left(\sum_{i=1}^{3} \alpha_i\right) - \sum_{i=4}^{9} \alpha_i + 2\alpha_{10} = 0 \, . \quad (6.7.9)$$

Then the 9 shape functions can be determined. Suppose in each element

$$u(x, y) = \sum_{k=1}^{3} (N_k u_k + \bar{N}_k u_{kx} + \tilde{N}_k u_{ky}) \, . \quad (6.7.10)$$

At node i

$$N_i = 1 \qquad \frac{\partial N_i}{\partial x} = 0 \qquad \frac{\partial N_i}{\partial y} = 0$$

$$\bar{N}_i = 0 \qquad \frac{\partial \bar{N}_i}{\partial x} = 1 \qquad \frac{\partial \bar{N}_i}{\partial y} = 0 \quad (6.7.11)$$

$$\tilde{N}_i = 0 \qquad \frac{\partial \tilde{N}_i}{\partial x} = 0 \qquad \frac{\partial \tilde{N}_i}{\partial y} = 1 \, .$$

At node j and m there are similar conditions as in Eq. (6.7.11). In local coordinates, Eq. (6.7.10) is then written as

$$u(\xi, \eta) = \sum_k [N_k(\xi, \eta) u_k(\xi, \eta) + \bar{N}_k(\xi, \eta) u_{k\xi}(\xi, \eta) + \tilde{N}_k(\xi, \eta) u_{k\eta}(\xi, \eta)] \quad (6.7.12)$$

where

$$u_\xi = \frac{\partial u}{\partial x}\frac{\partial x}{\partial \xi} + \frac{\partial u}{\partial y}\frac{\partial y}{\partial \xi} = u_x x_\xi + u_y y_\xi$$

$$u_\eta = \frac{\partial u}{\partial x}\frac{\partial x}{\partial \eta} + \frac{\partial u}{\partial y}\frac{\partial y}{\partial \eta} = u_x x_\eta + u_y y_\eta \quad (6.7.13)$$

$u_x, u_y, x_\xi, y_\xi, x_\eta$ and y_η are partial derivatives of u, x, y with respect to x, y and ξ, η, respectively. $\bar{N}, \tilde{N}$ are shape functions with respect to u_x and u_y. Substitution of Eq. (6.7.13) into Eq. (6.7.12), yields

$$u(\xi, \eta) = \sum_k [N_k u_k + (\bar{N}_k x_\xi + \tilde{N}_k x_\eta) u_x + (\bar{N}_k y_\xi + \tilde{N}_k y_\eta) u_y] \, . \quad (6.7.14)$$

Comparison of Eq. (6.7.14) and Eq. (6.7.10) shows that the relationships between the global coordinates and local coordinates are

$$N_k(x, y) = N_k(\xi, \eta)$$
$$\bar{N}_k(x, y) = \bar{N}_k(\xi, \eta) x_\xi + \tilde{N}_k(\xi, \eta) x_\eta \quad (k = i, j, m).$$
$$\tilde{N}_k(x, y) = \bar{N}_k(\xi, \eta) y_\xi + \tilde{N}_x(\xi, \eta) y_\eta \quad (6.7.15)$$

Substitution of the nodal value $u_i, u_j, u_m, \dfrac{\partial u_i}{\partial x}, \dfrac{\partial u_i}{\partial y}, \dfrac{\partial u_j}{\partial x}, \dfrac{\partial u_j}{\partial y}, \dfrac{\partial u_m}{\partial x}, \dfrac{\partial u_m}{\partial y}$ into Eq. (6.7.14) via the manipulation yields:

$$N_i = L_i^2(3 - 2L_i) + 2L_1 L_2 L_3 \quad (i = 1, 2, 3)$$
$$\bar{N}_i = L_i^2 L_2 + 0.5 L_1 L_2 L_3 \quad (i = 1, 3)$$
$$\bar{N}_2 = L_2^2(L_2 - 1) - L_1 L_2 L_3$$
$$\tilde{N}_i = L_i^2 L_3 + 0.5 L_1 L_2 L_3 \quad (i = 1, 2)$$
$$\tilde{N}_3 = L_3^2(L_3 - 1) - L_1 L_2 L_3. \quad (6.7.16)$$

These are the 9 shape functions of a 2-D first order Hermite element. In the above equations,

$$L_1 = 1 - \xi - \eta \qquad L_2 = \xi \qquad L_3 = \eta. \quad (6.7.17)$$

6.7.3 Evaluation of a Hermite element matrix

By following the same procedures as those used for deriving the Lagrange element matrix, $[N_k]$ is substituted by $[N_k]_H([N_k]_H$ which are composed of three terms, $N_k, \bar{N}_k$ and $\tilde{N}_k$), then Eq. (6.6.12) becomes:

$$\mathbf{k}_{ij}^e = \int_\Delta \varepsilon \left\{ \left[\left(\frac{\partial N_i}{\partial x} + \frac{\partial \bar{N}_i}{\partial x} + \frac{\partial \tilde{N}_i}{\partial x} \right) \frac{\partial N_j}{\partial x} + \left(\frac{\partial N_i}{\partial y} + \frac{\partial \bar{N}_i}{\partial y} + \frac{\partial \tilde{N}_i}{\partial y} \right) \frac{\partial N_j}{\partial y} \right] \right.$$

$$+ \left[\left(\frac{\partial N_i}{\partial x} + \frac{\partial \bar{N}_i}{\partial x} + \frac{\partial \tilde{N}_i}{\partial x} \right) \frac{\partial \bar{N}_j}{\partial x} + \left(\frac{\partial N_i}{\partial y} + \frac{\partial \bar{N}_i}{\partial y} + \frac{\partial \tilde{N}_i}{\partial y} \right) \frac{\partial \bar{N}_j}{\partial y} \right]$$

$$\left. + \left[\left(\frac{\partial N_i}{\partial x} + \frac{\partial \bar{N}_i}{\partial x} + \frac{\partial \tilde{N}_i}{\partial x} \right) \frac{\partial \tilde{N}_j}{\partial x} + \left(\frac{\partial N_i}{\partial y} + \frac{\partial \bar{N}_i}{\partial y} + \frac{\partial \tilde{N}_i}{\partial y} \right) \frac{\partial \tilde{N}_j}{\partial y} \right] \right\} dxdy$$

$$(6.7.18)$$

The complete formulation will be obtained by the following steps:

a) To develop the derivatives of $N_i \bar{N}_i \tilde{N}_i$ with respect to ξ and η.

6.7 Hermite shape function

b) Using the relationship between $\frac{\partial N_i}{\partial x}, \frac{\partial N_i}{\partial y}$ and $\frac{\partial N_i}{\partial \xi}, \frac{\partial N_i}{\partial \eta}$, then Eq. (6.7.18) can be expressed by

$$k_{ij}^e = \int_\Delta f(\xi, \eta)|J|d\xi d\eta. \tag{6.7.19}$$

The integration of Eq. (6.7.19) is carred out numerically. The efficiency of the triangular Hermite shape function is verified by the following example :

Example 6.7.1. Two coaxial cylindrical electrodes subject to Dirichlet boundary conditions are shown in Fig. 6.7.1(a) where $u_1|_{r=R_1} = 100$, $u_2|_{r=R_2} = 0$.

Solution. The region is subdivided by a Lagrange or Hermite triangular element as shown in Fig. 6.7.1(b). The relative errors of the field strength using Lagrange and Hermite shape functions are given in Table 6.7.1.

The advantage of the Hermite shape function is that not only a higher accuracy of the field strength obtained but while the Hermite interpolation is used, the symmetry conditions $\frac{\partial u}{\partial x}\bigg|_{x=0} = 0$, and $\frac{\partial u}{\partial y}\bigg|_{y=0} = 0$ are substituted in the matrix

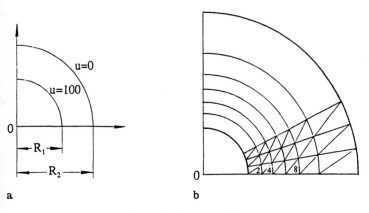

Fig. 6.7.1a, b. An example of Hermite element

Table 6.7.1. Comparison between a Lagrange element and a Hermite element

Shape function	No. of nodes and elements	Max. error of $E(\%)$
Lagrange	200 nodes, 342 elements	32.5
	600 nodes[†], 1062 elements	8.2
Hermite	200 nodes, 342 elements	8.0

[†] Near the inner conductor, the density of the elements are increased.

equation as a constraint condition, hence $E_y = 0$, and $E_x = 0$ on the x and y axis are satisfied. But these conditions cannot be guaranteed in Lagrange shape function. The problem is that the CPU time of the Hermite shape function is 3 or 4 times longer than the Lagrange shape function.

Hermite shape functions for a quadrilateral element are discussed in reference [11]. In the case of three-dimensional elements, satisfaction of C^1 continuity is quite difficult and practically no such element has been used in the references cited.

6.8 Application discussions

Isoparametric elements are widely used to solve various problems. The choice of the order of the interpolation function depends on the requirements. The linear element usually obtains good results for some purposes. Figure 6.8.1(a) shows the flux distribution around a pair of coils near a ferromagnetic plate. However, the flux density along the surface of the plate is discontinuous as shown in Fig. 6.8.1(b). This phenomenon is caused by using the linear interpolation function to approximate the vector potential. Then the derivative of the

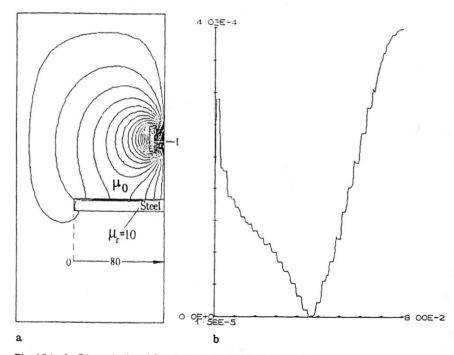

Fig. 6.8.1a, b. Discontinuity of flux density obtained by a linear element

6.8 Application discussions

potential within each element is a constant. If a more accurate result for the field strength is expected, higher order elements, are needed.

The accuracy of a quadratic element is higher than a linear element in general. An example of a one-dimensional case is given in reference [2]. That example also shows that the contribution of the cubic element has no significant effect. Usually the quadratic element is the better choice to obtain accurate results. A comparison of high order polynomial elements for solving a 2-D Laplacian problem is given in [11]. The comparative study of isoparametric and subparametric elements of a higher order element is given in [12].

In calculating the field strength of an axisymmetric problem, the 8-node quadrilateral element gives a more accurate solution than the 3-node triangular element. An example is shown in Table 6.8.1. The values given in the table are the field strengths on the conductor surfaces and on the three interfaces (on the right hand side of the interface) of the different conducting materials $(\gamma_1:\gamma_2:\gamma_3:\gamma_4 = 1000:100:10:1)$ shown in Fig. 6.8.2. The results show that the quadrilateral element is much more suitable than the 3-node triangular element.

Generally speaking, the increase in the number of elements will result in a more accurate solution for any given problems. However there will be a certain number of elements beyond which the accuracy cannot be improved by a significant amount. Reference [1] gives an example.

In some 3-D cases, if the hexahedron element is used to solve the problem in a cubic volume, the subparameter element (the function is approximated by quadratic function but the geometry is approximated by linear function) is preferred. It reduces the computation time and the good results are obtained. The comparison of subparametric and isoparametric elements to solve a nonlinear electromagnetic problem is discussed in [13].

In order to obtain more efficient results to solve some specific problems, several special elements are used. For example, the exponential shape function [14] and the spline function [15] are used to solve the eddy current problem. The edge element [16–20] is used to solve 3-dimensional vector problems. These new techniques are beyond of the purpose of this book. References are provided for the readers who are interested in these problems.

Table 6.8.1. A comparison of an 8-node quadrilateral element and a 3-node triangular element

Theoretical results	E (V/mm)	8.6707	0.2890	0.01734	0.001239	0.000963
Triangular element (160)	E (V/mm)	5.9467	0.2549	0.01622	0.001189	0.001049
	ΔE (%)	−31.4	−11.8	−6.2	−4.0	8.9
Triangular element (640)	E (V/mm)	6.9910	0.2689	0.01664	0.001205	0.000999
	ΔE (%)	−19.4	−7.0	−3.8	−2.7	3.7
Quadrilateral element (25)[†]	E (V/mm)	8.0101	0.2769	0.01703	0.001227	0.000954
	ΔE (%)	−7.6	−4.2	−1.6	−1.0	−0.9

[†] The region having the conductivity of γ_1 is divided into ten 8-node quadrilateral elements, each of the other regions is divided into five 8-node quadrilateral elements.

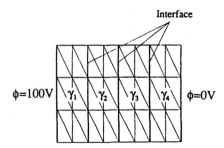

Fig. 6.8.2. Comparison of the accuracy of different elements for an axisymmetrical problem

6.9 Summary

In FEM the coefficients of the element matrix are dependent upon the formulation of the shape functions. Hence the construction of the shape function is one of the key steps in the use of the finite element method.

The formulations of the shape functions are based on the geometry of the element, the order of the interpolation function and the degrees of freedom of the nodes. The Lagrange shape function considers only the value of the function as unknowns, it is C^0-continuous. The Hermite shape function considers both the value of the function and its derivatives as unknowns, it is a C^1-continuous element. In this chapter both the construction of the shape function of C^0 and C^1-continuous elements have been discussed.

An universal recurrence formulation to construct Lagrange elements has been introduced. For these elements, the number of shape functions are identical to the number of nodes. For high-order triangular elements, the universal matrix is very useful to construct the element matrix. A great number of numerical integrations are eliminated.

Using the coordinates transformation, any curved element can be transformed into a regular shape in local coordinates. This technique ensures that any curved boundary can be modelled very well. The isoparametric elements may conveniently be used because for both the coordinates and the function variables the Jacobian is identical.

The choice of different kinds of elements are discussed in Sect. 6.8 and given in the references.

References

1. K.H.Huebner, E.A. Thornton, *The Finite Element Method for Engineers*, John Wiley & Sons, 1982
2. Frank L. Stasa, *Applied Finite Element Analysis for Engineers*, Holt, Rinehart and Winston, 1985
3. S.S. Rao, *The Finite Element Method in Engineering*, 2nd edn. Pergamon Press, 1989
4. H.R. Schwarz, *Finite Element Methods*, Academic Press, 1988

5. O.C. Zienkiewicz, R.L. Taylor, *The Finite Element Method*, 4th edn. Vol. 1, Chap. 7, McGraw-Hill, 1989
6. P.P. Silvester, High-Order Polynomial Triangular Finite Elements for Potential Problems, *Int. J. Engng. Sci.*, 7, 849–861, 1969
7. H. Hoole, *Computer-Aided Analysis and Design of Electromagnetic Device*, Elsevier: 1989
8. O.C. Zienkiewicz, *The Finite Element Method in Engineering Science*, McGraw-Hill, 1971
9. P.P. Silvester, R.L. Ferrari, *Finite Elements for Electrical Engineers*, 2nd Edn, Cambridge University Press, 1990
10. Zhou Peibai, Li Jiang, The Fast Algorithm for Calculating the Coefficient Matrix of High Order Finite Element, *ICEF'92, 0B1-4*, Hang Zhou, China, Oct. 1992
11. A.I. Zafrany, R.A. Cookson, Derivation of Lagrangian and Hermitian Shape Functions for Quadrilateral Elements, *Int. J. for Numerical Methods in Engineering*, vol. 23, 1939–1958, 1986
12. T. Nakata, N. Takahashi, Comparative Study of the Polynomial Orders of High Order Finite Elements, Electromagnetic Fields in Electrical Engineering, *Proc. of BISEF '88*, International Academic Publishers, pp. 449–451, 1989
13. Liu Nian, Jiang Kexun, Comparative Study of Two Kinds of High-order Triangular Finite Element Methods for Calculating 2-D Nonlinear Electromagnetic Field Problems, *Proc. of BISEF '88*, pp. 339–342, 1989
14. C.R.I. Emson, C. Greenough, *Special Finite Element Shape Function for Skin Depth Problems*, Rutherford Appleton Laboratory, RAL-84-066, 1984
15. Liang Xuebiao, Jian Baidun, Ni Guangzhen, The B-Spline Finite Element in Electromagnetic Field Numerical Analysis, *IEEE Trans. on Mag* 23(5), 2641–2643, 1987
16. M.L. Sarton, Z.J. Cendes, New Vector Finite Elements for Three-Dimensional Magnetic Field Computation, *J. Appl. Phys.* 61(15), 3919–3921, 1987
17. J.P. Webb, B. Forghani, A Single Scalar Potential Method for 3-D Magnetostatics Using Edge Element, *IEEE Trans. on Mag-*25(5), 4126–4129, 1989
18. A. Bossavit, I. Mayergoyz, Edge Elements for Scattering Problems, *IEEE Trans. Mag* 25(4), 2816–2821, 1989
19. M.L. Sarton, Z.J. Cendes, New Vector Finite Element for Three-Dimensional Magnetic Field Computation, *J. Appl. Phys.* 61(15) 3919–3921, 1987
20. Gerrit Mur, Optimum Choice of Finite Elements for Computing Three-Dimensional Electromagnetic Fields in Inhomogeneous Media, *IEEE Trans on Mag.* 24(1), 330–333, 1988

Appendix 6.1 Lagrangian shape functions for a 2-D case

Geometric figure	Interpolation function	Shape functions
3-node triangle	$u = \alpha_1 + \alpha_2 x \, \alpha_3 y$	$N_1 = 1 - \xi - \eta$ $N_2 = \xi$ $N_3 = \eta$
6-node triangle	$u = \alpha_1 + \alpha_2 + \alpha_3 y$ $\quad + \alpha_4 xy + \alpha_5 x^2$ $\quad + \alpha_6 y^2$	$N_1(\xi, \eta) = (1 - \xi - \eta)[2(1 - \xi - \eta) - 1]$ $N_2(\xi, \eta) = \xi(2\xi - 1)$ $N_3(\xi, \eta) = \eta(2\eta - 1)$ $N_4(\xi, \eta) = 4\xi(1 - \xi - \eta)$ $N_5(\xi, \eta) = 4\xi\eta$ $N_6(\xi, \eta) = 4\eta(1 - \xi - \eta)$

4-node quadrilateral $\quad u = \alpha_1 + \alpha_2 x + \alpha_3 xy$
$\qquad\qquad\qquad\qquad\quad + \alpha_4 y$

$N_1 = 1/4(1 - \xi)(1 - \eta)$
$N_2 = 1/4(1 + \xi)(1 - \eta)$
$N_3 = 1/4(1 + \xi)(1 + \eta)$
$N_4 = 1/4(1 - \xi)(1 + \eta)$

8-node quadrilateral $\quad u = \alpha_1 + \alpha_2 x + \alpha_3 y$
$\qquad\qquad\qquad\qquad + \alpha_4 xy + \alpha_5 x^2 + \alpha_6 y^2$
$\qquad\qquad\qquad\qquad + \alpha_7 x^2 y + \alpha_8 xy^2$

$N_i = 1/4(1 + \xi_i \xi)(1 + \eta_i \eta) \times$
$\qquad (\xi_i \xi + \eta_i \eta - 1) \quad i = 1, 2, 3, 4$
$N_i = 1/2(1 - \xi^2)(1 + \eta_i \eta) \quad i = 5,$
$N_i = 1/2(1 - \eta^2)(1 + \xi_i \xi) \quad i = 6, 8$
$[\xi_i] = [-1\ 1\ 1\ -1\ 0\ 1\ 0\ -1]^T$
$[\eta_i] = [-1\ -1\ 1\ 1\ -1\ 0\ 1\ 0]$

Appendix 6.2 Commonly used shape functions for 3-D cases

Geometric figure	Interpolation function	Shape function
4-node tetrahedron	$u = \alpha_1 + \alpha_2 x + \alpha_3 y$ $+ \alpha_4 xy$	$N_1 = 1 - \xi - \eta - \zeta$ $N_2 = \xi$ $N_3 = \eta$ $N_4 = \zeta$
10-node tetrahedron	$u = \alpha_1 + \alpha_2 x + \alpha_3 y + \alpha_4 z$ $+ \alpha_5 xy + \alpha_6 yz + \alpha_7 xz$ $+ \alpha_8 x^2 + \alpha_9 y^2 + \alpha_{10} z^2$	$N_1 = [2(1 - \xi - \eta - \zeta) - 1] \times$ $\quad (1 - \xi - \eta - \zeta)$ $N_2 = (2\xi - 1)\xi$ $N_3 = (2\eta - 1)\eta$ $N_4 = (2\zeta - 1)\zeta$ $N_5 = 4\xi\eta$ $N_6 = 4(1 - \xi - \eta - \zeta)\xi$ $N_7 = 4(1 - \xi - \eta - \zeta)\eta$ $N_8 = 4(1 - \xi - \eta - \zeta)\zeta$ $N_9 = 4\eta\zeta$ $N_{10} = 4\xi\zeta$
8-node hexahedron	$u = \alpha_1 + \alpha_2 x + \alpha_3 y$ $+ \alpha_4 z + \alpha_5 xy + \alpha_6 yz$ $+ \alpha_7 xz + \alpha_8 xyz$	$N_i = 1/8(1 + \xi_i \xi)(1 + \eta_i \eta) \times$ $\quad (1 + \zeta_i \zeta) \quad i = 1, 2, \ldots, 7,$

Appendix 6.3 The universal matrix of axisymmetric fields

20-node hexahedron

$u = \alpha_1 + \alpha_2 x + \alpha_3 y + \alpha_6 y^2$
$+ \alpha_4 z + \alpha_5 x^2 + \alpha_6 y^2$
$+ \ldots + \alpha_8 xy + \ldots$
$+ \alpha_{11} x^2 y + \alpha_{20} z^2 xy$

$N_i = 1/8(1 + \xi_i\xi)(1 + \eta_i\eta)(1 + \zeta_i\zeta)$
$i = 1, 3, 5, 7, 13, 15, 17, 1$

$N_i = 1/4(1 - \xi^2)(1 + \eta_i\eta)(1 + \zeta_i\zeta)$
$i = 2, 6, 14, 1$

$N_i = 1/4(1 - \zeta^2)(1 + \xi_i\xi)(1 + \eta_i\eta)$
$i = 9, 10, 11, 1$

$N_i = 1/4(1 - \eta^2)(1 + \xi_i\xi)(1 + \zeta_i\zeta)$
$i = 4, 8, 16, 2$

Appendix 6.3 The universal matrix of axisymmetric fields

The equivalent functional of scalar Helmholtz equation: $(\nabla^2 + k^2)u = g$ is:

$$I(u) = \int_\Omega \pi r(\beta \nabla u \cdot \nabla u - k^2 u^2 + 2ug) d\Omega \quad (A.6.3.1)$$

Because the coordinate r is included within the integration, the calculation of the coefficients of element matrices will be complicated than which of the 2-D case. Suppose the coordinates of r and z are approximated by:

$$\begin{cases} r = r_1 + (r_2 - r_1)\xi + (r_3 - r_1)\eta \\ z = z_1 + (z_2 - z_1)\xi + (z_3 - z_1)\eta \end{cases} \quad (A.6.3.2)$$

By using the same step as used in 2-D case, the following equation is obtained.

$$\pi \sum_{i=1}^{3} \int_0^1 \int_0^{1-r} r_i \left\{ \sum_{k=1}^{n} \beta \left(\frac{\partial N_k}{\partial r} \frac{\partial N_q}{\partial r} + \frac{\partial N_k}{\partial z} \frac{\partial N_q}{\partial z} \right) u_k - 2k^2 \sum_{k=1}^{n} N_k N_q u_k \right.$$

$$\left. + 2 \sum_{k=1}^{n} N_k N_q g_k \right\} \xi_i \, drdz = 0 \quad (A.6.3.3)$$

Substitute the terms of $\dfrac{\partial N_k}{\partial r}, \dfrac{\partial N_q}{\partial r}, \dfrac{\partial N_k}{\partial z}, \dfrac{\partial N_q}{\partial z}$ and the parameters $\cot \theta_i$ in to above equation yields:

$$\beta \sum_{k=1}^{n} \sum_{i=1}^{3} Q_{kq}^i \cot \theta_i u_q - 2k^2 \sum_{k=1}^{n} T_{kq} u_q + 2 \sum_{k=1}^{n} T_{kq} g_q = 0 \quad (A.6.3.4)$$

where

$$Q_{kq}^i = \sum_{i=1}^{3} r_i \int_0^1 \int_0^{1-\xi_1} \xi_i \left(\frac{\partial N_k}{\partial \xi_{i+1}} - \frac{\partial N_k}{\partial \xi_{i-1}} \right) \left(\frac{\partial N_q}{\partial \xi_{i+1}} - \frac{\partial N_q}{\partial \xi_{i-1}} \right) |J| d\xi_2 d\xi_3$$

$$(A.6.3.5)$$

$$T_{kq} = |J| \sum_{i=1}^{3} r_i \int_0^1 \int_0^{1-\xi_1} \xi_i N_k N_q \, d\xi_2 d\xi_3 \tag{A.6.3.6}$$

The matrix Q^1 of the first order triangle is:

$$Q^1 = \frac{1}{6} \begin{pmatrix} (0,0,0), & (0, & 0, & 0), & (0, & 0, & 0) \\ (0,0,0), & (1, & 1, & 1), & (-1, & -1, & -1) \\ (0,0,0), & (-1, & -1, & -1), & (1, & 1, & 1) \end{pmatrix}$$

It means $Q^1_{1,1} = (0 \cdot r_1 + 0 \cdot r_2 + 0 \cdot r_3)/6 = 0$

$$Q^1_{2,2} = (1 \cdot r_1 + 1 \cdot r_2 + 1 \cdot r_3)/6 = (r_1 + r_2 + r_3)/6$$

and so on. Where r_1, r_2, r_3 are the global coordinate of the vertices of the triangle. The universal matrix of Q^1, Q^2, Q^3 and T are summarized below:

$Q^1 = 1/6 \backslash Q^2 = 1/6$ $\quad$ (1, 1, 1) (0, 0,   0), (−1, −1, −1)
$\qquad\qquad\qquad\qquad\qquad\qquad\qquad$ (0, 0,   0), (0, 0, 0)
$\qquad\qquad\qquad\qquad$ (0, 0, 0), $\qquad\qquad\qquad\qquad$ (1, 1, 1)
$\qquad\qquad\qquad\qquad$ (0, 0, 0), (1, 1, 1),
$\qquad\qquad\qquad\qquad$ (0, 0, 0), (−1, −1, −1), (1, 1, 1)

$Q^3 = 1/6 \backslash T = 1/120$ $\quad$ (6, 2, 2), (2, 2, 1), (2, 1, 2)
$\qquad\qquad\qquad\qquad\qquad\qquad\qquad$ (2, 6, 2), (1, 2, 3)
$\qquad\qquad\qquad\qquad$ (1, 1, 1), $\qquad\qquad\qquad$ (2, 2, 6)
$\qquad\qquad\qquad\qquad$ (−1, −1, −1), (1, 1, 1),
$\qquad\qquad\qquad\qquad$ (0, 0, 0), (0, 0, 0), (0, 0, 0)

The universal matrices Q^1 and T of the second order triangular element of axisymmetric field are:

$Q^1 = 1/30 \cdot \backslash Q^2 = 1/30$

(9, 3, 3), (0, 0, 0), (2, 1,   2), (3, −2, −1), (−3, 2, 1), (−11, −4, −5)
$\qquad\qquad$(0, 0, 0), (0, 0,   0), (0, 0, 0), (0, 0, 0), (0, 0, 0)
(0, 0, 0), $\qquad\qquad$ (3, 3,   9), (1, 2, −3), (−1, −2, 3), (−5, −4, −11)
(0, 0, 0), (3, 9, 3), $\qquad\qquad$ (8, 24, 8), (−8, −24, −8), (−4, 0, 4)
(0, 0, 0), (1, 2, 2), (3, 3,   9), $\qquad\qquad$ (8, 24, 8), (4, 0, −4)
(0, 0, 0), (−2, 3, −1), (2, 1, −3), (24, 8, 8), $\qquad\qquad$ (16, 8, 16)
(0, 0, 0), (−4, −11, −5), (−4, −5, −11), (0, −4, 4), (8, 16, 16),
(0, 0, 0), (2, −3, 1), (−2, −1,   3), (−24, −8, −8), (0, 4, −4), (24, 8, 8)

Appendix 6.3 The universal matrix of axisymmetric fields

$$Q^3 = 1/30 \cdot \backslash T = 1/2520$$

```
 242,    6,    6), ( -4,   -4,    1), ( -4,    1,   -4), (  12,   -8,   -4), ( -4,  -12,  -12), (  12,   -4,   -8)
           (   6,  242,    6), (  1,   -4,   -4), ( -8,   12,   -4), ( -4,   12,   -8), (-12,   -4,  -12)
   9,    3,    3),                    (  6,    6,  242), (-12,  -12,   -4), ( -4,   -8,   12), ( -8,   -4,   12)
   2,    2,    1), (  3,    9,    3),                    ( 96,   96,   32), ( 32,   48,   32), ( 48,   32,   32)
   0,    0,    0), (  0,    0,    0), (  0,    0,    0),                    ( 32,   96,   96), ( 32,   32,   48)
 -11,  -5,   -4), ( -5,  -11,   -4), (  0,    0,    0), ( 16,   16,    8),                    ( 96,   32,   96)
  -3,    1,    2), ( -1,    3,   -2), (  0,    0,    0), (  4,   -4,    0), (  8,    8,   24),
   3,   -1,   -2), (  1,   -3,    2), (  0,    0,    0), ( -4,    4,    0), ( -8,   -8,  -24), (  8,    8,   24)
```

Part Three
Boundary Methods

Electromagnetic field problems can be handled both in different or in integral equatioins. Integral equations are very difficult to solve analytically. Hence different equations are now more familiar to have been in use for a long time. Now numerical methods are available to solve the integral equations for engineering applications. Integral equation methods may use volume integral equations or boundary integral equations. The most important advantages of boundary integral equation methods are as follows

(1) They reduce the dimension of the problem. The data preparation and storage are easier than that required for the differential equation methods.

(2) Errors caused by the approximation will be averaged out in the integral sense, but errors will be increased in the derivative process.

(3) Integral equation methods may obtain more accurate results of the field strength at the boundary.

(4) Integral equation methods are more convenient to solve open boundary problems and inverse problems.

Hence these methods have been developed more rapidly and are more widely used in recent years. However, there are a few problems with these methods, namely:

(1) Fundamental solutions are usually used in integral equation methods but these are hard to obtain in some cases.

(2) Numerical integrations are time consuming.

(3) Singular integrals must be taken care of.

(4) The solution of the integral equations may not converge as compared with variational techniques.

(5) The resulting matrices are dense and often ill-conditioned. Hence the solution of these matrix equations is not easy as compared with the matrix equations of the finite element method.

Here the emphasis is on boundary integral equation methods. Both the simple and more complicated cases are considered. Charge simulation and the surface charge simulation methods are given in the Chaps. 7 and 8. The generalized boundary element method is introduced in Chap. 9, and the method of moments is treated last as it includes all of the other methods.

Chapter 7

Charge Simulation Method

7.1 Introduction

According to the uniqueness theorem of electromagnetic fields, if a solution satisfies Laplace's equation or Poisson's equation and all the corresponding boundary conditions, no matter how that solution is obtained – even if guessed – it is the only solution of the specified boundary value problem. For example, the field distribution of an isolated charged spherical conductor equals the field distribution of a point charge if it is located at the centre of the sphere and its charge equals the total amount of surface charge of the sphere. This point charge is called the equivalent charge or simulated charge of the original charged conductor. Thus the distributed charge on the conductor surface is replaced by a lumped fictitious point charge. It should be noted that the region of interest is now the region outside the sphere. In other words, the fictitious simulated charges must be placed outside the space in which the field is under consideration.

Early in the 1950s, Loeb [1] used a set of lumped charges to analyse the field distribution of a rod-plane gap. At that time he obtained the solution without using a computer. Later on during 1968–1969 Abou-seada and Nasser [2, 3] used a digital computer to calculate the potential distribution of the same problem. Almost at the same time, Steinbigler published a more complete procedure which he called the charge simulation method (CSM) in his doctoral thesis [4]. The basic concept of the CSM is to replace the distributed charge of conductors and the polarization charges on the dielectric interfaces by a large number of fictitious discrete charges [5]. The magnitudes of these charges have to be calculated so that their integrated effect satisfies the boundary conditions exactly at a selected number of points on the boundary. From this point of view, the CSM is called a point matching method, it is one kind of equivalent source method. Starting with the 2nd ISH (International Symposium on High Voltage Engineering), in Zurich in 1975, many papers have been discussing the use of CSM. Now, it is regarded as an effective and simple method for solving Laplace's or Poisson's equations. It proved to be successful for many high-voltage field problems. It is applicable to any system that includes one or more homogeneous media. A special advantage of this method is its good applicability to three-dimensional field problems without axial symmetry. It is a practical method for engineering design [6] and for the optimum design of electrode contours [7, 8] and insulator interfaces [9]. Recently, the application of CSM has been extended to the analyse of two-dimensional elasticity problems [10], and to the Stefan problem which is a free boundary problem [11].

In this chapter, both the conventional charge simulation method and the optimized charge simulation method are discussed. The error analysis of this method is also classified in Sect. 7.7.

7.2 Matrix equations of simulated charges

The first step of the CSM is to find out the equivalent simulated charges by the charge simulation equation. These charges are always located outside the domain where the field distribution is calculated. Once these lumped charges are determined, the solution of potential and field strengths anywhere in the domain are computed by the analytical formulations and the superposition principle.

7.2.1 Matrix equation in homogeneous dielectrics

Matrix equations in the charge simulation method are discussed in the following cases.

7.2.1.1 Governing equation subject to Dirichlet boundary conditions

Figure 7.2.1 shows a rod-plane electrode, the potential distribution of this field is obtained by solving the partial differential equation subject to the given boundary conditions as follows

$$\nabla^2 \varphi = 0 \quad \text{in domain } \Omega$$
$$\varphi|_{\Gamma_1} = U_0 \quad \text{on boundary } \Gamma_1 \quad (7.2.1)$$
$$\varphi|_{\Gamma_2} = 0 \quad \text{on boundary } \Gamma_2.$$

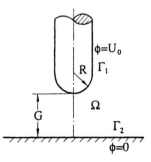

Fig. 7.2.1. Rod-plane electrode configuration

7.2 Matrix equations of simulated charges

One could also solve this problem by using the corresponding integral equation

$$\varphi = \frac{1}{4\pi\varepsilon} \int_{s'} \frac{\sigma(\mathbf{r}')}{R} ds' \qquad (7.2.2)$$

where $\sigma(\mathbf{r}')$ is the surface charge density on the electrode, s' is the area where surface charges are located, $R = |\mathbf{r} - \mathbf{r}'|$, $\mathbf{r}'$ is the position of the sources and $\mathbf{r}$ is the position of the observation point where the potential will be calculated.

Applying *CSM, the unknown distributed surface charges are replaced by a set of lumped simulated charges in such a way that the known potentials on the boundaries are satisfied.* Thus, if the types and the positions of the simulated charges are assumed, the values of these charges are determined by the known boundary conditions on the electrode surface. For example,

$$\varphi_1 = p_{11} Q_1 + p_{12} Q_2 + \cdots + p_{1j} Q_j + \cdots + p_{1N} Q_N \qquad (7.2.3)$$

where φ_1 is the known potential of point 1 on boundary Γ_1, p_{1j} is the potential coefficient between the boundary point 1 and the jth simulated charge Q_j, and $j = 1, 2, \ldots, N$ is the sequence of the simulated charges. N is the total number of simulated charges. In order to determine the magnitude of charges Q_j, N boundary points (collocation points or matching points) must be selected i.e.

$$\varphi_i = p_{i1} Q_1 + p_{i2} Q_2 + \cdots + p_{iN} Q_N \quad i = 1, 2, \ldots, N. \qquad (7.2.4)$$

If Q_1 equals 1, p_{11} is the potential of point 1 produced by charge Q_1, the total potential of point 1 is the sum of the potentials produced by N charges. Equation (7.2.4) is expressed as a matrix equation

$$\mathbf{P}^\dagger \{Q\} = \{\varphi\} \qquad (7.2.5)$$

This is a matrix equation of the simulated charges. Here $\{\varphi\}$ is a known column-matrix with N components, $\{Q\}$ is an unknown column matrix with N components and $\mathbf{P}$ is a square matrix of the order $N \times N$. When the types and the positions of simulated charges and the positions of collocation points have been selected, the potential coefficient matrix $\mathbf{P}$ can be calculated and then the unknown charges $\{Q\}$ are uniquely determined. The only requirement is that the type of simulated charges must be such that the potential produced by these charges can be expressed by an analytical formulation. In the case where the positions of the simulated charges or collocation points are changed, the magnitudes of the simulated charges will be different. Thus the substitution is not unique. A large number of different sets of solutions is possible.

---

† The bold character represents the matrix.

7.2.1.2 Governing equation subject to Neumann boundary conditions

If the problem is a Neumann boundary problem, then the surface charge distribution or the field strength along the boundary is known. In this case, the matrix equation of simulated charges is expressed as

$$\{E\} = [F]^\dagger \{Q\} \tag{7.2.6}$$

Equation (7.2.6) is a vector matrix equation where $\{E\}$ is a column vector and $[F]$ is a vector matrix of the order $N \times N$. It is defined as a field strength coefficient matrix with a similar meaning as the potential coefficient matrix P. In 2-dimensional rectangular coordinates, Eq. (7.2.6) can be separated into two scalar matrix equations

$$\begin{aligned} \{E_x\} &= \mathbf{F}_x\{Q\} \\ \{E_y\} &= \mathbf{F}_y\{Q\} \end{aligned} \tag{7.2.7}$$

This kind of formulation is suitable for use in the optimized contour design of electrodes.

7.2.1.3 Mixed boundary conditions and free potential conductors

In the problem with mixed boundary conditions, the Dirichlet boundary condition applies to one part and the Neumann boundary condition applies to a different part of the boundaries e.g.

$$\begin{cases} \nabla^2 \varphi = 0 & \text{in } \Omega \\ \varphi|_{\Gamma_1} = f_1(s) & \text{on } \Gamma_1 \\ \dfrac{\partial \varphi}{\partial n}\bigg|_{\Gamma_2} = f_2(s) & \text{on } \Gamma_2 . \end{cases} \tag{7.2.8}$$

In this case, the matrix equation is the combination of Eq. (7.2.5) and Eq. (7.2.6). They are

$$\begin{cases} \mathbf{P}\{Q\} = \{\varphi\} & \text{on } \Gamma_1 \\ [F]\{Q\} = \{E\} & \text{on } \Gamma_2 . \end{cases} \tag{7.2.9}$$

An especially important case in practical problems is the existence of free potential conductors. For instance, when an aeroplane flies acorss a charged cloud, the induced potential on the plane is unknown. In this case, the right hand side of Eq. (7.2.5) is an unknown constant. But it is obvious that the total

---

† In a 2-D case, the element of F consists of two components, hence the vector matrix is denoted by $[F]$.

7.2 Matrix equations of simulated charges

induced charge on the conductor surface is zero, i.e.

$$\int_s \frac{\partial \varphi}{\partial n} = 0.$$

Thus the sum of the simulated charges of this electrode must be zero. Hence an additional equation is required which is

$$\sum_{k=1}^{M} Q_k = 0 \tag{7.2.10}$$

where Q_k are the discrete simulated charges within the conductor with free potentials. This equation combined with Eq. (7.2.5) compensates the missing potential value in (7.2.5). The matrix equations for a free boundary problem are then extended to

$$\begin{cases} \mathbf{P}\{Q\} = \{\varphi\} \\ \sum Q_k = 0 \quad k = 1, \ldots, M \end{cases} \tag{7.2.11}$$

where M is the total number of conductors with free potentials.

7.2.1.4 Matrix form of Poisson's equation

The three cases discussed have yielded equivalent matrix equations of a Laplacian problem. If there are space charges in a linear medium and the distribution of the space charges is known, by using the principle of superposition, the matrix equation is

$$\mathbf{P}\{Q\} + \mathbf{P}_s\{Q_s\} = \{\varphi\} \tag{7.2.12}$$

where $\{Q\}$ is column vector of space charges and $\mathbf{P}_s$ is potential coefficient related to the space charges. As $\mathbf{P}_s$ and $\{Q_s\}$ are known, they can be multiplied and moved to the RHS of Eq. (7.2.12). It is important that matrix $\mathbf{P}$ should not be enlarged whether the space charge exists or not. The charge simulation method is convenient here.

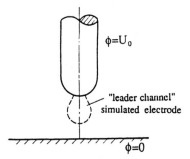

Fig. 7.2.2. Physical model of a breakdown discharge in a rod-plane gap

If the space charge distribution is not known, a reasonable physical model may be used. For example simulating a creepage discharge path when breakdown occurs in a long air gap, one could assume a constant potential gradient along the channel and the leading creepage path will then be considered as a quasi-electrode is shown in Fig. 7.2.2 by the line of dashes.

Once the contour points at the boundary of the electrode and quasi-electrode (with different potentials) are selected, then the matrix equation is established. Simulated charges having an identical number as the contour points are arranged at the electrode and quasi-electrode. Thus the matrix size of the potential coefficient P is enlarged.

7.2.3 Matrix equation in piece-wise homogeneous dielectrics

Figure 7.2.3 shows an axisymmetric problem with different dielectrics, each one being uniform. The region abcda is the cross section of a dielectric disk with a permittivity of ε_b. The permittivity of the other space is ε_a.

Due to the uniform polarization of the homogeneous dielectric, there are no volume polarization charges inside the dielectric, only surface polarization charges exist on the interface. In the charge simulation method, the surface polarization charges are considered as a single layer source. This causes the discontinuity of the field strength but the tangential component of the field strength and therefore the potential on both sides of the dielectrics are continuous. Based on the principle of CSM, the surface charge of the electrode is simulated by $Q(j), j = 1, \ldots, N$. The polarization charge of the interface of different dielectrics (the contour abcd) is simulated by $Q_a(j)$ ($j = N+1, \ldots, N+N_a$) and $Q_b(j)$ ($j = N+N_a+1, \ldots, N+N_a+n_b$) on both sides of the region with permittivities ε_b and ε_a, respectively. This is because to calculate the field in the region with permittivity ε_a, the simulated charges must be placed outside this region, hence the simulated charges $Q(j)$ and $Q_b(j)$ are used to calculate the field in the region with permittivity ε_a. Similarly, $Q(j)$ and $Q_a(j)$ are used to calculate the field in the region with permittivity ε_b. N, N_a and N_b are the total number of simulated charges within the conductor and on both sides of the interface. If there are N charges within the electrode, then the matching points on the electrode surface are also N. If the matching points on

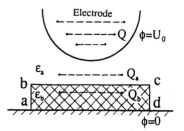

Fig. 7.2.3. Simulated charges in multi-dielectrics

7.3 Commonly used simulated charges

the interface are N_d, then the simulated charges on both sides near the interface are $2N_d$. The total unknown charges, $N + 2N_d$, are solved by the identical equations of $N + 2N_d$. To ensure the interfacial boundary conditions are satisfied, $2N_d$ equations are obtained by the following conditions:

$$\begin{cases} \varphi_a = \varphi_b \\ \varepsilon_a \frac{\partial \varphi_a}{\partial n} = \varepsilon_b \frac{\partial \varphi_b}{\partial n} . \end{cases} \tag{17.2.13}$$

Here the subscripts a and b indicate different regions with different permittivities of ε_a and ε_b, and n indicates the normal direction of the interface. Therefore, sufficient boundary conditions are provided to solve the unknown simulated charges. These equations are:

$$\begin{cases} \sum_{j=1}^{N} p_{ij} Q(j) + \sum_{j=N+1}^{N+N_b} p_{ij} Q_b(j) = \varphi_i \quad (i = 1, 2, \ldots, N) \\ \sum_{j=1}^{N} p_{ij} Q(j) + \sum_{j=N+1}^{N+N_a} p_{ij} Q_a(j) = \sum_{j=1}^{N} p_{ij} Q(j) + \sum_{j=N+N_a+1}^{N+N_a+N_b} p_{ij} Q_b(j) \\ \varepsilon_b \left[\sum_{j=1}^{N} (f_{ij})_n Q(j) + \sum_{j=N+1}^{N+N_a} (f_{ij})_n Q_a(j) \right] \\ = \varepsilon_a \left[\sum_{j=1}^{N} (f_{ij})_n Q(j) + \sum_{j=N+N_a+1}^{N+N_a+N_b} (f_{ij})_n Q_b(j) \right] . \end{cases} \tag{7.2.14}$$

To calculate the potential in either region a or b, the permittivity of material is ε_0, hence the second equation of (7.2.14) is simplified to:

$$\sum_{j=N+1}^{N+N_a} p_{ij} Q_a(j) - \sum_{j=N+N_a+1}^{N+N_a+N_b} p_{ij} Q_b(j) = 0 . \tag{7.2.15}$$

A typical example is presented in reference [12]. It shows a dielectric sphere with permittivity ε_a placed in an uniform electric field within a material with permittivity of ε_b. When $\varepsilon_a/\varepsilon_b < 6$, the error of the field strength is less than 0.1%. The drawback of this method is, that in case of more than two different materials, the number of simultaneous equations is increased rapidly, especially in the case of complicated shapes of the dielectric interface. In the latter case, more computer time is required. The surface charge simulation method (Chap. 8) is more convenient than the charge simulation method.

7.3 Commonly used simulated charges

The most commonly used simulated charges are point, line and ring charges. Combining these charges, a great number of different shapes of electrodes can be simulated. In some special cases, elliptic cylindrical or ellipsoidal surface charges can be used as simulated charges. The computation formulae of point, line, ring and elliptic cylinder charges are given in this section. Formulas for a disk charge and other shapes can be found in references [14] and [13], respectively.

As many engineering problems are axisymmetric, cylindrical coordinates are used to express the commonly used formulations. The influence of the ground or any other grounded conductor plane is shown by images. To make the contents simple only the formulation for calculating the potentials are listed in the contents. The formulations of the field strengths are given in Appendices A.7.1–A.7.4.

7.3.1 Point charge

In axisymmetric problems, a point charge Q is located at point (r', z') and its image $-Q$ is located at point $(r', -z')$, as shown in Fig. 7.3.1.

The potential at any point $P(r, z)$ is expressed as:

$$\varphi_P = \frac{Q}{4\pi\varepsilon} \left[\frac{1}{\sqrt{(r-r')^2 + (z-z')^2}} - \frac{1}{\sqrt{(r-r')^2 + (z+z')^2}} \right]. \quad (7.3.1)$$

If the source point and the field point are both located at the axis, i.e. $r = r' = 0$, then Eq. (7.3.1) is simplified as:

$$\varphi_P = \frac{Q}{4\pi\varepsilon} \left[\frac{1}{|z-z'|} - \frac{1}{|z+z'|} \right] \quad (7.3.2)$$

For a 3-D case,

$$\varphi_P = \frac{Q}{4\pi\varepsilon} \left[\frac{1}{\sqrt{(x-x')^2 + (y-y')^2 + (z-z')^2}} - \frac{1}{\sqrt{(x-x')^2 + (y-y')^2 + (z+z')^2}} \right]$$

where (x, y, z) and (x', y', z') are the field point and source point, respectively.

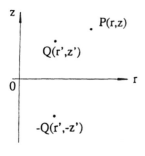

Fig. 7.3.1. A point charge Q and its image $-Q$

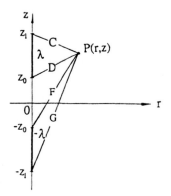

Fig. 7.3.2. A line charge and its image

7.3.2 Line charge

A line charge of length l lies on the z-axis of cylindrical coordinates shown in Fig. 7.3.2.

Assume the density of the line charge is $\lambda C/m$. Then the potential at any point $P(r, z)$ is

$$\varphi_P = \frac{\lambda}{4\pi\varepsilon}\left[\ln\frac{z_1 - z + C}{z_0 - z + D} + \ln\frac{z_0 + z + F}{z_1 + z + G}\right]. \tag{7.3.3}$$

If $l \to \infty$, then

$$\varphi_P = \frac{\lambda}{4\pi\varepsilon}\ln\frac{z_0 + z + F}{z_0 - z + D}. \tag{7.3.4}$$

The formulations of the constants C, D, F, G are listed in Appendix A.7.2

7.3.3 Ring charge

A filamentary circular loop with radius r_0 carrying a uniform charge density $\lambda\ C/m$ lies on the plane $z = 0$ shown in Fig. 7.3.3.

The potential at any point $P(r, z)$ is

$$\varphi_P = \frac{1}{4\pi\varepsilon}\oint\frac{\lambda dl}{R} + C \tag{7.3.5}$$

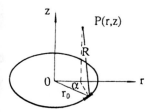

Fig. 7.3.3. A ring charge

where

$$R^2 = [(r_0 \cos \alpha - r)^2 + (r_0 \sin \alpha)^2 + z^2] \tag{7.3.6}$$

and C is a constant, it is determined by the position of the zero reference of the potential. Then the final result is

$$\varphi = \frac{Q}{4\pi\varepsilon} \frac{2}{\pi} \left[\frac{K(k)}{\beta} \right] + C \tag{7.3.7}$$

where

$$K(k) = \int_0^{\pi/2} \frac{d\theta}{\sqrt{1 - k^2 \sin^2\theta}} \tag{7.3.8}$$

$$\begin{cases} k^2 = \dfrac{4rr_0}{(r + r_0)^2 + z^2} \\ \beta^2 = (r_0 + r)^2 + z^2 \\ 2\theta = \pi - \alpha \,. \end{cases} \tag{7.3.9}$$

$K(k)$ is a complete elliptic integral of the first kind.

Consider now, that both the ring charge and its image are as shown in Fig. 7.3.4 and the zero potential is chosen on the symmetrical plane, then:

$$\varphi = \frac{Q}{4\pi\varepsilon} \frac{2}{\pi} \left[\frac{K(k_1)}{\alpha_1} - \frac{K(k_2)}{\alpha_2} \right] \tag{7.3.10}$$

where

$$\alpha_1 = [(r + r_0)^2 + (z - z_0)^2]^{1/2} \qquad \alpha_2 = [(r + r_0)^2 + (z + z_0)^2]^{1/2}$$

$$k_1 = \frac{2\sqrt{rr_0}}{\alpha_1} \qquad\qquad k_2 = \frac{2\sqrt{rr_0}}{\alpha_2} \,. \tag{7.3.11}$$

The approximate formulae for calculating elliptic integrals are given in Appendix A.7.5 [15].

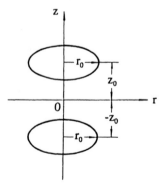

Fig. 7.3.4. A ring charge and its image

7.3.4 Charged elliptic cylinder

A charged elliptic cylinder is shown in Fig. 7.3.5(a).

The expansion form of Laplace's equation in elliptic cylindrical coordinates (Fig. 7.3.5(b)) is [16]

$$\nabla \varphi = \frac{1}{f^2(\cos^2\eta - \cos^2\psi)}\left[\frac{\partial^2\phi}{\partial\eta^2} + \frac{\partial^2\phi}{\partial\psi^2} + \frac{\partial^2\phi}{\partial z^2}\right] = 0. \tag{7.3.12}$$

If the cylinder is infinitely long and the surface of the ellipse is an equipotential surface, then the potential φ is independent of the variables z and the ψ. Equation (7.3.12) is simplified to

$$\frac{d^2\varphi}{d\eta^2} = 0 \tag{7.3.13}$$

then

$$\varphi = A\eta + B. \tag{7.3.14}$$

Let

$$\eta = \eta_a, \quad \phi = U_0$$
$$\eta = \eta_b, \quad \phi = 0$$

consequently,

$$\varphi = \frac{U_0}{\eta_a - \eta_b}(\eta - \eta_b) = \frac{\lambda}{2\pi\varepsilon}(\eta - \eta_b). \tag{7.3.15}$$

According to the coordinate transformations between Cartesian coordinates

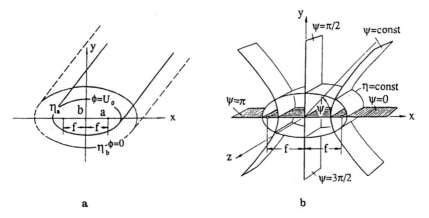

Fig. 7.3.5a, b. An elliptic cylinder in Cartesian coordinates (x, y, z) and in elliptic cylindrical coordinates (η, ψ, z)

and the elliptic cylindrical coordinates

$$\begin{cases} x = f\cosh\eta \cos\psi \\ y = f\sinh\eta \sin\psi \\ z = z \end{cases} \quad (7.3.16)$$

the following equation is obtained

$$\left(\frac{x}{f\cosh\eta}\right)^2 + \left(\frac{y}{f\sinh\eta}\right)^2 = 1 \quad (7.3.17)$$

where f is the focal length of the ellipse

$$\sinh\eta = \sqrt{\frac{(x^2 + y^2 - f^2) \pm \sqrt{(f^2 - x^2 - y^2)^2 + 4f^2 y^2}}{2f^2}} = A$$

Let
$$\eta = \sinh^{-1} A = \ln(A + \sqrt{A^2 + 1}). \quad (7.3.18)$$

then
$$R_1^2 = (x - f)^2 + y^2 \qquad R_2^2 = (x + f)^2 + y^2 \qquad s = \tfrac{1}{2}(R_1 + R_2)$$

hence
$$\eta = \ln[\tfrac{1}{2}(s + \sqrt{s^2 - f^2})] \quad (7.3.19)$$

$$\varphi = K - \frac{\lambda}{2\pi\varepsilon} \ln[\tfrac{1}{2}(s + (s^2 - f^2)^{1/2})] \quad (7.3.20)$$

where K is a constant depending on the reference. Consider that $f^2 = a^2 - b^2$, if $a = b$, i.e. $f = 0$, Eq. (7.3.20) represents the potential produced by a charged cylinder of infinite length. If $a = b = 0$, Eq. (7.3.20) represents the potential induced by a charged plate with a width of 2a and zero thickness which is reduced by an elliptic cylinder. The thin plate is used to simulate the field produced by the conductors with sharp edges.

7.4 Applications of the charge simulation method

General procedures of CSM are shown in Fig. 7.4.1.

In the fourth step of the diagram, Gauss's elimination method is used to solve the simultaneous equations. This is due to the fact that the coefficient matrix of the simulated charges is a full and asymmetric matrix.

According to the method used, some possible problems must be considered.
(1) Is the solution of the simulated charges $\{Q\}$ stable?

In CSM, the matrix of the simulated charge equations is easy to be ill-conditioned. For example, if the number of the simulated charges is too large or the relative distance between the simulated charges and the matching points is too small, then the coefficients of two parallel rows or columns of the matrix are very close. Thus the matrix could be ill-conditioned. If the matrix is ill-condi-

7.4 Applications of the charge simulation method

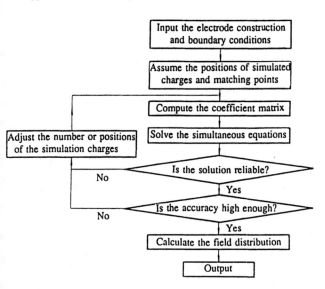

Fig. 7.4.1. Flow-chart of CSM

tioned then the solution is not the real solution of the problem. A simple way to check this possibility is by giving a small perturbation to the matrix. If the solution remains stable then it is an acceptable solution. Otherwise the positions or the number of the simulated charges have to be adjusted.

(2) Does the solution matrix $\{Q\}$ satisfy the whole boundary?

As the positions of the simulated charges are determined arbitrarily and the number of matching points are very limited compared with a continuous boundary it is possible that the solution matrix $\{Q\}$ does not match the whole boundary condition very well. In this case, the initial positions of the simulated charges or the matching points on the boundaries have to be adjusted or the number of the discretized charges have to be increased in such a way that the given boundary conditions are satisfied. A measure of the accuracy of the results may be shown as the potential error on the surface of the electrode or the continuity condition of the tangential component of the field strength on the interface of different dielectrics.

(3) In order to improve the accuracy of the solution or to overcome the slight ill-conditioned of the matrix, double precision computation is advised.

Once an adequate number of simulated charges have been selected the field distribution in the whole domain is computed by using analytical formulations

$$\varphi_i = \sum_{j=1}^{n} p_{ij} Q_j \tag{7.4.1}$$

$$E_i = \sum_{j=1}^{n} (f_{x,ij} Q_j \mathbf{i} + f_{y,ij} Q_j \mathbf{j}) \tag{7.4.2}$$

where the bold characters '**i**' and '**j**' are unit vectors of the x and y axes.

Example 7.4.1. Find the equivalent simulated charges of the rod-plane electrodes, shown in Fig. 7.2.1. Assume the radius of the cylindrical rod is 1 cm.

Solution. The simulated model is shown in Fig. 7.4.2. The influence of the ground is replaced by an image. The surface charge density is larger on the top of the rod and becomes smaller when the surface of the rod is far away from the ground. To simulate this field, a point charge is assumed at the centre of the hemi-sphere of the rod and a set of line charges are placed along the axis of the rod. The starting point of these line charges with half infinite length and different densities are located at different positions as the symbol '.' shows in Fig. 7.4.2.

Assume the starting positions of line charges are z_j and the charge densities are $\lambda_j (j = 1, 2, \ldots, N - 1)$. The matching points on the conductor surface are indicated by the symbol '×'. Then the simulated charge equations are expressed as:

$$\begin{bmatrix} p_{11} & p_{12} & \cdots & p_{1n} \\ p_{21} & p_{22} & \cdots & p_{2n} \\ \vdots & & & \vdots \\ p_{n1} & & & p_{nn} \end{bmatrix} \times \begin{Bmatrix} \lambda_1 \\ \lambda_2 \\ \vdots \\ \lambda_{n-1} \\ Q_0 \end{Bmatrix} = \begin{Bmatrix} \varphi_1 \\ \varphi_2 \\ \vdots \\ \varphi_n \end{Bmatrix} \quad (7.4.3)$$

where

$$p_{ij} = \ln \frac{(z_j + z_i) + \sqrt{r_i^2 + (z_j + z_i)^2}}{(z_j - z_i) + \sqrt{r_i^2 + (z_j - z_i)^2}} \quad \begin{pmatrix} i = 1, \ldots, n \\ j = 1, \ldots, n-1 \end{pmatrix}$$

$$p_{in} = \frac{1}{\sqrt{r_i^2 + (z_n - z_i)^2}} - \frac{1}{\sqrt{r_i^2 + (z_n + z_i)^2}} \quad (i = 1, \ldots, n) \quad (7.4.4)$$

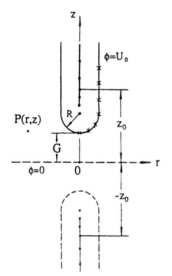

Fig. 7.4.2. Simulated model of a rod-plane gap

7.4 Applications of the charge simulation method

and r_i and z_i are the coordinates of matching points, z_n is the position of point charge. These two formulae include the influence of the ground by considering the image charges. The results of (7.4.3) multiplying the factor $4\pi\varepsilon_0$ is the real value of the simulated charges. The calculation results are shown in Tables 7.4.1 and 7.4.2.

Examining Table 7.4.1, one should note that:

(1) The line charges are alternatively 'positive' and 'negative' so that several semi-infinite line charges are equivalent to a line source having non-uniform charge density.

(2) The magnitudes of the simulated charges are changing as well as the different gap lengths between the rod and the grounded plane.

Table 7.4.2 shows that if the positions of the simulated charges are fixed but the gap length between the rod-plane is changed, then the accuracy is different. For example, as the ratio G/R is decreased, the maximum potential error on the surface of the rod is increased. This indicates that for different dimensions of the electrode the positions of the simulated charges should be changed. Desired accuracy could be obtained if a reasonable number of simulated charges and matching points are selected.

It is imperative that the matching points must cover the region of interest. Otherwise the simulation is meaningless.

Example 7.4.2. A charged sphere is close to a grounded plane as shown in Fig. 7.4.3, find the capacitance of this system.

Solution. Eight ring charges are located on the fictitious sphere surface with radius R_c and two point charges are placed on the two poles of the same sphere. Both the simulated ring and point charges are indicated by '$\cdot$—$\cdot$' and '$\cdot$' in Fig. 7.4.3. The matching points on the conductor surface are indicated by '×'. After running the program written according to the diagram shown in Fig. 7.4.1, the amplitude of each simulated charge is obtained. Assume $U_0 = 1$V, then the capacitance is the summation of these charges, i.e.

$$C = 4\pi\varepsilon_0 \sum_{i=1}^{10} Q_i$$

$$= 4\pi\varepsilon_0(-1.06647 + 1.04015 - 0.43989 + 0.36833 + 0.43692$$
$$-0.49344 + 1.13355 - 1.10882 + 0.60225 + 0.64069)$$

$$= 4\pi\varepsilon_0 \times (1.11127) = 1.23587 \text{pf}$$

The relative error is 9.39×10^{-2}% (the accurate result is calculated by image method). The relative errors of the potential and the field strength along the conductor sphere are shown in Fig. 7.4.4.

Example 7.4.3. Determine the breakdown voltage of a sphere-plane gap under impulse voltage.

Table 7.4.1. Simulated charges and matching points for different gap lengths

Number		1	2	3	4	5	6	7	8
Position of line charge		$G+1.0$	$G+1.1$	$G+1.2$	$G+1.5$	$G+2.0$	$G+10.0$	$G+25.0$	—
Position of point charge		—	—	—	—	—	—	—	$r=0.0$ $z=G+1.0$
Coordinates of matching points	r_i z_i	1.0 $G+1.0$	1.0 $G+2.0$	1.0 $G+5.0$	1.0 $G+15.0$	1.0 $G+40.0$	0 G	0.50 $G+0.134$	0.80 $G+0.40$
Magnitudes of simulated charges for different gap length, G/R	$G/R=5$ $G/R=20$ $G/R=200$	-80.6778101 -50.577172 -37.747776	140.772012 92.011102 69.244049	-71.471505 -50.104452 -38.167908	-13.402269 10.241929 7.886405	-1.849478 -1.428139 -1.097394	-0.0382239 -0.025475 -0.017328	-0.028444 -0.020458 -0.012356	5.2060099 3.314317 2.474494

7.4 Applications of the charge simulation method

Table 7.4.2. The calculated potential value along the electrode surface

Gap length (G/R)	5.0	20.0	200.0
Potential values on half hemi-sphere (V) (each point 0.2 radian apart)	0.99999997 0.99999423 0.99998317 0.99997959 0.99999078 1.00000151 0.99997106 0.99989623	1.00000000 0.99999529 0.99998642 0.99998449 0.99999409 0.99999919 0.99996412 0.99989541	1.00000000 0.99999927 0.99998962 0.99998791 0.99999559 0.99999947 0.99997186 0.99991749
max. error (%)	0.01	0.011	0.008
Potential values on rod (V) (The distance between each point is 2 cm)	0.99999997 1.00547751 0.99999995 1.02159402 1.02647452 0.99543609 0.99140996 0.99999991 1.01062371 1.02003660	1.00000000 1.00588382 0.99999998 1.01411101 1.01733291 0.99673632 0.99415117 1.00000000 1.00722890 1.01363741	1.00000000 1.00504270 1.00000070 1.00991230 1.01229780 0.99826051 0.99631070 1.00000000 1.00459242 1.00865760
Max. error (%)	2.6	1.7	1.2

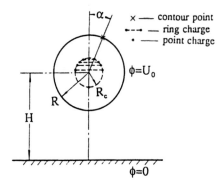

Fig. 7.4.3. Simulated charges in a charged sphere

Solution. When the voltage is applied to the sphere, the initial field strength is calculated by CSM. Assume that an electron exists at one point, according to the theorem of air-discharge, the equivalent number of electrons K is determined by:

$$K = e^{\int_{x_i}^{x_e} \alpha \, dx} \tag{7.4.5}$$

where α represents the coefficient of effective ionization. It is dependent on the field strength $E(\alpha/P = Ae^{-B(E/P)})$. x_i, x_e are the starting and ending positions of the ionization, for instance in Fig. 7.4.5, $K_G = e^{\int_0^G \alpha \, dx}$. Assume a set of ring charges is used to simulate the space charges, as shown in Fig. 7.4.5, the field strength is calculated repetitively. The results coincide well with the measured value.

When the number of electrons K at the end of the electron avalanche, e.g. the point W is larger than a given number K_0, the self-sustained discharge occurs., Then the breakdown voltage is obtained. The results are presented in Table 7.4.3 [17].

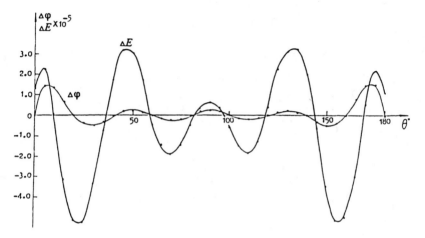

Fig. 7.4.4. Relative errors of potential and field strength distribution along the half sphere

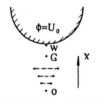

Fig. 7.4.5. Physical model of a breakdown discharge between the sphere-plane gap

Table 7.4.3. The breakdown voltage of the sphere-plane gap

Kind of gas	Gap distance (cm)	Radius of sphere (cm)	Calculated value (kV)	Measured value (kV)
Air	2.0	3.0	58.9	59.0
SF_σ	2.0	7.5	158.0	169.1

7.5 Coordinate transformations

In order to simplify computation, coordinate transformations are usually used in numerical calculation. Remember in finite element method, transformations between global and local coordinates are used so that the same formula is used

7.5 Coordinate transformations

to calculate the coefficients of every element matrix. In CSM, to simulate the complicated shape of the boundary, many line and ring charges are placed in different positions and ways. To use the identical formulation to the one in Eqs. (7.3.1) to (7.3.15), transformation of the coordinates is required.

7.5.1 Transformation matrix

First, a two-dimensional case as shown in Fig. 7.5.1 is considered.

The coordinates of an arbitrary point P in $x' - 0' - y'$ plane are x' and y', and x and y in the $x0y$ plane. If the axes of x' and y' are rotated by angle α in clockwise direction and move it parallel to the x and y axes, then these two coordiante systems are coincident. Hence, the relationship between (x, y) and (x', y') is

$$\begin{Bmatrix} x' \\ y' \end{Bmatrix} = T \begin{Bmatrix} x - x_0 \\ y - y_0 \end{Bmatrix} \quad \text{or} \quad \begin{Bmatrix} x - x_0 \\ y - y_0 \end{Bmatrix} = T^{-1} \begin{Bmatrix} x' \\ y' \end{Bmatrix} \qquad (7.5.1)$$

where α is the angle between x-axis and x'-axis in the $x0y$ plane,

$$T = \begin{bmatrix} \cos\alpha & \sin\alpha \\ -\sin\alpha & \cos\alpha \end{bmatrix} \qquad (7.5.2)$$

and T is a transformation matrix. It is a skewed[†] matrix. This relationship can be extended to the 3-dimensional case:

$$\begin{Bmatrix} x' \\ y' \\ z' \end{Bmatrix} = T \begin{Bmatrix} x - x_0 \\ y - y_0 \\ z - z_0 \end{Bmatrix} \qquad (7.5.3)$$

where

$$T = \begin{bmatrix} \cos\alpha\cos\beta & \sin\alpha\cos\beta & \sin\beta \\ -\sin\alpha & \cos\alpha & 0 \\ -\cos\alpha\sin\beta & -\sin\alpha\sin\beta & \cos\beta \end{bmatrix} \qquad (7.5.4)$$

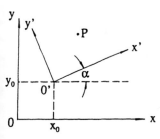

Fig. 7.5.1. Coordinate transformations

[†] A skewed matrix has elements which are symmetrical about the principal diagonal, but are opposite in sign.

Here, β is the angle between the x-axis and the x'-axis in the x0z plane. Equation (7.5.3) can also be expressed in an inverted form by $[\mathbf{T}]^{-1}$.

It should be noted that the transformation matrix is an orthogonal matrix. This is because

$$\mathbf{T}^{-1} = \mathbf{T}^{\mathrm{T}} = \begin{bmatrix} \cos\alpha & -\sin\alpha \\ \sin\alpha & \cos\alpha \end{bmatrix} \tag{7.5.5}$$

$$\mathbf{T}\mathbf{T}^{\mathrm{T}} = \begin{bmatrix} 1 & 0 \\ 0 & 1 \end{bmatrix} = I \tag{7.5.6}$$

Hence the inverse matrix $\mathbf{T}^{-1}$ is easy to be obtained.

7.5.2 Inverse transformation of the field strength

Figure 7.5.2 shows that a curved electrode surface is simulated by three short line charges. The length of each line charge is $l^{(1)}$, $l^{(2)}$, $l^{(3)}$, respectively. Due to coordinate transformations, each line charge is located on the axis of $y^{(1)}$, $y^{(2)}$, $y^{(3)}$, therefore the formulation given in Sect. 7.3.2 can be used directly, i.e.

$$\varphi^{(i)} = \frac{\lambda^{(i)}}{4\pi\varepsilon} \ln \frac{l^{(i)} - y^{(i)} + \sqrt{(x^{(i)})^2 + (l^{(i)} - y^{(i)})^2 + (z^{(i)})^2}}{-y^{(i)} + \sqrt{(x^{(i)})^2 + (y^{(i)})^2 + (z^{(i)})^2}}$$

$$(i = 1, 2, 3). \tag{7.5.7}$$

Notice that the coordinates of x, y and z must be transferred to the coordinates of $x^{(i)}$, $y^{(i)}$ and $z^{(i)}$. Then the potential at any point is given by the superposition of $\varphi^{(1)}$, $\varphi^{(2)}$, $\varphi^{(3)}$, i.e.

$$\varphi = \sum_{i=1}^{3} \varphi^{(i)} \tag{7.5.8}$$

However, the field strength is the derivative of

$$\frac{\partial \varphi^{(i)}}{\partial x^{(i)}}, \frac{\partial \varphi^{(i)}}{\partial y^{(i)}}, \frac{\partial \varphi^{(i)}}{\partial z^{(i)}}.$$

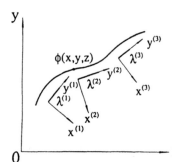

Fig. 7.5.2. The application of coordinate transformations

They are calculated in different coordinates; the total results of the field strength must be added vectorally in the same coordinates, i.e.

$$\begin{Bmatrix} E_x \\ E_y \\ E_z \end{Bmatrix} = \mathbf{T}^{(1)\mathrm{T}} \begin{Bmatrix} E_x^{(1)} \\ E_y^{(1)} \\ E_z^{(1)} \end{Bmatrix} + \mathbf{T}^{(2)\mathrm{T}} \begin{Bmatrix} E_x^{(2)} \\ E_y^{(2)} \\ E_z^{(2)} \end{Bmatrix}$$

$$+ \mathbf{T}^{(3)\mathrm{T}} \begin{Bmatrix} E_x^{(3)} \\ E_y^{(3)} \\ E_z^{(3)} \end{Bmatrix} \tag{7.5.9}$$

Here the superscript represents the number of different coordinates such as 1, 2, 3 shown in Fig. 7.5.2.

7.6 Optimized charge simulation method (OCSM)

It was stated in Sect. 7.2 that if the number of unknown fictitious simulated charges is N, then the number of matching points on the boundaries has to be the same number N in order to ensure the matrix is of the order $N \times N$ and a unique solution is obtained. If high accuracy is required, the number of matching points has a large value, thus the size of the matrix equation becomes large. The consequence is that more computer storage and computing time are required. The more serious problem is that the matrix might be ill-conditioned. *If the number of matching points are increased, but the simulated charges are not increased correspondingly, then superparametric equations are obtained. This means that the matrix equation has the order of $M \times N$ ($M > N$). The solution method used for superparametric equations is the one which finds the minimizer of the matrix equation.* This method will be discussed in Chap. 11. Another problem in using CSM is that the positions of the simulated charges are obtained by experience. If the shape of the boundary is simple, the simulated charges are easily placed and a good result is obtained. If the shape of boundary is complicated, the locations of the simulated charges are difficult to find. *It is desirable to determine the suitable positions and amplitudes of the simulated charges automatically. This method is called the optimized charge simulation method* (OCSM). It was provided by Yializes in 1978 [18]. This section concentrates on discussing this topic.

7.6.1 Objective function

The purpose of OCSM is to find the equivalent lumped sources which best replace the distributed surface charges. This means that the error between the real potential φ_a and the simulated value φ_c (which is calculated by the simulated charges), should be as small as possible. If the errors are zero, then the calculated value φ_c is the solution of the problem, e.g. if

$$\varphi_{ai} - \varphi_{ci} = 0 \quad i = 1, 2, \ldots, \infty \tag{7.6.1}$$

the simulation is completely successful. Actually it is impossible. Usually the average mean square error F of the potentials along the whole boundary approaching zero is considered as a criterion, i.e.

$$F(\mathbf{X}) = \frac{1}{S}\int_s (\varphi_a - \varphi_c)^2 \, ds \to 0 \tag{7.6.2}$$

If
$$F(\mathbf{X}) = \min \tag{7.6.3}$$

it means that the simulated potentials along the whole surface satisfy the governing equation and the given boundary conditions. By combining Eq. (7.6.2) and (7.6.3) and approximating the resultant equation in a discretized form, Eq. (7.6.3) leads to:

$$F(\mathbf{X}) = \sum_{i=1}^{M} (\varphi_{ai} - \varphi_{ci})^2 = \min \tag{7.6.4}$$

where M is the number of discretization points along the boundary. The function F is called the objective function, X is a column vector of design variables of the objective function.

According to Eq. (7.2.5)

$$\varphi_{ci} = \sum p_{ij} Q_j \tag{7.6.5}$$

after the type and the number of the simulated charges are chosen, the design variables of Eq. (7.6.4) could be either the positions or the amplitudes of the simulated charges or both of them. The positions of the matching points can also be chosen as design variables. Usually, the number of the simulated charges is less than the number of the matching points and the influence of the simulated charges is stronger than the influence of the matching points. Consequently, the positions or both the positions and the amplitudes of the simulated charges are chosen as the design variables.

In order to have a valid substitution, simulated charges must be located outside the region of interest. Hence Eq. (7.6.4) must be completed by the constrained conditions as follows

$$\begin{cases} F(\mathbf{X}) = \sum (\varphi_{ai} - \varphi_{ci})^2 = \min & i = 1, \ldots, M \\ f(\mathbf{r}) < g(\mathbf{r}) & j = 1, \ldots, N \end{cases} \tag{7.6.6}$$

The second equation of Eq. (7.6.6) is a constrained condition of the objective function $F(\mathbf{X})$, it limits the positions of the simulated charges to be inside a desired region. Here $\mathbf{r}$ is the position vector of the simulated charges. Therefore the optimized charge simulation method is a constrained optimization problem (see Chap. 11).

7.6.2 Transformation of constrained conditions

The solution of the constrained optimization problem is more complex than in unconstrained optimization. In the charge simulation method, the constrained condition is that the simulated charges must be placed inside the conductors or outside the region under consideration. Hence *a simple transformation function such as a sinusoidal function, which has the properties of* $-1 \leq \sin x \leq 1$ *and* $0 \leq \sin^2 x \leq 1$ *can be used to transform the constrained optimization problem into an unconstrained optimization problem.*

For example, there is a charged spherical conductor over a grounded plane, as shown in Fig. 7.7.1. Two point charges Q_1, Q_2 are used to simulate the surface charge of the sphere. The design variables of the objective function are positions and magnitudes of these two point charges. They are H_1, H_2 and Q_1, Q_2. If a new variable X is introduced and used to express H_1, H_2 and Q_1, Q_2, i.e.

$$\begin{cases} H_1 = H + 0.9R \sin x_1 \\ H_2 = H + 0.9R \sin x_2 \\ Q_1 = K(\sin x_3) \\ Q_2 = K(\sin x_4) \end{cases} \tag{7.6.7}$$

then the design variables of the objective function F are x_1, x_2, x_3 and x_4, i.e.

$$\begin{cases} F(\mathbf{X}) = \sum_{i=1}^{M} (\varphi_{ai} - \varphi_{ci})^2 = \min \\ \{\mathbf{X}\} = \{x_1 \ x_2 \ x_3 \ x_4\}^T \end{cases} \tag{7.6.8}$$

where M is the number of matching points on the conductor surface. Equation (7.6.8) is an unconstrained optimization problem, the positions of the simulated charges are limited to be inside the sphere by the first two equations of Eq. (7.6.7). In Eq. (7.6.7), K is an arbitrary constant.

7.6.3 Examples

Example 7.6.1. Calculate the capacitance between the sphere-plate electrode as given in Example 7.4.2 by OCSM.

Solution. In Fig. 7.6.1, assume that $H = 5$ cm, $R = 1$ cm, and $U_0 = 1$ V. The influence of the grounded plane is taken into account by the symmetric image charges $-Q_1$ and $-Q_2$, the potential of the arbitrary point on the spherical surface is

$$\varphi(r, z) = \frac{1}{4\pi\varepsilon_0} \times \left[Q_1 \left(\frac{1}{\sqrt{r^2 + (z - H_1)^2}} + \frac{1}{\sqrt{r^2 + (z + H_1)^2}} \right) \right.$$
$$\left. + Q_2 \left(\frac{1}{\sqrt{r^2 + (z - H_2)^2}} + \frac{1}{\sqrt{r^2 + (z + H_2)^2}} \right) \right]. \tag{7.6.9}$$

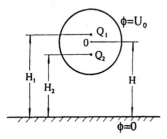

Fig. 7.6.1. An example of OCSM

Substitution of Eq. (7.6.9) and (7.6.7) into Eq. (7.6.8) yields the formulation to calculate the objective function $F(X)$. The gradient of this objective function $F(X)$ can be expressed in analytical form. If the initial value of the design variable is assumed as

$$\{X\} = \{0.012 \quad -0.2 \quad 0.05 \quad 0.01\}^T$$

when $M = 25$, after 6 iterations the objective function F decreases to 1.43842×10^{-6}. In this case

$$\begin{cases} H_1 = 5.016549 \text{ cm} \\ H_2 = 4.821025 \text{ cm} \\ Q_1 = 4\pi\varepsilon_0 \times 1.074916 \text{ C} \\ Q_2 = 4\pi\varepsilon_0 \times 0.036231 \text{ C} \end{cases} \quad (7.6.10)$$

Based on the solution of eqn. (7.6.10), the capacitance of this system is:

$$C = Q_1 + Q_2 = 1.235733 \text{ pf.}$$

In comparison with the classical result, the relative error is -8.02×10^{-2}%. During the process of optimization, an unconstrained optimization method, called DEP method is used. Details of the DFP method are given in Chap. 11.

Example 7.6.2. Field distribution of a transformer bushing.

Solution. A simplified figure of a transformer bushing is shown in Fig. 7.6.2 by the heavy solid line.

Suppose that the potential along the bushing decreases linearly. The grouping of the simulated charges in each part of this problem is shown in Table 7.6.1. There are 96 simulated charges and 172 design variables.
The objective function F is

$$F = \sum_{i=1}^{M} \left[\sum_{j=1}^{N_1} (p_{i,j})_l Q_j + \sum_{j=N_1+1}^{N_1+N_2} (p_{i,j})_r Q_j - \varphi_{ai} \right]^2 = \sum_{i=1}^{M} (f_i)^2 \quad (7.6.11)$$

where N_1 is the number of line charges, N_2 is the number of ring charges. $(p_{i,j})_l$ and $(p_{i,j})_r$ are potential coefficients of line and ring charges, respectively. φ_{ai} is the known potential on the boundary. The derivative of the objective function (it

7.6 Optimized charge simulation method

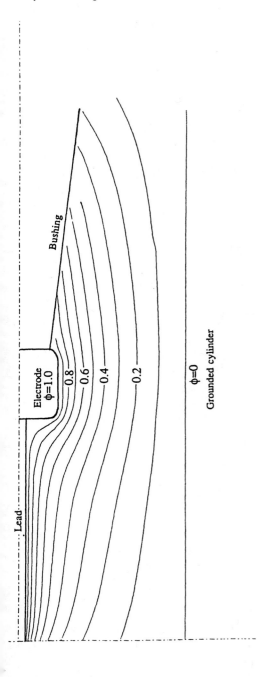

Fig. 7.6.2. Potential distribution of a transformer bushing

Table 7.6.1. The grouping of simulated charges and matching points

Component	Contour points	Simulated charges	Design variables
Lead	41	20 line charges	20
Electrode	30	14 ring charges	28
Bushing	41	21 ring charges	42
Grounded cylinder	81	41 ring charges	82

is needed in the process of optimization) can be obtained analytically

$$G = \nabla F = -2 \sum_{i=1}^{M} f_i (\nabla f_i) \tag{7.6.12}$$

$$f_i = \sum_{j=1}^{N_1} (p_{i,j})_l \, Q_j + \sum_{j=N+1}^{N_1+N_2} (p_{i,j})_r \, Q_j - \varphi_{ai} = \varphi_{ci} - \varphi_{ai} \tag{7.6.13}$$

Using Eqs. (7.3.4) and (7.3.10), a set of partial derivative terms is obtained.

$$\frac{\partial \varphi_i}{\partial \lambda_j} = \frac{1}{4\pi\varepsilon} \sum_{i=1}^{M} \ln \frac{z_{1,j} - z_i + C}{z_{0,j} - z_i + D},$$

$$\frac{\partial \varphi_i}{\partial z_{1,j}} = \frac{\lambda_i}{4\pi\varepsilon} \sum_{i=1}^{M} \frac{1}{C}, \quad \frac{\partial \varphi_i}{\partial z_{0,j}} = \frac{-\lambda_i}{4\pi\varepsilon} \sum_{i=1}^{M} \frac{1}{D}$$

$$\frac{\partial \varphi_i}{\partial Q_{r,j}} = \frac{1}{2\pi^2 \varepsilon} \sum_{i=1}^{M} \frac{K(k)}{\alpha_1},$$

$$\frac{\partial \varphi_i}{\partial z_{r,j}} = \frac{Q_{r,j}}{2\pi^2 \varepsilon} \sum_{i=1}^{M} \frac{(z_i - z_{r,j}) E(k)}{\alpha_1 \beta_1^2}$$

$$\frac{\partial \varphi_i}{\partial r_{r,j}} = \frac{Q_{r,j}}{4\pi^2 \varepsilon r_{0,j}} \sum_{i=1}^{M} \frac{[r_i^2 - r_{0,j}^2 + (z_i - z_{r,j})^2] E(k) - \beta_1 K(k)}{\alpha_1 \beta_1^2} \tag{7.6.14}$$

The meaning of each constant in Eq. (7.6.14) is the same as that defined in Sects. 7.3.2 and 7.3.3. In eq. (7.6.11), the ring charges are constrained inside the electrode by the transformation function

$$R_j = R_e \sin x_j \tag{7.6.15}$$

where R_e is the radius of the electrode.

If all the positions and the values of the simulated charges are chosen as the design variables, and all the initial values of these variables are assumed to be zero, the initial value of the objective function is 0.81×10^6. When the objective function decrease to 78.09 the errors on the 180 contour points are less than 1%. If the iteration process is continuous, the value of the objective function will be decreased continuously. If the iteration is stopped, the potential distribution around the electrode is shown in Fig. 7.6.2.

In this case, because the number of design variables are very large, Cholesky's decomposition (see Chap. 11) is introduced to improve the DFP program. The computer program called VA09A [19] is used.

7.7 Error analysis in the charge simulation method

7.7.1 Properties of the errors

If a Laplacian problem with Dirichlet boundary conditions

$$\nabla^2 \varphi = 0$$
$$\varphi|_s = f(s) \tag{7.7.1}$$

is solved by CSM, the solution $\varphi_c(r)$ must satisfy Laplace's equation, i.e.

$$\nabla^2 \varphi_c = 0. \tag{7.7.2}$$

Let $e_\varphi(r)$ represent the potential error, i.e.

$$e_\varphi(r) = \varphi_a(r) - \varphi_c(r) \tag{7.7.3}$$

where $\varphi_a(r)$ is the accurate solution of Eq. (7.7.1). After taking the gradient and divergence operation of both sides of Eq. (7.7.3), one obtains:

$$\nabla \cdot \nabla e_\varphi(r) = \nabla \cdot \nabla [\varphi_a(r) - \varphi_c(r)] \tag{7.7.4}$$

Substitution of Eq. (7.7.1) and (7.7.2) into Eq. (7.7.4), yields

$$\nabla^2 e_\varphi(r) = 0. \tag{7.7.5}$$

This equation demonstrates that the error function of the potential still satisfies Laplace's equation, hence the error function is a harmonic function. It has the following properties:

(a) The maximum or minimum value always appears on the boundaries.
(b) The derivative of any order of the harmonic function exists and is continuous.

Thus the error distribution is stable and smooth. Furthermore, consider that the fundamental solution of two-dimensional or three-dimensional Laplace's equations are $\ln 1/r$ or $1/4\pi r$, respectively. Since the error distribution in region Ω varies as the function of $\ln 1/r$ or $1/r$. Consequently the error inside the region satisfies the requirement if the errors on the boundaries are controlled.

It is well known that the field strength of a static and quasi-static electric field satisfies Laplace's equation

$$\nabla^2 E = 0. \tag{7.7.6}$$

The properties of each component of Eq. (7.7.6) in Cartesian coordinates are similar to the error properties of the potential.

According to Eq. (7.7.5), if the error distribution is known along the boundary, the error function can be calculated for the whole domain [20].

7.7.2 Error distribution pattern along the electrode contour

On matching boundaries, the maximum error appears in the middle between two matching points. The reason is that the potential error at matching points is zero. For a smooth contour, the error distribution along the boundary could be dependent on

$$e = e_m \sin^2 nd \tag{7.7.7}$$

or

$$e = e_m \sin nd \tag{7.7.8}$$

as shown in Fig. 7.7.1, where n is the number of discretization points, d is the distance between two matching points and e_m is the maximum error in each period. This result has been verified by obtaining the error distribution of the sphere-plane electrode as shown in Fig. 7.4.4. This figure also shows that the error of the potential is smaller than the error of the field strength. Let us define an average error e_{av} in a mean root square sense, e.g.

$$e_{av} = \sqrt{\frac{1}{M} \sum_{i=1}^{M} e_i^2} \tag{7.7.9}$$

where e_i are relative errors at every point. The average error of the potentials and field intensities along a half circular path are $(e_\varphi)_{av} = 5.48 \times 10^{-6}$ and $(e_E)_{av} = 25.52 \times 10^{-6}$, respectively. The ratio of $(e_E)_{av}/(e_\varphi)_{av}$ is 4.65. It shows that the error of field strength is several times that of the error of potentials. These results are obtained when the radius of the ring charges are various from $\eta(R_c/R) = 0.2$–0.5. If $\eta = 0.8$, and the matching points are arranged as before and the errors of the potentials and field strengths increase rapidly. The maximum error of the potential is 2.983×10^{-3}, the maximum error of the field strength is 2.36×10^{-2}. *In general, it is possible that the error of field strength is 10 times that of the potential* [5].

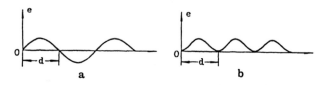

Fig. 7.7.1a, b. Two different patterns of error distribution

7.7.3 Factors influencing the errors

The number and position of simulated charges have a significant influence on the accuracy, as mentioned in Sect. 7.7.2. The positions of the matching points also play an important role. In Example 7.4.2, 8 ring charges are arranged on a circle with radius $R_c = 0.2 R$. If the matching points are not located on the same radial line as the simulated charges (see Fig. (7.7.2(a))), then the potential error distribution is shown by curve 'a' in Fig. 7.7.3. The largest errors appear at the point M and A. The potential error at two poles, N and S, is zero, as they are matching points. If the matching points and the simulated charges are located on the same radial

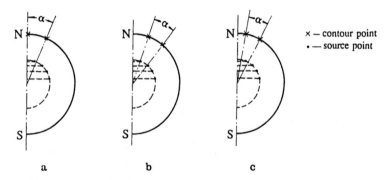

Fig. 7.7.2a–c. Different arrangements of the matching points versus the simulated charges

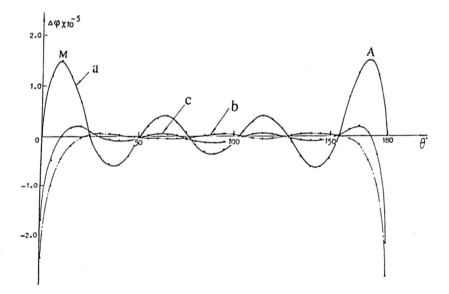

Fig. 7.7.3. Potential error distribution versus the different arrangement of the matching points

lines as shown in Fig. 7.7.2(b), the error distribution is shown by curve 'b' in Fig. 7.7.3. The errors along the semicircle are very small except in the region near the two poles. The reason is that these points are out of the range of the matching points. If the matching points are not located on the same radial line as the simulated charges, as shown in Fig. 7.7.2(c), the error distribution is pictured by curve 'c' in Fig. 7.7.3. The error is larger than in the case of Fig. 7.7.2(b).

Based on the above numerical results, usually the simulation charges and the matching points are placed on the same orthogonal line of the electrode contour, and usually the factor F is selected between 1.2 to 2.5[†]. In Example 7.4.2, when $R_c = 0.2R$ and $R_c = 0.5R$, the factor F is 2.3 and 1.4. If $R_c = 0.8$, the factor F is 0.57. The factor F is defined as

$$F = \frac{OA}{AB} \qquad (7.7.10)$$

where OA is the distance between the simulated charge and the contour point, and AB is the distance between two adjacent matching points as shown in Fig. 7.7.4. Reference [12] indicates that the value of factor F is influenced by the number of simulated charges. If the number of simulated charges is large, the factor F should be higher.

For OCSM, the error distribution of the potential and the field strength of the sphere-plane electrodes are shown in Fig. 7.7.5. This case illustrates that the error distribution is more even than the one obtained from the general CSM. In Fig. 7.7.5, another important phenomena is that the positions of the maximum error of the potential and field strength are almost at the same place. The ratio of the average error of the field strength and the potential is:

$$\frac{(e_E)_{av}}{(e_\varphi)_{av}} = \frac{8.507 \times 10^{-4}}{2.384 \times 10^{-4}} = 3.568 \ .$$

It is smaller than the ratio obtained by using the conventional charge simulation method.

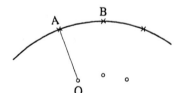

Fig. 7.7.4 Relative positions of matching points and simulated charges

---

[†] This value depends upon the geometry; Reference [12] suggests from 0.2–1.5.

7.7 Error analysis in the charge simulation method

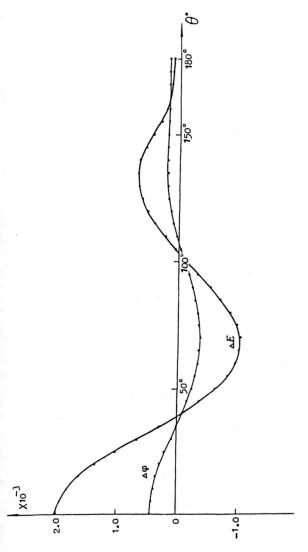

Fig. 7.7.5. Error distribution obtained by optimized charge simulation

7.8 Summary

CSM is efficient for calculating the electric field with fairly simple programs and little computing time. The surface charges of the electrodes and the polarization charges on the interface of different dielectrics are replaced by a set of discrete simulated charges. The types and positions of the simulated charges are predetermined. The mangitudes of these equivalent charges are determined by the boundary conditions on the collocation points of the boundary. Hence CSM is one of the collocation methods and can be classified as an equivalent source method.

The main advantages of this method are:

(1) It can be used to solve open boundary problems and is easily applied to three-dimensional problems.

(2) The solution domain does not need to be discretized, the troublesome of pre- and post-data processing is avoided. It is easy to calculate the potential and the field strength at any point of interest.

(3) Accurate results are obtained, as there is no truncation error. Higher accuracy is obtained in the field strength especially on the electrode surface than with FDM and FEM as no approximate numerical derivative is computed. If the continuous condition of the tangential component of the electric intensity of interfacial surface is introduced, the accuracy will be further increased [21].

(4) Compared to other boundary methods, such as the boundary element method, numerical integration and singularization of the integrand are not required, so fast economical calculation is obtained. Reference [22] gives comparison of a CSM and BEM, one example shows that the CPU time of BEM is 5.6 times than that of CSM.

(5) The distinctive advantage of CSM is that the estimation of error is simple. Only the errors on the boundaries need to be examined.

(6) The method can be extended to solve static and quasi-static magnetic field problems by using the equivalent current source [23].

A comparison between CSM, FEM, and FDM to solve static electric field problems is presented in reference [24].

One disadvantage of CSM is that errors are dependent on the location of the simulation charges especially for problems with a complicated shape. Except for OCSM, an additional charge system [25] is suggested to improve accuracy. To capitalize on the advantages of various methods, a combination of FEM and CSM may be applied effectively in some cases [26].

References

1. L.B. Loeb et al., The Choice of Suitable Gap Forms for the Study of Corona Breakdown and the Field Along the Axis of a Hemispherically Copped Cylinderical Point-to-Plane Gap. *Review of Scientific Instruments*, 21, 42–47, 1950
2. M.S. Abou-Seada, E. Nasser, Digital Computer Calculation of the Electric Potential and Field of a Rod Gap, *Proc. IEEE*, 56(5), 813–820, 1968
3. M.S. Abou-Seada, E. Nasser, Digital Computer Calculation of the Potential and its Gradient of a Twin Cylinderical Conductor, *IEEE Trans.*, 88(12), 1802–1814, 1969
4. H. Steinbigler, Anfangsfeldstärken und Ausnutzungsfaktoren Rotation-symmetrischer Elektrodenanordnungen in Luft (Dissertation TU Munich, 1969)
5. H. Singer, H. Steinbigler, P. Weiß, A Charge Simulation Method for the Calculation of High Voltage Fields, *IEEE Trans. on PAS*, 93, 1660–1667, 1974
6. M. Mathis, B. Bachmann et al., Potential Profile Calculation in Earthing System Using the Charge Simulation Method, *ETZ Archiv*, Bd. 9, 21–27, 1987
7. H. Singer, P. Grafoner, Optimization of Electrode and Insulator Contours, *Proc. of International Symposium on High Voltage Engineering (ISH)*, Zurich, 111–117, 1975
8. H. Klement, B. Bachmann, H. Gronewald, H. Singer, Layout of HV Components by Computer Design, *Cigre Paper*, 2, 33-04, 1980.
9. H.H. Daumling, Optimization of Insulators, *5th ISH*, Braunschweig, 31.05, 1987
10. S. Murashima, Y. Nonaka, H. Nieda, The Charge Simulation Method and Its Application to Two-Dimensional Elasticity. In *Boundary Elements. Procedings of the Fifth International Conference* (Ed. C.A. Brebbia), 339–344, Japan, 1983
11. Y. Nonaka, A Charge Simulation Method for the Stefan Problems, *Engineering Analysis*, 2(1), 16–19, 1985
12. T. Kouno, T. Takuma, *Numerical Calculation of Electric Field* (In Japanese), 1980
13. S. Sato, Electric Field Calculation by Charge Simulation Method Using Axi-Spheroidal Charge, *3rd ISH*, Milan 11.07, 1979
14. T. Takuma, Charge Simulation Method Including Disk Charge (*in Japanese*), *IEE Trans. of Japan*, 97A(8), 411–416, 1977
15. M. Abramowitz, I.A. Stegun (eds.), *Handbook of Mathematical Function*, Dover Publications, Inc., New York, 1964, 1970
16. P. Moon, D.E. Spencer, *Field Theory for Engineering*, Princeton, 1961
17. Zhong Liang Kun, The Develop of A 220/128 KV Gas Switch, Master's Thesis, Xi' an Jiao-tong University, 1985
18. A. Yializis, E. Kuffel and P.H. Alexander, An Optimized Charge Simulation Method for the Calculation of High Voltage Fields, *IEEE, Trans. PAS*, 97, 2434–2440, 1987
19. R. Fletcher, *A FORTRAN Subroutine VA09A for Minimization When the Function and First Derivative can be Calculated Explicitly*, Harwell Library
20. S. Kato, An Estimation Method for the Electric Field Error of a Charge Simulation Method, *3rd ISH*, Milan, 11.09, 1979
21. J.P. Chan, S.J. Lin, The Error Distribution of CSM and Improvement of the Local Accuracy, *5th ISH*, 33.05, 1987
22. Z. Andjelic, H. Steinbiglar et al., Comparative Analysis of the Boundary Element Method and the Charge Simulation Method in 2-D and 3-D Electrostatic Field Calculation, *Proc. of 6th ISH*, New Orleans, 24.09, 1989
23. P.K. Mukherjee, A Simulation Method for Computation of Magnetic Field, *IEEE PES*, Winter Meeting, 2 (C-74), 213–215, 1974
24. M.D.R. Beasley et al., Comparative Study of Three Methods for Computing Electric Fields, *Proc. IEE*, 126(1), 126–134, 1979
25. Surgen Hamschen, Paul Wiess, Charge Simulation Method with Additional Charge System, *Proc. of 6th ISH*, New Orleans, 24.01, 1989
26. H. Steinbiglar, Combined Application of Finite Element Method and Charge Simulation Method for the Computation of Electric Field, *3rd ISH*, Milan, 1979
27. Shanghai Computational Institute, *Handbook of Computer Algorithms* (in Chinese), Shanghai Educational Press, 1982

Appendix 7.1 Formulations for a point charge

The field strength at any point $P(r, z)$ produced by a point charge is:

$$E_r = -\frac{\partial \varphi}{\partial r} = \frac{Q}{4\pi\varepsilon}\left[\frac{r - r'}{\alpha} - \frac{r - r'}{\beta}\right] \qquad \text{(A.7.1.1)}$$

$$E_z = -\frac{\partial \varphi}{\partial z} = \frac{Q}{4\pi\varepsilon}\left[\frac{z - z'}{\alpha} - \frac{z - z'}{\beta}\right] \qquad \text{(A.7.1.2)}$$

$$\alpha = [(r - r')^2 + (z - z')^2]^{3/2} \qquad \beta = [(r - r')^2 + (z + z')^2]^{3/2}.$$

If $r = r' = 0$, then

$$E_r = 0 \qquad \text{(A.7.1.3)}$$

$$E_z = \frac{Q}{4\pi\varepsilon}\left[\frac{1}{(z - z')^2} - \frac{1}{(z + z')^2}\right]. \qquad \text{(A.7.1.4)}$$

Appendix 7.2 Formulations for a line charge

$$E_r = \frac{\lambda}{4\pi\varepsilon(z_l - z_0)}\left[\frac{z_l - z}{Cr} - \frac{z_0 - z}{Dr} + \frac{z_0 + z}{Fr} - \frac{z_l - z}{Gr}\right] \qquad \text{(A.7.2.1)}$$

$$E_z = \frac{\lambda}{4\pi\varepsilon(z_l - z_0)}\left[\frac{1}{C} - \frac{1}{D} - \frac{1}{F} + \frac{1}{G}\right] \qquad \text{(A.7.2.2)}$$

$$C = [r^2 + (z_l - z)^2]^{1/2}$$
$$D = [r^2 + (z_0 - z)^2]^{1/2}$$
$$F = [r^2 + (z_0 + z)^2]^{1/2}$$
$$G = [r^2 + (z_l + z)^2]^{1/2}. \qquad \text{(A.7.2.3)}$$

If $l \to \infty$, then

$$E_r = \frac{\lambda}{4\pi\varepsilon}\left[\frac{r}{D(z_0 - z + D)} - \frac{r}{F(z_0 + z + F)}\right] \qquad \text{(A.7.2.4)}$$

$$E_z = -\frac{\lambda}{4\pi\varepsilon}\left[\frac{1}{F} + \frac{1}{D}\right]. \qquad \text{(A.7.2.5)}$$

Appendix 7.4 Formulations for a charged elliptic cylinder   249

Appendix 7.3 Formulations for a ring charge

$$E_r = -\frac{Q}{4\pi\varepsilon}\frac{1}{\pi r}\left\{\frac{[r_0^2 - r^2 + (z-z_0)^2]E(k_1) - \beta_1^2 K(k_1)}{\alpha_1 \beta_1^2}\right.$$
$$\left. - \frac{[r_0^2 - r^2 + (z-z_0)^2]E(k_2) - \beta_2^2 K(k_2)}{\alpha_2 \beta_2^2}\right\} \quad (A.7.3.1)$$

$$E_z = \frac{Q}{4\pi\varepsilon}\frac{2}{\pi}\left[\frac{(z-z_0)E(k_1)}{\alpha_1 \beta_1^2} - \frac{(z+z_0)E(k_2)}{\alpha_2 \beta_2^2}\right] \quad (A.7.3.2)$$

where

$$E(k) = \int_0^{\pi/2} \sqrt{1 - k^2 \sin^2\theta}\, d\theta \quad (A.7.3.3)$$

$$\begin{cases} \alpha_1 = [(r+r_0)^2 + (z-z_0)^2]^{1/2} \quad \alpha_2 = [(r+r_0)^2 + (z+z_0)^2]^{1/2} \\ \beta_1^2 = (r-r_0)^2 + (z-z_0)^2 \quad \beta_2^2 = (r-r_0)^2 + (z+z_0)^2 \\ k_1 = \frac{2\sqrt{rr_0}}{\alpha_1} \quad k_2 = \frac{2\sqrt{rr_0}}{\alpha_2} \end{cases} \quad (A.7.3.4)$$

$E(k)$ is a complete elliptic integral of the second kind.

Appendix 7.4 Formulations for a charged elliptic cylinder

$$E = -\nabla\varphi = \frac{1}{f\sqrt{\cosh^2\eta - \cos^2\psi}}\frac{U_0}{\eta - \eta_b}\mathbf{a}_\eta \quad (A.7.4.1)$$

$$E_x = \frac{\lambda}{4\pi\varepsilon\sqrt{s^2 - f^2}}\left(\frac{x-f}{R_1} + \frac{x+f}{R_2}\right) \quad (A.7.4.2)$$

$$E_y = \frac{\lambda}{4\pi\varepsilon\sqrt{s^2 - f^2}}\, y\left(\frac{1}{R_1} + \frac{1}{R_2}\right) \quad (A.7.4.3)$$

where

$$\begin{cases} R_1^2 = (x-f)^2 + y^2 \quad R_2^2 = (x+f)^2 + y^2 \\ s = \tfrac{1}{2}(R_1 + R_2). \end{cases} \quad (A.7.4.4)$$

Appendix 7.5 Approximate formulations for calculating $K(k)$ and $E(k)$ [15, 27]

$$K(k) = \int_0^{\pi/2} \frac{d\theta}{\sqrt{1 - k^2 \sin^2 \theta}} \qquad E(k) = \int_0^{\pi/2} \frac{d\theta}{\sqrt{1 - k^2 \sin^2 \theta}}$$

Let $k_1 = 1 - k^2$, the Hastings polynomial approximations are

$$K(k) = \sum_{t=0}^{n} a_t k_1^t - \ln k_1 \sum_{t=0}^{n} b_t k_1^t + \varepsilon_1(k) \qquad (A.7.5.1)$$

$$E(k) = 1 + \sum_{t=0}^{n} c_t k_1^t - \ln k_1 \sum_{t=0}^{n} d_t k_1^t + \varepsilon_2(k). \qquad (A.7.5.2)$$

When $n = 4$, the constants a_t, b_t, c_t and d_t are given in the following table.

t	a	b	c	d
0	1.38629436112	0.5		
1	0.09666344259	0.12498593597	0.44325141463	0.24998368310
2	0.03590092383	0.06880248576	0.06260601220	0.09200180037
3	0.03742563713	0.03328355346	0.04757383546	0.04069697526
4	0.01451196212	0.00441787012	0.01736506451	0.00526449639

In Eqs. (A.7.5.1) and (A.7.5.2), $|\varepsilon(k)| \le 2 \times 10^{-8}$. These approximations are very easy and fast to calculate the $K(k)$ and $E(k)$, the drawback is that the error is not uniform for the different variable of 'k'. More accurate results can be obtained if the arithmetic-geometrical mean polynomials are used, especialy when k is greater than 0.9. The formulations are

Let
$$k_1 = 1 - k^2, \qquad a_0 = 1,$$

then
$$b_0 = k_1^{1/2}, \qquad c_0^2 = a_0^2 - b_0^2,$$
$$a_n = \tfrac{1}{2}(a_{n-1} + b_{n-1}), \qquad b_n = (a_{n-1} \times b_{n-1})^{1/2}, \qquad c_n = \tfrac{1}{2}(a_{n-1} - b_{n-1})$$

when $c_n \to 0$, then

$$K(k) = \frac{\pi}{2 a_n}$$

$$E(k) = K(k) \left[1 - \tfrac{1}{2}(c_0^2 + 2c_1^2 + 2^2 c_2^2 + \ldots + 2^n c_n^2) \right].$$

Chapter 8

Surface Charge Simulaton Method (SSM)

8.1 Introduction

The surface charge simulation method (SSM) is similar to the charge simulation method (CSM). In these methods, the real source distribution is simulated by a great number of accumulated or surface charges. They are both convenient for solving open boundary and 3-D problems. *SSM can solve field problems that cannot be solved by CSM and also those that can.* Using SSM one can obtain solutions to problems containing a number of dielectric materials and problems having thin electrodes. Hence SSM is a more general method than CSM. One could also consider it to be a kind of boundary element method, or a method of equivalent source.

The basic concept of SSM is to simulate the real charges distributed on the surface of the electrode and the polarization charges on the interfaces of dielectrics by the equivalent sources of a single or double layer on the surface. The equivalent surface charge density is determined by known boundary conditions. After the equivalent surface charge density is found, the potential and the field strength at any point can be calculated. As the unknown distributed charges are located on the surface of the electrodes and the interfaces of the dielectrics, it is not necessary to find the locations and characteristics of the simulated charges as is the case when using CSM. The disadvantages of SSM are that there are more unknowns and consequently CPU-time is increased. It should be also noted that SSM requires faster methods to calculate the many integrals and singular integrals involved in integral equations.

The original idea of SSM might even have been known to Maxwell [1] when he manually calculated the capacitance of an isolated square metal plate with the dimensions 1 m × 1 m by solving the corresponding integral equation. He divided the plate into 36 square elements and assumed the charge density to be constant in each element. He found that the capacitance of the plate to be 40.13 pf. This is an accurate result compared with today's results [2]. Since Maxwell's work, no significant publications then appeared for about 90 years. Early in the 1970s, Singer used the surface charge simulation method to calculate the field distribution of a high voltage insulator [3]. Recently another doctoral thesis was written by Shuji Sato [2]. He deals with 3-D problems by using SSM.

Now the SSM is well known and is used to solve 2-D and 3-D electric [4] and magnetic [5] problems. It is especially useful in electron optics [6] where many thin electrodes are used for focusing. In optimal shape design of electric apparatus, the SSM is more suitable as the tool for field analysis [7-9].

8.1.1 Example

Consider an isolated and very thin plate with dimensions $2a \times 2b$ as shown in Fig. 8.1.1. When the plate is charged to a potential of 1 V, the charge density distribution $\sigma(x', y')$ is not uniform. According to Coulomb's law, the potential of any point $P(\mathbf{r})$ is

$$\varphi(\mathbf{r}) = \int_{-b}^{b} dy' \int_{-a}^{a} \frac{\sigma(x', y')}{4\pi\varepsilon R} dx'. \tag{8.1.1}$$

This is the Fredholm integral equation of the first kind, where $\sigma(x', y')$ is not known, R is the distance between the source point $\mathbf{r}'$ and field point $\mathbf{r}$, i.e. $R^2 = [(x - x')^2 + (y - y')^2 + (z - z')^2]$.

Subdivide the plate into M square elements (Fig. 8.1.1(b)) with the dimensions $2e \times 2e$, shown in Fig. 8.1.1(c). The continuous distributed charge density of each element is assumed as a constant (the potential of each element is also assumed constant and is represented by the potential of the centre point), thus

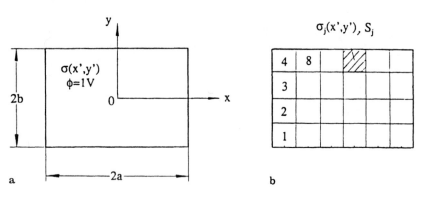

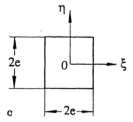

Fig. 8.1.1a–c. An isolated plate charged to 1 V

8.1 Introduction

$\sigma(x', y')$ can be taken out of the integral operator, then Eq. (8.1.1) becomes

$$\varphi(\mathbf{r}) = \sum_{j=1}^{M} \sigma_j(x', y') \int_e \frac{1}{4\pi\varepsilon R} dS_j . \tag{8.1.2}$$

In consideration of the boundary condition that the potential of the plate is 1 V, Eq. (8.1.2) can be expressed as:

$$\varphi(x, y, 0) = \sum_{j=1}^{M} \sigma_j(x_j, y_j) \int_{-e}^{e} \int_{-e}^{e} \frac{dx\,dy}{4\pi\varepsilon R} = 1 \quad \begin{array}{l} -a \le x \le a \\ -b \le y \le b. \end{array} \tag{8.1.3}$$

Equation (8.1.3) can be written in a matrix form

$$\mathbf{P}\{\sigma\} = \{\varphi\} \tag{8.1.4}$$

where

$$p_{ij} = \int_{-e}^{e} \int_{-e}^{e} \frac{dx\,dy}{4\pi\varepsilon R_{ij}} \tag{8.1.5}$$

$$p_{ii} = \int_{-e}^{e} \int_{-e}^{e} \frac{dx\,dy}{4\pi\varepsilon R_{ii}} . \tag{8.1.6}$$

The subscripts i, j represent the field point and the source point, respectively. p_{ij} represents the potential of element 'i' induced by an unit charge density located at the element 'j'. As the charge is assumed to be concentrated at the centre of element 'j', and the potential of element 'i' is represented by the potential at the centre point of the sub-element, then

$$p_{ij} = \frac{S_j}{4\pi\varepsilon R_{ij}} \tag{8.1.7}$$

where S_j is the area of element 'j'. R_{ij} is the distance between the centre point of element 'i' and 'j'.

To calculate the coefficient p_{ii}, it is considered that the charge is distributed on a square surface. The coefficient p_{ii} is evaluated analytically, i.e.

$$p_{ii} = \int_{-e}^{e} \int_{-e}^{e} \frac{dx\,dy}{4\pi\varepsilon_0(x^2 + y^2)^{1/2}} = \frac{2e}{\pi\varepsilon_0} \ln(1 + 2^{1/2}) = \frac{0.2806}{\varepsilon_0}\sqrt{S_i} . \tag{8.1.8}$$

If the element is not a square, an equivalent circle can be used to calculate the coefficient p_{ii}, i.e. the potential at the centre of the circle is

$$p_{ii} = \frac{1}{4\pi\varepsilon} \int_0^a \int_0^{2\pi} \frac{r\,d\theta\,dr}{r} = \frac{0.2821}{\varepsilon_0}\sqrt{S_i} . \tag{8.1.9}$$

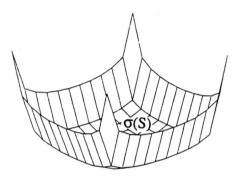

Fig. 8.1.2. Charge distribution of an isolated plate

After programming the above equations, the solution of Eq. (8.1.4) yields the surface charge distribution of a square isolated plate as shown in Fig. 8.1.2. The charge densities on the four corners of the plate are much higher than elsewhere.

8.2 Surface integral equations

The potential integral equation of a Poisson's equation has been derived in Sect. 1.4. In this section, the surface integral equations of interfacial boundaries are derived.

8.2.1 Single layer or double layer integral equations

The surface charge simulation method or the magnetic surface charge simulation method is based on surface integral equations. They may be described by single or double layer source. For a single layer source, the potentials on both sides of the layer are continuous, but the normal derivative of the potential suffers an abrupt change, i.e.

$$\varphi_+ = \varphi_- \tag{8.2.1}$$

$$\frac{\partial \varphi_+}{\partial n} - \frac{\partial \varphi_-}{\partial n} = -\sigma/\varepsilon_0 . \tag{8.2.2}$$

This abrupt change is considered on both sides of the inter-surface S, i.e.

$$\frac{1}{2}\left[\frac{\partial \varphi_+}{\partial n} + \frac{\partial \varphi_-}{\partial n}\right] = \left.\frac{\partial \varphi}{\partial n}\right|_s$$

thus

$$\frac{\partial \varphi_+}{\partial n} = -\frac{\sigma}{2\varepsilon_0} + \left.\frac{\partial \varphi}{\partial n}\right|_s \tag{8.2.3}$$

$$\frac{\partial \varphi_-}{\partial n} = +\frac{\sigma}{2\varepsilon_0} + \left.\frac{\partial \varphi}{\partial n}\right|_s \tag{8.2.4}$$

8.2 Surface integral equations

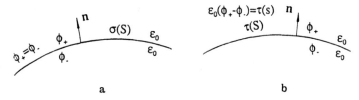

Fig. 8.2.1a, b. Single and double layer source

As the point **r** approaches the surface and considering the Eq. (1.3.1)

$$\varphi(\mathbf{r}) = \int_{s'} \frac{\sigma(\mathbf{r'})}{4\pi\varepsilon_0 R} ds' = \int_{s'} \frac{\sigma(\mathbf{r'})}{\varepsilon_0} G(\mathbf{r},\mathbf{r'}) ds'$$

one obtains

$$\left.\frac{\partial \varphi}{\partial n}\right|_{\pm} = \mp \frac{\sigma(\mathbf{r})}{2\varepsilon_0} + \int_{s'} \frac{\sigma(\mathbf{r'})}{\varepsilon_0} \left.\frac{\partial G(\mathbf{r},\mathbf{r'})}{\partial n}\right|_s ds' . \qquad (8.2.5)$$

Note that the subscript '$\pm$' represents the point which approaches the surface from the opposite side as shown in Fig. 8.2.1.

For a double layer source, as discussed in Sect. 1.3.2, the potential is discontinuous at surface S but the normal derivative of the potential is continuous, i.e.

$$\varphi_+ - \varphi_- = \frac{\tau(\mathbf{r})}{\varepsilon_0} \qquad (8.2.6)$$

$$\frac{\partial \varphi_+}{\partial n} = \frac{\partial \varphi_-}{\partial n} = \left.\frac{\partial \varphi}{\partial n}\right|_s . \qquad (8.2.7)$$

By using a similar procedure, the corresponding equation of eq. (8.2.5) is

$$\varphi(\mathbf{r})|_{\pm} = \pm \frac{\tau(\mathbf{r})}{2\varepsilon_0} + \int_{s'} \frac{\tau(\mathbf{r'})}{\varepsilon_0} \left.\frac{\partial G(\mathbf{r},\mathbf{r'})}{\partial n}\right|_s ds' . \qquad (8.2.8)$$

Equations (8.2.5) and (8.2.8) are Fredholm integral equations of the second kind. In these equations, the different sign of +ve, −ve before each term of the RHS is due to the different normal directions of both the interior and exterior region.

8.2.2 Integral equations of the interfacial surface [10, 11]

Recalling Eqs. (1.2.30) and (1.4.9) and consider both the interior and the exterior (represented by −ve and +ve) problems, these are

$$\varphi_-(\mathbf{r}) = \int_{s'} \left[G(\mathbf{r},\mathbf{r}') \frac{\partial \varphi_-}{\partial n} - \varphi_- \frac{\partial G(\mathbf{r},\mathbf{r}')}{\partial n} \right] ds' \tag{8.2.9}$$

$$\varphi_+(\mathbf{r}) = \int_{v'} G(\mathbf{r},\mathbf{r}') f(\mathbf{r}') dv' - \int_{s'} \left[G(\mathbf{r},\mathbf{r}') \frac{\partial \varphi_+}{\partial n} - \varphi_+ \frac{\partial G(\mathbf{r},\mathbf{r}')}{\partial n} \right] ds' . \tag{8.2.10}$$

In eq. (8.2.10), the first term of the RHS is the potential produced by source density inside the surface, while the second term represents the single layer source and the third term is the effect of the double layer source. For Eq. (8.2.9), there is no direct contribution from the source density, it occurs indirectly by interface conditions

In the exterior region when $\mathbf{r} \to \mathbf{r}_s$, as discussed in Sect. 1.3.2, the contributions to the potential φ_+ due to the impressed source density and the equivalent single layer source remain continuous, i.e.

$$\lim_{\mathbf{r} \to \mathbf{r}_s} \varphi'(\mathbf{r}_s) = \int_{s'} \frac{\rho(\mathbf{r}')}{\varepsilon_0} G(\mathbf{r}_s,\mathbf{r}') ds' \tag{8.2.11}$$

$$\lim_{\mathbf{r} \to \mathbf{r}_s} \int_{s'} G(\mathbf{r},\mathbf{r}') \frac{\partial \varphi_+}{\partial n} ds = \int_{s'} G(\mathbf{r}_s,\mathbf{r}') \frac{\partial \varphi_+}{\partial n} \bigg|_s ds \tag{8.2.12}$$

where the term $\varphi'(\mathbf{r})$ denotes the potential produced by the impressed source.

However, the contribution to $\varphi_+(\mathbf{r})$ due to the equivalent double layer source is discontinuous, i.e.

$$\lim_{\mathbf{r} \to \mathbf{r}_s} \int_{s'} \varphi_+(\mathbf{r}) \frac{\partial G(\mathbf{r},\mathbf{r}')}{\partial n} ds' = \frac{\varphi_+(\mathbf{r}_s)}{2} + \int_{s'} \varphi_+(s') \frac{\partial G(\mathbf{r}_s,\mathbf{r}')}{\partial n} ds' . \tag{8.2.13}$$

Therefore, in a limited case such as $\mathbf{r} \to \mathbf{r}_s$

$$\frac{\varphi_+(\mathbf{r}_s)}{2} = \varphi'_+(\mathbf{r}_s) + \int_{s'} \varphi_+(s') \frac{\partial G(\mathbf{r}_s,\mathbf{r}')}{\partial n} ds' - \int_{s'} G(\mathbf{r}_s,\mathbf{r}') \frac{\partial \varphi_+}{\partial n} \bigg|_s ds' . \tag{8.2.14}$$

In a similar case, if the point P moves from the domain Ω_i to the boundary,

$$\lim_{\mathbf{r} \to \mathbf{r}_s} \varphi_-(\mathbf{r}) = \varphi_-(\mathbf{r}_s) \tag{8.2.15}$$

$$\lim_{\mathbf{r} \to \mathbf{r}_s} \int_{s'} \varphi_-(\mathbf{r}') \frac{\partial G(\mathbf{r}_s,\mathbf{r}')}{\partial n} ds' = -\frac{\varphi_-(\mathbf{r}_s)}{2} + \int_{s'} \varphi_-(s') \frac{\partial G(\mathbf{r}_s,\mathbf{r}')}{\partial n} ds' \tag{8.2.16}$$

$$\lim_{\mathbf{r} \to \mathbf{r}_s} \int_{s'} G(\mathbf{r},\mathbf{r}') \frac{\partial \varphi_-}{\partial n} ds' = \int_{s'} G(\mathbf{r}_s,\mathbf{r}') \frac{\partial \varphi_-(s)}{\partial n} ds' . \tag{8.2.17}$$

The resultant equation is

$$\frac{\varphi_-(\mathbf{r}_s)}{2} = -\int_{s'} \varphi_-(\mathbf{r}') \frac{\partial G(\mathbf{r}_s, \mathbf{r}')}{\partial n} ds' + \int_{s'} G(\mathbf{r}_s, \mathbf{r}') \frac{\partial \varphi_-(s)}{\partial n} ds'. \quad (8.2.18)$$

Using the interface boundary conditions

$$\varphi_+(\mathbf{r})|_s = \varphi_-(\mathbf{r})|_s = \varphi(\mathbf{r})|_s \quad (8.2.19)$$

$$\varepsilon_+ \frac{\partial \varphi_+}{\partial n}\bigg|_s = \varepsilon_- \frac{\partial \varphi_-}{\partial n}\bigg|_s \quad (8.2.20)$$

multiply Eq. (8.2.14) by ε_+ and Eq. (8.2.18) by ε_-, respectively, then add these two equations and by considering the Eqs. (8.2.19) and (8.2.20), one obtains

$$\varphi(\mathbf{r}_s) = \frac{2\varepsilon_+}{\varepsilon_+ + \varepsilon_-} \varphi'(\mathbf{r}_s) + \frac{(\varepsilon_+ - \varepsilon_-)}{(\varepsilon_+ + \varepsilon_-)} \int_{s'} \varphi(\mathbf{r}) \frac{\partial G(\mathbf{r}_s, \mathbf{r}')}{\partial n} ds' \quad (8.2.21)$$

and

$$\sigma(\mathbf{r}_s) = \frac{2}{\varepsilon_+ + \varepsilon_-} \sigma_f(\mathbf{r}_s) + \frac{(\varepsilon_- - \varepsilon_+)}{(\varepsilon_+ + \varepsilon_-)} \int_{s'} \sigma(\mathbf{r}) \frac{\partial G(\mathbf{r}_s, \mathbf{r}')}{\partial n} ds'. \quad (8.2.22)$$

In magnetostatic fields,

$$\varphi_m(\mathbf{r}_s) = \frac{2\mu_-}{\mu_+ + \mu_-} \varphi'_m(\mathbf{r}_s) + \frac{(\mu_- - \mu_+)}{(\mu_+ + \mu_-)} \int_{s'} \varphi_m(\mathbf{r}) \frac{\partial G(\mathbf{r}_s, \mathbf{r}')}{\partial n} ds' \quad (8.2.23)$$

these are Fredholm integral equations of the second kind. These equations are very useful for solving the problems containing multiple materials as used in references [12, 13]. Where σ_f is an imposed charge density, and $\varphi'(\mathbf{r}_s)$, and $\varphi'_m(\mathbf{r}_s)$ are potentials induced by imposed source terms.

8.3 Types of surface boundary elements and surface charge densities

8.3.1 Representations of boundary and charge density

Two different kinds of discretization are required. One deals with the discretization of the boundary surface and the other deals with the distributed charge density. In a 2-D case, the boundary could be subdivided by thin plate or an arced thin plate with infinite length. The cross-section of this element is a short line or arc as shown in Fig. 8.3.1. These are called linear or circular elements. For an axisymmetric problem, the subelement consists of filament ring charges.

In 3-D problems, the surface of the boundary could be a plane or a part of a cylinder or any other curved surface [4]. The charge density of these subelements can also be constant or represented by any other functions. Here, to avoid the many special functions, only constant and linear charge distributions on linear and circular elements in a 2-D case are discussed. Other functions like the Fourier-type and Tchebycheff-type expansions are described in references [2, 14].

8.3.2 Potential and field strength coefficients for 2-D and axisymmetrical problems

In 2-D cases, the sub-element of the boundary could be a thin plate or an arced thin plate of infinite length. The cross-section of a thin narrow plate and of an arced plate in the x-y plane are shown in Fig. 8.3.1.

The potential at any point $P(x_i, y_i)$ produced by these elements is

$$\varphi_{ij} = \frac{1}{4\pi\varepsilon} \int_A^B \lambda(x_j, y_j) \ln \frac{1}{(x_i - x_j)^2 + (y_i - y_j)^2} dl + C' \qquad (8.3.1)$$

where $\lambda(x_j, y_j)$ represents the line charge density. The constant C' in Eq. (8.3.1) is determined by the potential reference point. If the image source is considered and the symmetric plane is chosen as a potential reference point, then Eq. (8.3.1) is altered to

$$\varphi_{ij} = \frac{1}{4\pi\varepsilon} \int_A^B \lambda(x_j, y_j) \ln \frac{(x_i - x_j)^2 + (y_i + y_j)^2}{(x_i - x_j)^2 + (y_i - y_j)^2} dl \,. \qquad (8.3.2)$$

For an axisymmetric field, the sub-element could be a frustum of a cone or a spherical segment as shown in Fig. 8.3.2(a) and (b). For simplicity, in this chapter, these elements are called ring elements with a linear or arced lateral surface.

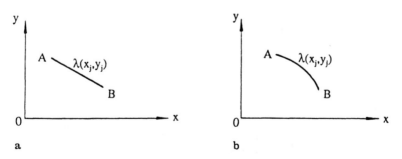

Fig. 8.3.1a, b. Cross-section of subelements in translational symmetric cases

8.3 Surface boundary elements and surface charge densities

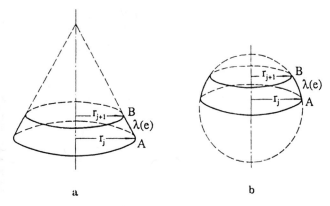

Fig. 8.3.2a, b. A ring element

The potential induced by these elements is

$$\varphi_{ij} = \frac{1}{\pi\varepsilon} \int_A^B \frac{\lambda(e_j) r_j K(k_j)}{[(r_i - r_j)^2 + (z_i - z_j)^2]^{1/2}} dl + C' \tag{8.3.3}$$

where

$$k_j = \left[\frac{4 r_i r_j}{(r_i + r_j)^2 + (z_i - z_j)^2}\right]^{1/2}.$$

$K(k_j)$ is the elliptic integral of the first kind [15].

8.3.2.1 Planar element with constant or linear charge density

If $\lambda(e)$ is constant, Eq. (8.3.1) is simplified to:

$$\varphi_{ij} = \frac{\lambda(e_j)}{4\pi\varepsilon} \int_A^B \ln \frac{1}{(x_i - x_j)^2 + (y_i - y_j)^2} dl + C'. \tag{8.3.4}$$

To obtain a generalized formula to calculate the integral, the coordinate transformations are used and the relationship between x-y and x''-y'' (as shown in Fig. 8.3.3) is,

$$\begin{Bmatrix} x'' \\ y'' \end{Bmatrix} = \begin{bmatrix} \cos\alpha & \sin\alpha \\ -\sin\alpha & \cos\alpha \end{bmatrix} \begin{Bmatrix} x' \\ y' \end{Bmatrix} \tag{8.3.5}$$

$$\cos\alpha = \frac{x_2 - x_1}{l} \qquad \sin\alpha = \frac{y_2 - y_1}{l}$$

$$x' = x - \frac{x_1 + x_2}{2} \qquad y' = y - \frac{y_1 + y_2}{2},$$

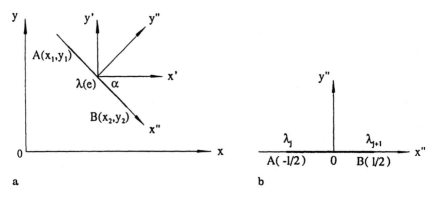

Fig. 8.3.3a, b. Transformation of coordinates

where the axis x'' is along the length of the element, and l is the length of the element, $l^2 = (x_2 - x_1)^2 + (y_2 - y_1)^2$. As

$$dl = \sqrt{1 + \left(\frac{dy}{dx}\right)^2}\, dx = \sqrt{1 + tg^2\alpha}\, dx = \sec\alpha\, dx$$

$$dx'' = \cos\alpha\, dx + \sin\alpha\, dx = \sec\alpha\, dx = dl$$

then

$$\varphi_{ij} = \frac{\lambda(e)}{4\pi\varepsilon} \int_{-l/2}^{l/2} \ln \frac{1}{(x'' - x_i'')^2 + (y'' - y_i'')^2}\, dx'' + C'\,. \tag{8.3.6}$$

The distance between the source point (x_j, y_j) and the observation point (x_i, y_i) is unchanged during the transformation of coordinates, i.e.

$$(x - x_i)^2 + (y - y_i)^2 = (x'' - x_i'') + (y'' - y_i'')^2\,. \tag{8.3.7}$$

The integration of Eq. (8.3.6) yields

$$\varphi_{ij} = \frac{\lambda_j(e)}{4\pi\varepsilon} p_{ij}\,. \tag{8.3.8}$$

The components of field strength in the coordinates $x'' \circ y''$ are

$$\begin{cases} E''_{x_{ij}} = -\dfrac{\partial \varphi_i}{\partial x_i''} = -\dfrac{\lambda_j}{4\pi\varepsilon} f''_{x_{ij}} \\ E''_{y_{ij}} = -\dfrac{\partial \varphi_i}{\partial y_i''} = -\dfrac{\lambda_j}{4\pi\varepsilon} f''_{y_{ij}} \end{cases} \tag{8.3.9}$$

then

$$\begin{Bmatrix} E_{x_{ij}} \\ E_{y_{ij}} \end{Bmatrix} = \begin{bmatrix} \cos\alpha & -\sin\alpha \\ \sin\alpha & \cos\alpha \end{bmatrix} \begin{Bmatrix} E''_{x_{ij}} \\ E''_{y_{ij}} \end{Bmatrix}\,. \tag{8.3.10}$$

8.3 Surface boundary elements and surface charge densities

The formulations for evaluating the coefficients of p_{ij}, $f''_{x_{ij}}$ and $f''_{y_{ij}}$ are given in Appendix A.8.1, Eq. (A.8.1.1) to (A.8.1.4).

To avoid discontinuities of the charge density between adjacent elements as shown in Fig. 8.3.4(a), a linear charge (density) distribution is used as shown in Fig. 8.3.4(b).

In Fig. 8.3.4(b), let

$$\lambda(e) = \frac{l_{j+1} - l}{l_{j+1} - l_j} \lambda_j + \frac{l - l_j}{l_{j+1} - l_j} \lambda_{j+1} . \tag{8.3.11}$$

Substitute Eq. (8.3.11) into Eq. (8.3.1) and consider Eq. (8.3.7), this yields:

$$\begin{aligned}
\varphi_{ij} &= -\frac{\lambda_j}{4\pi\varepsilon} \int_{-l/2}^{l/2} \frac{x''_2 - x''}{x''_2 - x''_1} \ln\left[(x'' - x''_i)^2 + y''^2_i\right] dx'' \\
&\quad - \frac{\lambda_{j+1}}{4\pi\varepsilon} \int_{-l/2}^{l/2} \frac{x'' - x''_1}{x''_2 - x''_1} \ln\left[(x'' - x''_i)^2 + y''^2_i\right] dx'' + C' \\
&= \frac{\lambda_j}{4\pi\varepsilon} p^{(1)}_{ij} + \frac{\lambda_{j+1}}{4\pi\varepsilon} p^{(2)}_{ij} .
\end{aligned} \tag{8.3.12}$$

Using the derivation of Eq. (8.3.9), the components of the field strength in the $x''-y''$ plane are:

$$\begin{cases} E''_{x_{ij}} = \dfrac{\lambda_j}{4\pi\varepsilon} f''^{(1)}_{x_{ij}} + \dfrac{\lambda_{j+1}}{4\pi\varepsilon} f''^{(2)}_{x_{ij}} \\ E''_{y_{ij}} = \dfrac{\lambda_j}{4\pi\varepsilon} f''^{(1)}_{y_{ij}} + \dfrac{\lambda_{j+1}}{4\pi\varepsilon} f''^{(2)}_{x_{ij}} . \end{cases} \tag{8.3.13}$$

The formulations of $p^{(1)}_{ij}$, $p^{(2)}_{ij}$, $f''^{(1)}_{x_{ij}}$, $f''^{(2)}_{x_{ij}}$, $f''^{(1)}_{y_{ij}}$ and $f''^{(2)}_{y_{ij}}$ are given in Appendix A.8.1, Eqs. (A.8.1.5), (A.8.1.6), and (A.8.1.9) to (A.8.1.12).

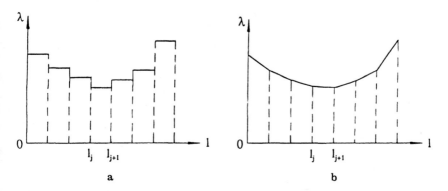

Fig. 8.3.4a, b. Distribution of constant and linear charge density

8.3.2.2 Arced element with constant or linear charge density

An arced element in the x–y and x'–y' planes are shown in Fig. 8.3.5(a) and (b). The formulation of coordinate transformations are given in Eqs. (8.3.14) to (8.3.16)

$$\begin{cases} x' = x - x_0 \\ y' = y - y_0 \end{cases} \tag{8.3.14}$$

$$\begin{cases} x_0 = \dfrac{(x_1^2 - x_3^2)(y_2 - y_3) - (x_2^2 - x_3^2)(y_1 - y_3) - (y_1 - y_2)(y_2 - y_3)(y_3 - y_1)}{2(x_1 - x_3)(y_2 - y_3) - 2(x_2 - x_3)(y_1 - y_3)} \\ y_0 = \dfrac{(y_2^2 - y_3^2)(x_1 - x_3) - (y_1^2 - y_3^2)(x_2 - x_3) - (x_1 - x_2)(x_2 - x_3)(x_3 - x_1)}{2(x_1 - x_3)(y_2 - y_3) - 2(x_2 - x_3)(y_1 - y_3)} \end{cases} \tag{8.3.15}$$

Let

$$\begin{cases} R_0^2 = (x_k - x_0)^2 + (y_k - y_0)^2 \quad k = 1, 2, 3 \\ \psi_1 = \operatorname{arctg} \dfrac{y_1'}{x_1'} = \operatorname{arctg} \dfrac{y_1 - y_0}{x_1 - x_0} \\ \psi_2 = \operatorname{arctg} \dfrac{y_2'}{x_2'} = \operatorname{arctg} \dfrac{y_2 - y_0}{x_2 - x_0} . \end{cases} \tag{8.3.16}$$

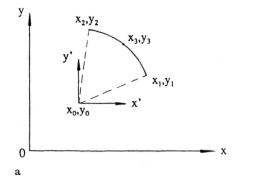
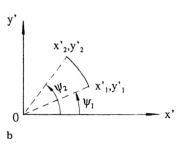

Fig. 8.3.5a–c. Coordinate transformations of an arced element

8.3 Surface boundary elements and surface charge densities

If $\lambda(e)$ is a constant, then

$$\varphi_{ij} = \frac{\lambda_j(e)}{4\pi\varepsilon} \int_{\psi_1}^{\psi_2} \ln[(x'-x_i')^2 + (y'-y_i')^2] R_0 d\psi + C' = -\frac{\lambda_j(e)}{4\pi\varepsilon} p_{ij}. \quad (8.3.17)$$

It is easy to integrate if the local coordinate (ξ, η) are introduced, i.e. suppose

$$\begin{cases} x = \sum_{k=1}^{2} N_k x_k \quad y = \sum_{k=1}^{2} N_k y_k \quad k=1,2 \\ N_1 = \frac{1}{2}(1-\xi) \quad N_2 = \frac{1}{2}(1+\xi) \end{cases} \quad (8.3.18)$$

$$\xi = \frac{\psi - \frac{\psi_1 + \psi_2}{2}}{\frac{\psi_1 - \psi_2}{2}} \quad \xi \in (-1, +1) \quad (8.3.19)$$

then

$$\varphi_{ij} = \frac{\lambda}{4\pi\varepsilon} \int_{-1}^{1} \ln\left[\left(\sum_{k=1}^{2} N_k x_k - x_i\right)^2 + \left(\sum_{k=1}^{2} N_k y_k - y_i\right)^2\right] R_0 d\xi + C'$$

$$= \frac{\lambda}{4\pi\varepsilon} p_{ij} \quad (8.3.20)$$

$$\begin{cases} E_{x_{ij}} = \frac{\lambda}{4\pi\varepsilon} f_{x_{ij}} \\ E_{y_{ij}} = \frac{\lambda}{4\pi\varepsilon} f_{y_{ij}}. \end{cases} \quad (8.3.21)$$

The integration of Eq. (8.3.20) is evaluated either numerically or analytically. The analytical results are given in Appendix A.8.2, Eqs. (A.8.2.4), (A.8.2.7) and (A.8.2.8).

If the charge distribution is linear, then

$$\lambda_j(e) = \sum N_k \lambda_k \quad k=1,2 \quad (8.3.22)$$

$$\varphi_{ij} = -\frac{\lambda_j}{4\pi\varepsilon} \int_{-1}^{1} N_1 \ln[(\sum N_k x_k - x_i)^2 + (\sum N_k y_k - y_i)^2] R_0 d\xi$$

$$- \frac{\lambda_{j+1}}{4\pi\varepsilon} \int_{-1}^{1} N_2 \ln[(\sum N_k x_k - x_i)^2 + (\sum N_k y_k - y_i)^2] R_0 d\xi + C'.$$

$$= \frac{\lambda_j}{4\pi\varepsilon} p_{ij}^{(1)} + \frac{\lambda_{j+1}}{4\pi\varepsilon} p_{ij}^{(2)}. \quad (8.3.23)$$

The formulations of $p_{ij}^{(1)}$, $p_{ij}^{(2)}$ are found in Appendix A.8.2, Eqs. (A.8.2.9), and (A.8.2.10).

8.3.2.3 Ring element with linear charge density

For a axisymmetric field, the boundary surface is subdivided by a number of ring elements. The base of the ring element is a filamentary ring charge. When calculating the potential induced by the ring element, the elliptic integral has to be solved. As the integration path dl is a curve as shown in Fig. 8.3.6(a), these integrals are cumbersome. To reduce the CPU-time for the numerical integration, parabolic curves are recommended to replace the curved contour [3, 16]. The variables of the integration are reduced by using the following equations

$$z = a + br + cr^2 \quad \theta \leq 45^0 \tag{8.3.24}$$

$$r = a + bz + cz^2 \quad \theta \geq 45^0 \tag{8.3.25}$$

$$\theta = \arctan\left|\frac{z_{j+1} - z_j}{r_{j+1} - r_j}\right|. \tag{8.3.26}$$

By using Eqs. (8.3.24) and (8.3.25), the variable of integration in φ_{ij} is only 'r' or 'z'.

Assuming that the charge density is linearly distributed, i.e.

$$\lambda_j(e) = \frac{l_{j+1} - l}{l_{j+1} - l_j}\lambda_j + \frac{l - l_j}{l_{j+1} - l_j}\lambda_{j+1} \tag{8.3.27}$$

substitution of Eqs. (8.3.24) and (8.3.27) into Eq. (8.3.3), leads to

$$\varphi_{ij} = \frac{1}{\pi\varepsilon} \int_{l_j}^{l_{j+1}} \lambda_j(e) r_j \frac{K(k_1)}{\alpha_1} dl + C$$

$$= \frac{\lambda_j}{\pi\varepsilon(l_{j+1} - l_j)} \int_{l_j}^{l_{j+1}} r_j \frac{K(k_1)(l_{j+1} - l)}{\alpha_1} dl$$

$$+ \frac{\lambda_{j+1}}{\pi\varepsilon(l_{j+1} - l_j)} \int_{l_j}^{l_{j+1}} r_j \frac{K(k_1)(l - l_j)}{\alpha_1} dl + C. \tag{8.3.28}$$

Since

$$l^2 = (r_{j+1} - r_j)^2 + (z_{j+1} - z_j)^2$$

$$dl = [1 + (b + 2cr)^2]^{1/2} dr$$

or

$$dl = [1 + (b + 2cz)^2]^{1/2} dz$$

8.3 Surface boundary elements and surface charge densities

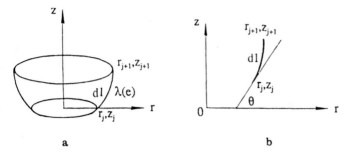

Fig. 8.3.6a, b. A parabolic curve of a ring element

when the symmetric plane is chosen as a potential reference, then $C = 0$,

$$\varphi_{ij} = \frac{\lambda_j}{\pi\varepsilon(z_{j+1} - z_j)} \int_{z_j}^{z_{j+1}} \left[\frac{K(k_1)}{\alpha_1} - \frac{K(k_2)}{\alpha_2}\right]$$
$$\times (a + bz + cz^2)[1 + (b + 2cz)^2]^{1/2}(z_{j+1} - z)dz$$
$$+ \frac{\lambda_{j+1}}{\pi\varepsilon(z_{j+1} - z_j)} \int_{z_j}^{z_{j+1}} \left[\frac{K(k_1)}{\alpha_1} - \frac{K(k_2)}{\alpha_2}\right]$$
$$\times (a + bz + cz^2)[1 + (b + 2cz)^2]^{1/2}(z - z_j)dz \quad (8.3.29)$$

where

$$\begin{cases} \alpha_1^2 = (r_i + r_j)^2 + (z_i - z_j)^2 & \alpha_2^2 = (r_i + r_j)^2 + (z_i + z_j)^2 \\ k_1^2 = \dfrac{4r_i r_j}{\alpha_1^2} & k_2^2 = \dfrac{4r_i r_j}{\alpha_2^2} \end{cases} \quad (8.3.30)$$

In Eq. (8.3.29) $K(k_1)$ and $K(k_2)$ are elliptic integrals of the first kind, the parameters a, b and c are determined by the following equations, e.g.

$$\begin{cases} r_j = a + bz_j + cz_j^2 \\ r_{j+1/2} = a + bz_{j+1/2} + cz_{j+1/2}^2 \\ r_{j+1} = a + bz_{j+1} + cz_{j+1}^2 \end{cases} \quad (8.3.31)$$

where r_j, z_j, $r_{j+1/2}$, $z_{j+1/2}$, r_{j+1} and z_{j+1} are coordinates of each element.

Let

$$z = \frac{z_{j+1} - z_j}{2}x + \frac{z_{i+1} + z_j}{2} \quad (8.3.32)$$

then

$$x = (2z - z_{j+1} - z_j)/(z_{j+1} - z_j) . \quad (8.3.33)$$

Equation (8.3.29) is rewritten to

$$\varphi_{ij} = \frac{\lambda_j}{2\pi\varepsilon} \int_{-1}^{1} g_1[\xi(x)]dx + \frac{\lambda_{j+1}}{2\pi\varepsilon} \int_{-1}^{1} g_2[\xi(x)]dx . \tag{8.3.34}$$

The range of the integration is changed to -1 to $+1$ where

$$g_1[\xi(x)] = g_1(z) = \left[\frac{K(k_1)}{\alpha_1} - \frac{K(k_2)}{\alpha_2}\right]$$
$$\times (a + bz + cz^2)[1 + (b + 2cz)^2]^{1/2}(z_{j+1} - z) \tag{8.3.35}$$

$$g_2[\xi(x)] = g_2(z) = \left[\frac{K(k_1)}{\alpha_1} - \frac{K(k_2)}{\alpha_2}\right]$$
$$\times (a + bz + cz^2)[1 + (b + 2cz)^2]^{1/2}(z - z_j) . \tag{8.3.36}$$

The field strength is:

$$E_{r_{ij}} = -\frac{\partial \varphi}{\partial r}$$

$$= -\frac{\lambda_i}{4\pi\varepsilon r} \int_{-1}^{1} [H_1 - H_2](a + bz + cz^2)[1 + (b + 2cz)^2]^{1/2}$$

$$\times (z_{j+1} - z)dx - \frac{\lambda_{i+1}}{4\pi\varepsilon r} \int_{-1}^{1} [H_1 - H_2](a + bz + cz^2)$$

$$\times [1 + (b + 2cz)^2]^{1/2} (z - z_j)dx \tag{8.3.37}$$

$$E_{z_{ij}} = -\frac{\partial \varphi}{\partial z}$$

$$= -\frac{\lambda_j}{2\pi\varepsilon} \int_{-1}^{1} [H_3 - H_4](a + bz + cz^2)[1 + (b + 2cz)^2]^{1/2}$$

$$\times (z_{j+1} - z)dx - \frac{\lambda_{j+1}}{2\pi\varepsilon} \int_{-1}^{1} [H_3 - H_4](a + bz + cz^2)$$

$$\times [1 + (b + 2cz)^2]^{1/2} (z - z_j)dx . \tag{8.3.38}$$

The integrations of Eqs. (8.3.34) (8.3.37) and (8.3.38) are calculated numerically. All the formulations of the constants a, b, c and H_1 to H_4 are listed in Appendix A.8.3, Eqs. (A.8.3.1) to (A.8.3.8).

8.3.3 Elements for 3-D problems

In 3-D cases, the curved boundary is discretized by a great number of surface elements, these may be a planar triangle, curved triangle, cylindrical element, or any quadrangular element [16].

8.3.3.1 Planar triangular element

The simplest element for 3-D problems is the 3-node planar triangle. Assume the surface charge density is expressed by a linear function of the x and y coordinates, e.g.

$$\sigma(e) = a + bx + cy \tag{8.3.39}$$

then

$$\varphi(\mathbf{r}) = \int_{S_e} \frac{\sigma(e)}{4\pi\varepsilon} \mathrm{d}x\mathrm{d}y \ . \tag{8.3.40}$$

In Eq. (8.3.39), the constants a, b and c are determined by

$$\begin{Bmatrix} a \\ b \\ c \end{Bmatrix} = \begin{bmatrix} 1 & x_i & y_i \\ 1 & x_j & y_j \\ 1 & x_m & y_m \end{bmatrix}^{-1} \begin{Bmatrix} \sigma_i \\ \sigma_j \\ \sigma_m \end{Bmatrix} \tag{8.3.41}$$

where (x_i, y_i), (x_j, y_j), (x_m, y_m) and σ_i, σ_j, σ_m are coordinates and charge densities of the three vertices of the triangle. Equation (8.3.40) can be evaluated numerically or analytically. The analytical integration formulae are given in Appendix 6 of reference [16].

8.3.3.2 Cylindrical tetragonal bilinear element

In order to avoid discontinuity of the charge density along the boundary of the element and the resulting jump phenomenon, the charge density is assumed to be linear along the boundary. For a cylindrical element, as shown in Fig. 8.3.7,

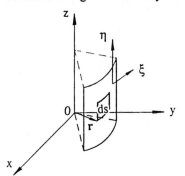

Fig. 8.3.7. Cylindrical tetragonal element

the surface charge density has a nonlinear distribution at the surface of the element e.g.

$$\sigma_j(z, \psi) = \alpha_0 + \alpha_1 z + \alpha_2 \psi + \alpha_3 z\psi . \quad (8.3.42)$$

Using the notations $\sigma(z_1, \psi_1)$, $\sigma(z_1, \psi_2)$, $\sigma(z_2, \psi_2)$, $\sigma(z_2, \psi_1)$ to denote the charge densities at the vertices of the element, the parameters $\alpha_0, \alpha_1, \alpha_2, \alpha_3$ are determined by knowing the charge densities and the coordinates of the four vertices of the element, i.e.

$$\begin{aligned}\sigma(z, \psi) = \frac{1}{N} \{ &\sigma(z_1, \psi_1)[z_2\psi_2 - z_2\psi - \psi_2 z + \psi z] \\ &+ \sigma(z_1, \psi_2)[-z_2\psi_2 - z_2\psi - \psi_1 z - \psi z] \\ &+ \sigma(z_2, \psi_2)[z_1\psi_1 - z_1\psi - \psi_1 z + \psi z] \\ &+ \sigma(z_2, \psi_1)[-z_1\psi_2 + z_1\psi + \psi_2 z - \psi z]\} \end{aligned} \quad (8.3.43)$$

where $N = (z_2 - z_1)(\psi_2 - \psi_1)$. Substitution of these equations into Eq. (8.3.40) leads to

$$\begin{aligned}\varphi_{ij} &= \frac{1}{4\pi\varepsilon} \int_{z_1}^{z_2} \int_{\psi_1}^{\psi_2} \frac{\sigma_j(z, \psi)}{[(R\cos\psi - x_i)^2 + R^2 \sin^2\psi + (z - z_i)^2]^{1/2}} R\,dz\,d\psi \\ &= \frac{1}{4\pi\varepsilon} \{\alpha_3[I_{\psi z}(z_2) - I_{\psi z}(z_1)] + (\alpha_3 z_0 + \alpha_2)[I_\psi(z_2) - I_\psi(z_1)] \\ &\quad + \alpha_1[I_z(z_2) - I_z(z_1)] + (\alpha_1 z_0 + \alpha_0)[I_c(z_2) - I_c(z_1)]\} . \end{aligned} \quad (8.3.44)$$

In Eq. (8.3.44), $I_c, I_z, I_\psi, I_{\psi z}$ represent the four integrals. These integrals have no analytical solution. Reference [14] states that if point P is situated near the surface, an approximate solution is possible as shown in the Appendix of reference [14].

8.3.3.3 Isoparametric high order element

For problems with a curved boundary, curved triangular and tetragonal elements with a high order interpolation function may be used in the same way as in FEM. These elements are used to the optimum design of insulators in reference [9].

Both the charge density and coordinates of a point are expressed by the local coordinates

$$\begin{cases} \sigma(\xi, \eta) = \sum N_j(\xi, \eta)\sigma_j \\ x(\xi, \eta) = \sum N_j(\xi, \eta)x_j \\ y(\xi, \eta) = \sum N_j(\xi, \eta)y_j \\ z(\xi, \eta) = \sum N_j(\xi, \eta)z_j \end{cases} \quad (8.3.45)$$

where $N_j(\xi, \eta)$ are shape functions as was defined in Chap. 6 and 'j' is the

8.3 Surface boundary elements and surface charge densities

sequential number of the nodes. The potential at any observation point $P(x_i, y_i)$ produced by one element is

$$\varphi_{ij} = \frac{1}{4\pi\varepsilon} \int_{S^e} \frac{\sigma_j \, ds}{[(x_i - x_j)^2 + (y_i - y_j)^2 + (z_i - z_j)^2]^{1/2}}$$

$$= \frac{1}{4\pi\varepsilon} \int_{-1}^{1} \int_{-1}^{1} \frac{\sum N_j \sigma_j |J| \, d\xi \, d\eta}{F(\xi, \eta)}$$

$$= \sum K_{ij} \sigma_j \qquad (8.3.46)$$

where $ds = \left| \frac{\partial \mathbf{r}}{\partial \xi} \times \frac{\partial \mathbf{r}}{\partial \eta} \right| d\xi \cdot d\eta = |J| \, d\xi \, d\eta$ (refer to Fig. 6.6.1). In eq. (8.3.46)

$$K_{ij} = \frac{1}{4\pi\varepsilon} \int_{-1}^{1} \int_{-1}^{1} \frac{\sum N_j |J| \, d\xi \, d\eta}{F(\xi, \eta)} \qquad (8.3.47)$$

$$F(\xi, \eta) = [(\sum N_j x_j - x_i)^2 + (\sum N_j y_j - y_i)^2 + (\sum N_j z_j - z_i)^2]^{1/2} \qquad (8.3.48)$$

$$\mathbf{J} = \begin{bmatrix} \frac{\partial x}{\partial \xi} & \frac{\partial y}{\partial \xi} & \frac{\partial z}{\partial \xi} \\ \frac{\partial x}{\partial \eta} & \frac{\partial y}{\partial \eta} & \frac{\partial z}{\partial \eta} \\ 1 & 1 & 1 \end{bmatrix} \qquad (8.3.49)$$

By using Gauss integration

$$K_{ij} = \frac{1}{4\pi\varepsilon} \sum_{m=1}^{I_m} \sum_{n=1}^{I_n} W_m W_n \frac{N_j(\xi_m, \eta_n) |J|}{F(\xi_m, \eta_n)} \qquad (8.3.50)$$

where ξ_m, η_n are coordinates along the path of integration, W_m and W_n are weighted coefficients of the integration. I_m and I_n are the numbers of the integration points.

8.3.3.4 Spline function element

Reference [17] provides a spline function to approximate both the shape of the contour and the charge density, i.e.

$$\begin{cases} \sigma = f(t) = a_\sigma t^3 + b_\sigma t^2 + c_\sigma t + d_\sigma \\ x = f(t) = a_x t^3 + b_x t^2 + c_x t + d_x \\ y = f(t) = a_y t^3 + b_y t^2 + c_y t + d_y \end{cases} \qquad (8.3.51)$$

where t is a parameter. Using a high order spline function, the complex contours can be represented. A more detailed application is given in reference [17].

8.4 Magnetic surface charge simulation method

The presence of magnetization of magnetic materials can be simulated by an equivalent magnetic surface charge, even if it is impractical. It has been derived in Sect. 1.3, that the scalar magnetic potential produced by the magnetization of the material is

$$\varphi_m(\mathbf{r}) = \frac{1}{4\pi} \int_{\Omega'} -\frac{\nabla' \cdot \mathbf{M}}{R} d\Omega' + \frac{1}{4\pi} \oint_{s'} \frac{\mathbf{M} \cdot \mathbf{n}}{R} ds'$$

$$= \frac{1}{4\pi} \int_{\Omega'} \frac{\rho_m}{R} d\Omega' + \frac{1}{4\pi} \oint_{s'} \frac{\sigma_m}{R} ds' \qquad (8.4.1)$$

where
$$R = |\mathbf{r} - \mathbf{r}'|. \qquad (8.4.2)$$

$\mathbf{M}$ is the magnetization vector, σ_m is the magnetic surface charge density and $\mathbf{n}$ is the positive normal direction of the surface. Suppose that the magnetization of the material is uniform, then $\rho_m = 0$. The magnetic field caused by the magnetization is

$$\mathbf{H}_m = -\operatorname{grad} \varphi_m = -\nabla \left(\frac{1}{4\pi} \oint_s \frac{\sigma_m}{R} ds' \right). \qquad (8.4.3)$$

The total magnetic strength $\mathbf{H}$ is composed of two parts, i.e.

$$\mathbf{H} = \mathbf{H}_c + \mathbf{H}_m = \mathbf{H}_c - \nabla \left(\frac{1}{4\pi} \oint_{s'} \frac{\sigma_m}{R} ds' \right). \qquad (8.4.4)$$

$\mathbf{H}_c$ is the magnetic field strength induced by the impressed current. After discretization of the surface, Eq. (8.4.4) can be approximated as

$$\mathbf{H}_i = \mathbf{H}_{c_i} - \frac{1}{4\pi} \sum \int_s \nabla \left(\frac{\sigma_{mj}}{R} \right) ds' \qquad (8.4.5)$$

where the subscript 'i' denotes the field points. Multiplying both sides of Eq. (8.4.4) by $\mathbf{n}$, the normal direction of the interface, one obtains

$$H_{in} = H_{cin} - \frac{1}{4\pi} \sum_{j=1}^{N} \int_{s'} \mathbf{n} \cdot \nabla \left(\frac{\sigma_{mj}}{R} \right) ds'. \qquad (8.4.6)$$

8.4 Magnetic surface charge simulation method

The second term of the RHS of Eq. (8.4.6) includes the fields H_{mns} and H_{mno}, they are due to the magnetization of element 'i' itself and due to the magnetization of all the other elements except the element 'i', i.e.

$$H_{in} = H_{cin} + H_{mns} + H_{mno} . \qquad (8.4.7)$$

The field strength H_c and the H_{mno} are continuous on both sides of the interface. However the values of H_{mns} on both sides of the interface are equal but in the opposite directions.

By using the interfacial boundary conditions

$$\mathbf{n} \cdot (\mathbf{B}_2 - \mathbf{B}_1) = 0 \qquad (8.4.8)$$

$$\mathbf{n} \times (\mathbf{H}_2 - \mathbf{H}_1) = 0 \qquad (8.4.9)$$

i.e.

$$\mu_0 \mathbf{H}_1 \cdot \mathbf{n} = \mu_0 (\mathbf{H}_2 + \mathbf{M}) \cdot \mathbf{n} = \mu_0 \mathbf{H}_2 \cdot \mathbf{n} + \mu_0 \sigma_m$$

thus

$$\sigma_m = (\mathbf{H}_1 - \mathbf{H}_2) \cdot \mathbf{n} . \qquad (8.4.10)$$

For the component H_{mns}, the corresponding magnetization charge density is

$$\sigma_{mi} = \mathbf{n} \cdot (\mathbf{H}_{1s} - \mathbf{H}_{2s}) = 2 H_{1ns} = -2 H_{2ns}$$

$$= 2 \left(\frac{1}{4\pi} \sigma_{mi} \int_{s_i} \mathbf{n}_i \cdot \nabla \left(\frac{1}{R} \right) ds \right) \qquad (8.4.11)$$

therefore

$$\frac{1}{4\pi} \int_{s_i} \mathbf{n}_i \cdot \nabla \left(\frac{1}{R} \right) ds = \frac{1}{2} .$$

Introducing the susceptibility χ_m to express magnetic field strength yields

$$H_{in} = \frac{1}{\chi_m} M_{in} = \frac{1}{\chi_m} \sigma_{mi} \qquad (8.4.12)$$

hence

$$\sigma_{mi} \left(\frac{1}{\chi_m} + \frac{1}{2} \right) + \sum_{i \neq j} \frac{\sigma_{mj}}{4\pi} \int_{s_j} \mathbf{n}_i \cdot \nabla \left(\frac{1}{R} \right) ds' = H_{cin} . \qquad (8.4.13)$$

Consider that $\chi_m = \mu_r - 1$, Eq. (8.4.13) is written to

$$\frac{\mu_2 + \mu_1}{2(\mu_2 - \mu_1)} \sigma_m + \sum_{i \neq j} \frac{\partial}{\partial n} \int_s \frac{\sigma_{mj} ds}{4\pi R} = H_{cn} \qquad (8.4.14)$$

this is the Fredholm integral equation of the second kind. It is used to calculate the magnetic field due to the magnetization of the material. This result can also be obtained from Eq. (8.2.22) by using the analogy between the electric and magnetic field. After using the technique of discretization, Eq. (8.4.14) is transformed to

$$\mathbf{P}\{\sigma_m\} = \{H_{cn}\} \tag{8.4.15}$$

The elements of matrix $\mathbf{P}$ are

$$p_{ii} = \frac{\mu_2 + \mu_1}{2(\mu_2 - \mu_1)} \tag{8.4.16}$$

$$p_{ij} = \frac{\partial}{\partial n} \int_s \frac{ds}{4\pi R} . \tag{8.4.17}$$

If the 3-node triangular element is used, i.e.

$$\sigma_m(\xi, \eta) = \alpha_1 + \alpha_2 \xi + \alpha_3 \eta = \sum N_k(\xi, \eta) \sigma_{mk} \tag{8.4.18}$$

then

$$p_{ij} = \frac{1}{4\pi} \frac{\partial}{\partial n} \int_s \frac{1}{R} N_k(\xi, \eta) d\xi \, d\eta . \tag{8.4.19}$$

Solve Eq. (8.4.15) to obtain $\{\sigma_m\}$, then the magnetic flux density is calculated by:

$$\begin{cases} B_x = \mu_0 \left(H_{c_x} - \frac{1}{4\pi} \frac{\partial}{\partial x} \sum_{j=1}^{N} \int_{s_j} \frac{\sigma_m}{R} ds \right) \\ B_y = \mu_0 \left(H_{c_y} - \frac{1}{4\pi} \frac{\partial}{\partial x} \sum_{j=1}^{N} \int_{s_j} \frac{\sigma_m}{R} ds \right) \\ B_z = \mu_0 \left(H_{c_z} - \frac{1}{4\pi} \frac{\partial}{\partial x} \sum_{j=1}^{N} \int_{s_j} \frac{\sigma_m}{R} ds \right) \end{cases} \tag{8.4.20}$$

An example of this method is given in [18]. Another application of this method is shown in Example 8.6.4.

8.5 Evaluation of singular integrals

One of the drawbacks of the methods of integral equations including SSM is the presence of singular integrals. The use of SSM requires the solution of improper numerical integrals. The numerical solution of improper integrals is either inaccurate or requires considerable CPU time. The calculation of a singular integral depends on the property of the kernel of the integral. The kernels of an integral equation are composed of two kinds: either indicating weak singularity or strong singularity. *If the singularity can be removed by transformation of the function or by changing the variables, it is called weak singularity; otherwise it is*

8.5 Evaluation of singular integrals

a strong singularity. For instance, the kernel ln(r) includes a weak singularity and the kernel 1/r includes a strong singularity. In 2-D electromagnetic problems, the singularities are usually of a logarithmic type. The integration over these singularities is generally evaluated analytically or at least by partially analytical procedures which require tedious series expansions. Generally speaking, known techniques are used depending on the degree of singularity of the integrand. Several methods are introduced in this section.

8.5.1 The semi-analytical technique

If the integrand $f(x)$ of the following equation

$$F = \int_0^1 f(x)dx \tag{8.5.1}$$

is singular when $x = 0$, rewrite the integral as follows:

$$F = \int_0^1 f(x)dx = \int_0^1 [f(x) - h(x)]dx + \int_0^1 h(x)dx \tag{8.5.2}$$

where

$$\lim_{x \to 0} |f(x) - h(x)| = M < \infty . \tag{8.5.3}$$

The first integral of Eq. (8.5.2) is a regular integral, it can be evaluated by Gaussian quadrature. The second integral of Eq. (8.5.2) must be an analytical integral. Here $h(x)$ is a function added to remove the singularity. Thus the technique removes the essential singularity and separates the integration into two parts. For example,

$$F = \int_0^1 \log\left[\sin\left(\frac{\pi}{2}x\right)\right]dx$$

$$= \int_0^1 \left\{\log\left[\sin\left(\frac{\pi}{2}x\right)\right] - \log\left(\frac{\pi}{2}x\right)\right\}dx + H \tag{8.5.4}$$

where

$$H = \int_0^1 \log\left[\left(\frac{\pi}{2}x\right)\right]dx = \log\left(\frac{\pi}{2}\right) - 1 . \tag{8.5.5}$$

The first integration on the RHS of Eq. (8.5.4) is computed by Gaussian quadrature, the second part is analytically integrable as shown in Eq. (8.5.5).

If there are a number of possible functions for a given $f(x)$ to remove singularities, it is recommended to choose the one for which the value M in Eq. (8.5.3) becomes zero. This method requires the evaluation of fewer functions than the method using geometrical intervals. This technique was used to handle the singularity of Green's function in a 2-D case in reference [7]. Another example treats the singularities of a Fourier-type charge distribution composed of linear and circular elements [2]. It is also used in the example given in Sect. 10.4.2.

8.5.2 Method using coordinate transformations

The aim of coordinate transformations is to alternate the singular integral into an analytically integrable function. When the observation point P_i is located on the source segment j, as shown in Fig. 8.5.1(a), Eq. (8.3.47) cannot be evaluated numerically. Let the observation point P_i be at the origin of the polar coordinates $(\rho - \theta)$, shown in Fig. 8.5.1(b), then the integration of the area 1234 is subdivided into eight elements as shown in Fig. 8.5.1(c). The integration of each element can be calculated analytically in $\rho - \theta$ coordinates.

The coordinates transformation methods are usually used to remove any singularity, one example is shown in Example 3 of Sect. 8.6, another example can be seen in [19].

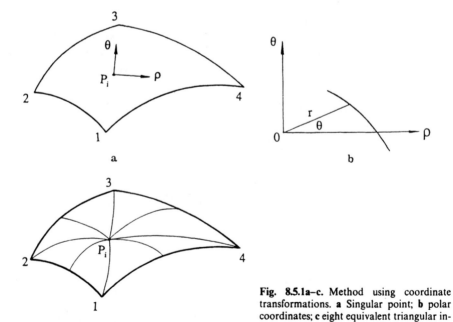

Fig. 8.5.1a–c. Method using coordinate transformations. **a** Singular point; **b** polar coordinates; **c** eight equivalent triangular integration

8.5.3 Numerical technique

One of the numerical integrations of logarithmic singularity is given in [15]. Another method was provided by Kasper [20]. If the integrand can be written as

$$F(x) = f(x) \ln \frac{1}{|x|} + g(x) \tag{8.5.6}$$

where $f(x)$, $g(x)$ remains regular, then the integral can be calculated by updated Gauss numerical integration as below

$$\int_{-h}^{h} F(x) dx = h \sum_{i=1}^{3} \mu_i [F(p_i h) + F(-p_i h)] \tag{8.5.7}$$

where the parameters are

$p_1 = 0.067\ 2230\ 4110 \qquad \mu_1 = 0.211\ 6130\ 9257$

$p_2 = 0.441\ 8550\ 6300 \qquad \mu_2 = 0.470\ 7574\ 6449$

$p_3 = 0.870\ 0987\ 6121 \qquad \mu_3 = 0.317\ 6294\ 4294$.

8.5.4 Combine the analytical integral and Gaussian quadrature

Separate the singular integrand into two parts, one part containing the removable singularity which can be calculated analytically and the other by regular integration. An example is given in Example 8.6.1.

In general, the methods for dealing with integrations around singularities requires experience. Many authors have their own method to handle these problems as shown in references [10, 20–24]. Only the bases are introduced in this section.

8.6 Applications

Example 8.6.1 The field distribution of a pair of spherical electrodes

A pair of charged spherical electrodes is chosen as an example to examine the accuracy of SSM. In Fig. 8.6.1, $S = 2$ cm, $R = 0.524194$ cm. Each spherical electrode is subdivided into six ring elements. Assume for each element that the charge distribution is linear. The matrix equation of $\{\sigma\}$ is:

$$\mathbf{P}\{\sigma\} = \{\pm U_0/2\}. \tag{8.6.1}$$

The elements of $\mathbf{P}$ are calculated using Eq. (8.3.29). The singular integration included in p_{ii} is handled as follows

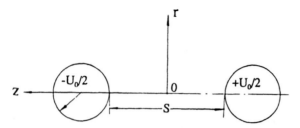

Fig. 8.6.1. A pair of charged spherical conductors

Subdivide the ring element into two parts, A and B, as shown in Fig. 8.6.2. For this singularity, the small area A containing the point 'i' and 'j' is approximated as a rectangular surface with dimensions of $2\beta R \times dl$. The potential at point 'j' produced by the charge density σ of this small area can be calculated analytically. If the charge density of this small part is a constant, then the results of integration are

$$\varphi_A = \frac{\sigma}{4\pi\varepsilon} \int_{-a}^{a} \int_{-b}^{b} \frac{d\xi d\eta}{[\xi^2 + \eta^2]^{1/2}}$$

$$= \frac{\sigma}{\pi\varepsilon}\left[a\ln\left(\frac{b + [a^2 + b^2]^{1/2}}{a}\right) + b\ln\left(\frac{a + [a^2 + b^2]^{1/2}}{b}\right)\right] \quad (8.6.2)$$

where

$$a = \beta R \qquad b = dl/2 . \tag{8.6.3}$$

The potential at point 'j' produced by the charge of the remaining part B is:

$$\varphi_B = \frac{1}{\pi\varepsilon_0} \int_{l_j}^{l_{j+1}} \frac{F(k,\vartheta)}{\alpha_1} r\lambda(e)dl \tag{8.6.4}$$

where $F(k,\vartheta)$ is an incomplete elliptic integral

$$F\left(\frac{\pi - \beta}{2}, k\right) = \int_{0}^{(\pi-\beta)/2} \frac{d\alpha}{[1 - k^2 \sin^2\alpha]^{1/2}} . \tag{8.6.5}$$

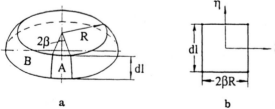

Fig. 8.6.2a, b. Integration around singularity, **a** Subdivision of an element into parts A and B; **b** Local coordinates of area A

8.6 Applications

Table 8.6.1. Validation of SSM

Position		SSM		Analytical solution		Errors
r (m)	z (cm)	E_r (kV/cm)	E_z (kV/cm)	E_r (kV/cm)	E_z (kV/cm)	of E_z (%)
0	0	0	5.564980	0	5.555555	0.169
2.0	0	0	1.214454	0	1.200000	1.203
4.0	0	0	0.244404	0	0.240495	1.624
0	0.4	0	6.891321	0	6.896591	0.076

The approximate evaluation of the incopmplete elliptic is referred in [15]. The comparative results of SSM and the analytical solution are shown in Table 8.6.1.

Example 8.6.2 Field distribution in a vacuum switchgear

Figure 8.6.3(a) shows a model of a vacuum switchgear r-axis is the axis of the symmetry, where 0 is the main electrode, 5 is the insulating envelope and 6 is the endplate. The thin electrodes 1, 2, 3, 4 are shielding electrodes, these are used to equalize the field distribution in the chamber. The shape of these electrodes are

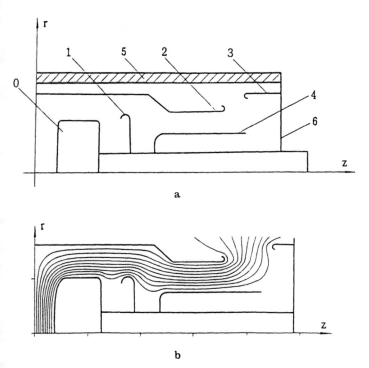

Fig. 8.6.3. **a** Electrodes of a 35-kV vacuum switchgear, **b** equipotential lines of the switchgear

optimized so that a uniform field distribution is obtained in the vacuum chamber. Due to the shielding electrodes being very thin, CSM is not suitable to be used. The surface simulation charges help the solution, the resultant equipotential lines of the switchgear shown in Fig. 8.6.3(b) [25].

Example 8.6.3 Calculation of the capacitance of an isolated plate

A plate with dimensions 1 m × 1 m is subdivided into 8-node isoparametric tetragonal elements. The integrations around singularities are handled using coordinate transformations. The results are shown in Table 8.6.2. The result shows that if only one 8-node tetragonal element is used, the result is very close to the accurate result. If the number of the elements is increased from 48 to 81, the result does not change. The values of Table 8.6.2 shows that the accuracy of an 8-node tetragonal element is quite good.

For comparison, the results of different methods given in reference [2] are listed in Table 8.6.3.

Table 8.6.2. Capacitance C versus the number of sub-elements

Number of element	C (pf)	CPU-Time(s)[†]
1	40.623	0.507813
9	40.825	1.742188
25	40.814	6.554688
49	40.808	25.984375
81	40.808	81.375000

[†] The macrosuper computer ELXSI is used.

Table 8.6.3. Capacitance C of a 1×1 m^2 plate calculated by different methods

Name of the author of the method	Year	Value of C (ppf)
Maxwell	1879	40.13
Reitan[†]	1957	40.2
Harrington[‡]	1970	39.5
Ruehli[§]	1973	40.8
Takuma[¶]	1980	41.11

[†] *AIEE Trans. Comput. Electrons*, **75**, 761–766, 1957
[‡] *Proc. IEE*, **55**, 136–149, 1967
[§] *IEEE Trans.*, MIT, **21**(2), 76–82, 1973
[¶] *CRIEPI Report No. 180029*, Dec. 1980

8.6 Applications

Example 8.6.4 Calculate the magnetic field strength in a ferromagnetic trough

A direct current of 150 A is passing through a long conductor with the cross section of 0.6×0.6 cm^2. It is inserted into the middle of a rectangular ferromagnetic trough as shown in Fig. 8.6.4. The magnetic field strength in the trough is caused by the impressed current and the magnetization of the ferromagnetic material of the trough. Assuming that the current is not strong enough, the

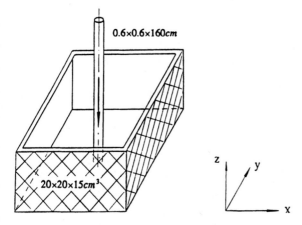

Fig. 8.6.4. The model of the ferromagnetic trough

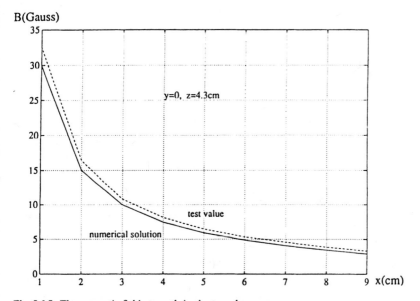

Fig. 8.6.5. The magnetic field strength in the trough

magnetic material is considered to be working in the linear part of the B–H curve. The influence of the magnetization is simulated by using the magnetic surface charge as introduced in Sect. 8.4. The singular integration is solved by coordinate transformations. The calculated and the measured results are shown in Fig. 8.6.5. In Fig. 8.6.5, the origin of the coordinates is assumed in the middle of the bottom of the trough.

8.7 Summary

SSM is one of the direct integral equation methods. The surface of the electrode or the interfacial boundaries are discretized by elements and the surface charge distribution of the subelement is approximated by a specific function. After the boundary is discretized and the approximate function of the charge distribution is chosen, a great number of integral expressions have to be evaluated. Then the solution of the matrix equation is shown by the values of the surface charges.

Compared to CSM, due to the flexibility of the approximate function of the surface chage, it is more suitable for problems with complex geometry and complex interfaces associated with dielectrics. It is more convenient for the optimum design of the electrodes and insulators [26]. The accuracy of SSM due to the discretization has been discussed in [27].

More general methods to deal with the singular integral have been introduced in this chapter.

The magnetic surface charge simulation is suitable for solving the problems contained in ferromagnetic materials.

References

1. J. C. Maxwell, The Scientific Papers of the Honourable Henry Cavendish, Vol. 1, *The Electrical Researchers*, Cambridge University Press, 1921. (Reprint of *The Electrical Researchers*, First edn, 1889)
2. S. Sato, Effective Three-Dimensional Electric Field Analysis by Surface Charge Simulation Method, Diss. *ETZ* Nr. 8221, 1987
3. H. Singer, Berechung von Hochspannungsfeldern mit Hilfe von Flachenladungen, Habilitation, Schrift, 1973
4. B. Krstajic, Z. Andjelic, S. Milojkovic, An Improvement In 3-D Electrostatic Field Calculation, *Proc. of 5th ISH*, Braunschweig, 31.02, 1987
5. Qui Jie, 3-D Magnetic Field Computation in Aluminimum Electrolyser by Using Boundary Element Method, MSc thesis, Xian Jiaotong University, 1987
6. Yoshiki Uchikawa et al., Electron Optics of Point Cathode Electron Gun, *J. Electron Microsc.*, 32(2), 85–98, 1983
7. T. Misaki, et al., Computation of Three-Dimensional Electric Field Problems by a Surface Charge Method and its Application to Optimum Insulator Design, *IEEE Trans. PAS* 101(3), 627–634, 1982
8. Z. Andjelic, B. Krstajic et al., A Procedure for Automatic Optimal Shape Investigation of Interfaces Between Different Media, *IEEE Trans. on Mag*, 24(1) 415–418, 1988

9. H. Tsuboi, T. Misaki, The Optimum Design of Electrode and Insulator Contours by Nonlinear Programming Using the Surface Charge Simulation Method, *IEEE Trans. on Mag*, **24**(1), 35–38, 1988
10. B.H. McDonald, M. Friedman, A. Wexler, Variational Solution of Integral Equations, *IEEE Trans. on MTT*, **22**(3), 237–248, 1974
11. J.D. Lavers, Lecture Notes on Application of Approximate Methods to Field Problems, University of Toronto, 1989
12. M.H. Lean, A. Wexler, Accurate Field Computation with the Boundary Element Method, *IEEE Trans. on Mag*, **18**(2), 331–335, 1982
13. S. Sato, W. Zaengl, Effective 3-Dimensional Electric Field Calculation by Surface Charge Simulation Method, *IEE, Proc.* **133**, Pt A(2), 77–83, 1986
14. F. Gutfleisch, Calculation of the Electric Field by the Boundary Element Method with Different Surface Elements, *5th ISH*, Braunschweig, 31.01, 1987
15. W. Abramowitz, I.A. Stegun (Ed), *Handbook of Mathematical Functions*, Dover Publications, Inc., New York, 1972
16. T. Kouno, T. Takuma, *Numerical Calculation Method of Electric Field* (in Japanese), 1980
17. D. Sautter, A. Schwab, Field Calculation with Surface-Charge Simulation Using Spline Function, *Proc. of 4th ISH*, Athens, 11.04, 1983
18. Masanori Kobayashi, Noboru Takachio, Surface Magnetic Charge Distribution of Cylindrical Cores (In Japanese), *IEE Trans. of Japan* A105, pp. 111–117, 1986
19. Qiu Jie, Feng Cizhang, Qian Xiuying, Treatment of Singularity and Numerical Integral in 3-D Magnetic Computation with BEM, Electromagnetic Fields in Electrical Engineering, *Proc. of BISEF'88*, International Academic Publishers, pp. 317–320, 1989
20. E. Kasper, On the Solution of Integral Equation Arising in Electron Electrical Field Computation, *Optik*, **64**(2), 157–169, 1983
21. S.K. Chow, M.M. Schaefer, Electromagnetic Field Computations Using the Method of Singularities, *IEEE Trans. on Mag*, **15**, 1674–1676, 1979
22. S. Kalaichelvan, J.D. Lavers, Singularity Evaluation in Boundary Integral Equation of Minimum Order for 3-D Eddy Currents, *IEEE Trans. on Mag*, **23**(5), 3053–3055, 1987
23. Jean-Lous Migeot, Numerical Integration in the Vicinity of a Logarithmic Singularity, *Engineering Analysis*, **2**(2), 92–94, 1985
24. Liu Jun, G. Beer and J.L. Meek, Efficient Evaluation of Integrals of Order $1/r$, $1/r^2$, $1/r^3$. Using Gauss Quadrature, *Engineering Analysis*, **2**(3), 118–123, 1985
25. Chen Meijuan, Wang Jimei, Zhou Peibai, Optimum Field Distribution in Vacuum Switchgears Using Integral Equation Method, Electromagnetic Fields in Electrical Engineering, *Proc. of BISEF'88*, International Academic Publishers, pp. 45–48, 1989
26. T. Misaki, H. Tsuboi, et al., Optimization of Three-Dimensional Electrode Contour Based on Surface Charge Method and Its Application to Insulation Design, *IEEE Trans. on PAS.*, **102**, 1687–1692, 1983
27. Y. Kuno, A. Yagi and Y. Uchikawa, 3-D Surface Charge Method — Accuracy Limit Due to Discretization, Electromagnetic Fields in Electrical Engineering, *Proc. of BISEF'88*, Beijing. International Academic Publishers, pp. 293–296, 1988

Appendix 8.1 Potential and field strength coefficients of 2-D planar elements with constant and linear charge density

For constant elements

$$p_{ij} = -\left[A_i \ln C_i + B_i \ln D_i + 2y_i'' \arctg\frac{A_i}{y_i''} + 2y_i'' \arctg\frac{B_i}{y_i''} - 2l\right] + C$$

(A.8.1.1)

$$\begin{cases} A_i = l/2 - x_i'' & C_i = A_i^2 + y_i''^2 \\ B_i = l/2 + x_i'' & D_i = B_i^2 + y_i''^2 \\ C' = \dfrac{\tau(e)}{4\pi\varepsilon} C \end{cases} \quad (A.8.1.2)$$

$$f_{x_{ij}}'' = \ln\frac{D_i}{C_i} + 2\left(\frac{B_i^2}{D_i} - \frac{A_i^2}{C_i}\right) + 2y_i''^2\left[\frac{1}{B_i^2 + y_i''^2} - \frac{1}{A_i^2 + y_i''^2}\right] \quad (A.8.1.3)$$

$$f_{y_{ij}}'' = 2y_i''\left(\frac{A_i}{C_i} + \frac{B_i}{D_i}\right) + 2\mathrm{arctg}\frac{A_i}{y_i''}$$

$$\qquad + 2\mathrm{arctg}\frac{B_i}{y_i''} 2y_i''\left[\frac{A_i}{A_i^2 + y_i''^2} + \frac{B_i}{B_i^2 + y_i''^2}\right] \quad (A.8.1.4)$$

For linear element

$$p_{ij}^{(1)} = -\frac{A_i}{l}\left[A_i \ln C_i + B_i \ln D_i + 2y_i'' \mathrm{arctg}\frac{A_i}{y_i''} + 2y_i'' \mathrm{arctg}\frac{B_i}{y_i''} - 2l\right]$$

$$\qquad + \frac{1}{2l}\left[C_i \ln C_i - D_i \ln D_i - A_i^2 + B_i^2\right] + C^{(1)} \quad (A.8.1.5)$$

$$p_{ij}^{(2)} = -\frac{B_i}{l}\left[A_i \ln C_i + B_i \ln D_i + 2y_i'' \mathrm{arctg}\frac{A_i}{y_i''} + 2y'' \mathrm{arctg}\frac{B_i}{y_i''} - 2l\right]$$

$$\qquad + \frac{1}{2l}\left[C_i \ln C_i - D_i \ln D_i - A_i^2 + B_i^2\right] + C^{(2)} \quad (A.8.1.6)$$

Let $\bar{p}_{ij}$ represent the potential coefficient of a constant element, then the relationshp between $\bar{p}_{ij}$ and p_{ij} is:

$$p_{ij}^{(1)} = \frac{A_i}{l}\bar{p}_{ij} + \frac{1}{2l}[C_i \ln C_i - D_i \ln D_i - A_i^2 + B_i^2] + C^{(1)} \quad (A.8.1.7)$$

$$p_{ij}^{(2)} = \frac{B_i}{l}\bar{p}_{ij} - \frac{1}{2l}[C_i \ln C_i - D_i \ln D_i - A_i^2 + B_i^2] + C^{(2)} \quad (A.8.1.8)$$

$$f_{x_{ij}}''^{(1)} = -\frac{1}{l}\left[2y_i'' \mathrm{arctg}\frac{A_i}{y_i''} + 2y'' \mathrm{arctg}\frac{B_i}{y_i''} 2l\right]$$

$$\qquad + \frac{A_i}{l}\left[\ln\frac{D_i}{C_i} + \frac{2B_i^2}{D_i} - \frac{2A_i^2}{C_i} + \frac{2y_i''}{B_i^2 + y_i''^2} - \frac{2y_i''}{A_i^2 + y_i''^2}\right] \quad (A.8.1.9)$$

$$f_{x_{ij}}''^{(2)} = -\frac{1}{l}\left[2y_i'' \mathrm{arctg}\frac{A_i}{y_i''} + 2y'' \mathrm{arctg}\frac{B_i}{y_i''} 2l\right]$$

$$\qquad + \frac{B_i}{l}\left[\ln\frac{D_i}{C_i} + \frac{2B_i^2}{D_i} - \frac{2A_i^2}{C_i} + \frac{2y_i''}{B_i^2 + y_i''^2} - \frac{2y_i''}{A_i^2 + y_i''^2}\right] \quad (A.8.1.10)$$

Appendix 8.2 Potential and field strength coefficients of 2-D arced elements

$$f'''^{(1)}_{y_{ij}} = \frac{A_i}{l}\left[\frac{2A_i y_i''}{C_i} + \frac{2B_i y_i''}{D_i} + 2\text{arctg}\frac{A_i}{y_i''}\right.$$
$$\left. + 2\text{arctg}\frac{B_i}{y_i''} - \frac{2A_i y_i''}{A_i^2 + y_i''^2} - \frac{2B_i y_i''}{B_i^2 + y_i''^2}\right] - \frac{1}{l}[y_i'' \ln C_i - Y_i'' \ln D_i]$$

(A.8.1.11)

$$f'''^{(2)}_{y_{ij}} = \frac{B_i}{l}\left[\frac{2A_i y_i''}{C_i} + \frac{2B_i y_i''}{D_i} + 2\text{arctg}\frac{A_i}{y_i''}\right.$$
$$\left. + 2\text{arctg}\frac{B_i}{y_i''} - \frac{2A_i y_i''}{A_i^2 + y_i''^2} - \frac{2B_i y_i''}{B_i^2 + y_i''^2}\right] + \frac{1}{l}[y_i'' \ln C_i - y_i'' \ln D_i]$$

(A.8.1.12)

In Eqs. (A.8.1.5) and (A.8.1.6), the coefficients A_i, B_i, C_i and D_i are identical to those of Eq. (A.8.1.2). The constants $C^{(1)}$ and $C^{(2)}$ are determined by reference to the potential.

Appendix 8.2 Potential and field strength coefficients of 2-D arced elements with constant and linear charge density

For constant charge distribution

$$p_{ij} = -\int_{-1}^{1} \ln\left[(\tfrac{1}{2}(1-\xi)x_1 + \tfrac{1}{2}(1+\xi)x_2 - x_i)^2\right.$$
$$\left. + (\tfrac{1}{2}(1-\xi)y_1 + \tfrac{1}{2}(1+\xi)y_2 - y_i)^2\right] R_0 d\xi + C''$$
$$= -R_0 \int_{-1}^{1} \ln\left[(\tfrac{1}{2}(x_2 - x_1)\xi + \tfrac{1}{2}(x_2 + x_1) - x_i)^2\right.$$
$$\left. + (\tfrac{1}{2}(y_2 - y_1)\xi + \tfrac{1}{2}(y_2 + y_1) - y_i)^2\right] d\xi + C''. \quad \text{(A.8.2.1)}$$

Let

$$\tfrac{1}{2}(x_2 - x_1) = D_x \qquad \tfrac{1}{2}(x_2 + x_1) = H_x + x_i$$
$$\tfrac{1}{2}(y_2 - y_1) = D_y \qquad \tfrac{1}{2}(y_2 + y_1) = H_y + y_i \quad \text{(A.8.2.2)}$$

then

$$p_{ij} = -R_0 \int_{-1}^{1} \ln[(D_x \xi + H_x)^2 + (D_y \xi + H_y)^2] d\xi + C''$$
$$= -R_0 \int_{-1}^{1} \ln[c\xi^2 + b\xi + a] d\xi + C''. \quad \text{(A.8.2.3)}$$

The integration of Eq. (A.8.2.3) yields

$$p_{ij} = -\frac{R_0}{S}[(D_x G_x + D_y G_y)\ln(G_x^2 + G_y^2)$$
$$- (D_x Q_x + D_y Q_y)\ln(Q_x^2 + Q_y^2) - 4S]$$
$$- \frac{2R_0}{S}\arctg\frac{2T}{G_x Q_x + G_y Q_y} + C'' \qquad (A.8.2.4)$$

where

$$\begin{cases} G_x = H_x + D_x = x_2 - x_i & G_y = H_y + D_y = y_2 - y_i \\ Q_x = H_x - D_x = x_1 - x_i & Q_y = H_y - D_y = y_1 - y_i \\ S = D_y^2 + D_y^2 \end{cases} \qquad (A.8.2.5)$$

$$T = |H_x D_y - H_y D_x| . \qquad (A.8.2.6)$$

The coefficients of field strength are

$$f_{x_{ij}} = -\frac{R_0}{S}\left[D_x \ln\left(\frac{G_x^2 + G_y^2}{Q_x^2 + Q_y^2}\right) + \frac{2G_x(D_x G_x + D_y G_y)}{G_x^2 + G_y^2}\right.$$
$$\left. + \frac{2Q_x(D_x Q_x + D_y Q_y)}{Q_x^2 + Q_y^2}\right] + (\pm D_y)\frac{2R_0}{S}\arctg\frac{2T}{(G_x Q_x + G_y Q_y)}$$
$$+ \frac{4R_0 T}{S}\frac{[T(Q_x + G_x) + (\pm D_y)(G_x Q_x + G_y Q_y)]}{[(G_x Q_x + G_y Q_y)^2 + 4T^2]} \qquad (A.8.2.7)$$

If $H_x D_y - H_y D_x \geq 0$, the sign of D_y is '$-$ve'. Otherwise the sign of D_y is '$+$ve'

$$f_{y_{ij}} = -\frac{R_0}{S}\left[D_y \ln\left(\frac{G_x^2 + G_y^2}{Q_x^2 + Q_y^2}\right)\right.$$
$$\left. + \frac{2G_y(D_x G_x + D_y G_y)}{G_x^2 + G_y^2} + \frac{2Q_y(D_x Q_x + D_y Q_y)}{Q_x^2 + Q_y^2}\right]$$
$$+ (\pm D_x)\frac{2R_0}{S}\arctg\frac{2T}{(G_x Q_x + G_y Q_y)}$$
$$+ \frac{4R_0 T}{S}\frac{[T(Q_y + G_y) + (\pm D_x)(G_x Q_x + G_y Q_y)]}{[(G_x Q_x + G_y Q_y)^2 + 4T^2]}. \qquad (A.8.2.8)$$

if $H_x D_y - H_y D_x \geq 0$, the sign of D_x is $+$ ve. Otherwise the sign of D_x is $-$ ve.

If the charge distribution is linear, then

$$p_{ij}^{(1)} = -R_0 \int_{-1}^{1} N_1 \ln[(\sum N_k x_k - x_i)^2 + (\sum N_k y_k - y_i)^2]\,d\xi + C^{(1)}$$
$$= -R_0 \int_{-1}^{1} \frac{1}{2}(1-\xi)\ln[(D_x\xi + H_x)^2 + (D_y\xi + H_y)^2]\,d\xi + C^{(1)}$$

$$= -\frac{R_0}{2S}[(D_xG_x + D_yG_y)\ln(G_x^2 + G_y^2)$$

$$- (D_xQ_x + D_yQ_y)\ln(Q_x^2 + Q_y^2) - 4S]$$

$$+ \frac{bR_0}{2C} + \frac{(S^2 - F^2 + T^2)R_0}{4S^2}\ln\frac{G_x^2 + G_y^2}{Q_x^2 + Q_y^2}\frac{R_0T}{S}\left(1 + \frac{F}{S}\right)$$

$$\text{arctg}\frac{2T}{G_xQ_x + G_yQ_y} + C^{(1)} \qquad (A.8.2.9)$$

$$p_{ij}^{(2)} = -R_0 \int_{-1}^{1} N_2 \ln[(\sum N_k x_k - x_i)^2 + (\sum N_k y_k - y_i)^2]\,d\xi + C^{(2)}$$

$$= -R_0 \int_{-1}^{1} \frac{1}{2}(1 + \xi)\ln[(D_x\xi + H_x)^2 + (D_y\xi + H_y)^2]\,d\xi + C^{(2)}$$

$$= -\frac{R_0}{2S}[(D_xG_x + D_yG_y)\ln(G_x^2 + G_y^2)$$

$$- (D_xQ_x + D_yQ_y)\ln(Q_x^2 + Q_y^2) - 4S]$$

$$- \frac{bR_0}{2C} + \frac{(S^2 - F^2 + T^2)R_0}{4S^2}\ln\frac{G_x^2 + G_y^2}{Q_x^2 + Q_y^2}\frac{R_0T}{S}\left(1 - \frac{F}{S}\right)$$

$$\text{arctg}\frac{2T}{G_xQ_x + G_yQ_y} + C^{(2)} \qquad (A.8.2.10)$$

where

$$F = D_xH_x + D_yH_y \qquad (A.8.2.11)$$

the meanings of $D_x, D_y, H_x, H_y, Q_x, Q_y, G_x, G_y$ and T are the same as before.

Appendix 8.3 Coefficients of ring elements with linear charge density

$$a = \frac{1}{\Delta}\begin{vmatrix} r_j & z_j & z_j^2 \\ r_{j+1/2} & z_{j+1/2} & z_{j+1/2}^2 \\ r_{j+1} & z_{j+1} & z_{j+1}^2 \end{vmatrix} \qquad (A.8.3.1)$$

$$b = \frac{1}{\Delta}\begin{vmatrix} 1 & r_j & z_j^2 \\ 1 & r_{j+1/2} & z_{j+1/2}^2 \\ 1 & r_{j+1} & z_{j+1}^2 \end{vmatrix} \qquad (A.8.3.2)$$

$$c = \frac{1}{\Delta} \begin{vmatrix} 1 & z_j & r_j \\ 1 & z_{j+1/2} & r_{j+1/2} \\ 1 & z_{j+1} & r_{j+1} \end{vmatrix} \quad \text{(A.8.3.3)}$$

$$\Delta = \begin{vmatrix} 1 & z_j & z_j^2 \\ 1 & z_{j+1/2} & z_{j+1/2}^2 \\ 1 & z_{j+1} & z_{j+1}^2 \end{vmatrix} \quad \text{(A.8.3.4)}$$

$$H_1 = \frac{[r_i^2 - r_j^2 + (z_i - z_j)^2] E(k_1) - \beta_1^2 K(k_1)}{\alpha_1 \beta_1^2} \quad \text{(A.8.3.5)}$$

$$H_2 = \frac{[r_i^2 - r_j^2 + (z_i - z_j)^2] E(k_2) - \beta_1^2 K(k_2)}{\alpha_2 \beta_1^2} \quad \text{(A.8.3.6)}$$

$$H_3 = \frac{(z_i - z_j) E(k_1)}{\alpha_1 \beta_1^2} \quad \text{(A.8.3.7)}$$

$$H_4 = \frac{(z_i + z_j) E(k_2)}{\alpha_2 \beta_2^2}. \quad \text{(A.8.3.8)}$$

Where $E(k_1)$, $E(k_2)$ are elliptic integrals of the second kind, β_1, β_2 were defined in Eq. (7.3.17).

Chapter 9

Boundary Element Method

9.1 Introduction

It is difficult to say who was the pioneer of the boundary element method (BEM). In Brebbia's opinion [1], the work started in 1960s. The first book entitled *Boundary Elements* was published in 1978 [2]. After that BEM developed rapidly. It has been expanded so as to include time-dependent and non-linear problems [3, 4]. During this time many papers [5, 6], theses [7–9] and books [10–12] have been published. The method is now regarded as important as FEM. An international conference to discuss BEM is held every year and the edited proceedings are valuable references.

The Boundary element method is based on the boundary integral equation and the principle of weighted residuals, where the fundamental solution is chosen as the weighting function. The value of the function and its normal derivative along the boundary are assumed to be the unknows. By using discretization, similar to that used in the finite element method, the boundary integral equation is transformed into a set of algebraic equations at the nodes of the boundary. Then the value of the function and its normal derivative are obtained simultaneously by solving the matrix equation.

The other kind of BEM creates an equivalent surface source either a single layer or a double layer [13] distribution to replace the effects of the source inside or outside the boundary. Such kinds of methods are the indirect boundary element methods. The surface charge simulation method, which was discussed in Chap. 8 is one of the indirect boundary element methods. The direct BEM is discussed in this chapter.

The method contains the following steps.

(1) The boundary Γ is discretized into a number of elements over which the unknown function and its normal derivative are assumed by the interpolation functions.

(2) According to the error minimization principle of weighted residuals, the fundamental solution as the weighting function is chosen to form the matrix equation.

(3) After the integrals over each element are evaluated analytically or numerically, the coefficients of the matrix equation are evaluated.

(4) Setting the proper boundary conditions to the given nodes, a set of linear algebraic equations are then obtained. The solutions of these equations result in the boundary value of the potential and its normal derivatives. Hence the field strength of most interest on the boundary is computed directly from the matrix equation.

(5) The value of the function at any interior point can be calculated once all the function values and their normal derivatives on the boundary are known.

The main characteristic of BEM is that it reduces the dimensions of the problem by one. For a 3-D problem, only the surface of the domain needs to be discretized hence it produces a much smaller number of algebraic equations. It is especially attractive that the data preparation is simple because the tedious domain descritization is avoid. The post-processing of data is also simpler than in the domain methods. Only the required values are calculated. The method is well suited to solve problems with boundaries at infinity. Finally, the solution of the derivatives of the unknown function are as accurate as the function itself.

Disadvantages of BEM are:

(1) A great number of integrations are required and the singularities of the integral must be considered. Hence the calculation of the coefficient matrix requires more time than for FEM.

(2) The fundamental solution of the governing equation is difficult in some problems.

(3) The method cannot be used directly for non-linear problems.

In this chapter the general form of the boundary integral equation is derived.

Formulations to calculate the coefficients of the matrix for a potential problem are derived explicitly. The application of the boundary element method is illustrated by an eddy current problem.

9.2 Boundary element equations

The general integral of the operator equation and therefore the boundary integral equation may be derived by the principle of weighted residuals, using Green's theorem or by the variational method. In this section, the integral equation will be derived by using these different ways and the volume integral equation to the boundary integral equation deduced.

9.2.1 Method of weighted residuals

The variational principle requires that the operator $\mathscr{L}$ is positive definite and self-adjoint. The method of weighted residuals may be applied for arbitrary

9.2 Boundary element equations

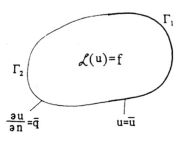

Fig. 9.2.1. A potential problem

operators. It requires knowledge of the governing equation and the corresponding boundary conditions. For a boundary value problem,

$$\begin{cases} \mathscr{L}(u) = f & \text{in } \Omega \\ u|_{\Gamma_1} = \bar{u} & \text{on } \Gamma_1 \\ \dfrac{\partial u}{\partial n}\bigg|_{\Gamma_2} = \bar{q} & \text{on } \Gamma_2 \end{cases} \quad (9.2.1)$$

assume that the solution of the governing equation is approximated by a function as shown in Eq. (9.2.2), i.e.

$$u = \sum_{k=1}^{N} \alpha_k \psi_k \quad (9.2.2)$$

where α_k are the unknown parameters and ψ_k are linearly independent functions taken from a complete sequence of functions, such as

$$\psi_1(x), \psi_2(x), \ldots, \psi_n(x) .$$

These functions are usualy chosen in such a way as to satisfy certain given conditions, called admissible conditions. Consider that the functions belong to a linear space, and are linearly independent, i.e.

$$\alpha_1 \psi_1 + \alpha_2 \psi_2 + \ldots + \alpha_n \psi_n = 0 \quad (9.2.3)$$

only if

$$\alpha_1 = \alpha_2 = \alpha_3 = \ldots = \alpha_n = 0 . \quad (9.2.4)$$

Then they can be combined linearly, e.g.

$$\psi = \alpha_1 \psi_1 + \alpha_2 \psi_2 . \quad (9.2.5)$$

Substituting Eq. (9.2.2) into Eq. (9.2.1), residuals, or called errors are unavoidable, i.e.

$$\begin{cases} R(u) = \mathscr{L}(u) - f & \text{in } \Omega \\ R_1(u) = u - \bar{u} & \text{on } \Gamma_1 \\ R_2(u) = \dfrac{\partial u}{\partial n} - \bar{q} & \text{on } \Gamma_2 . \end{cases} \quad (9.2.6)$$

In order to make the errors approach zero, the average error distribution principle is used i.e. let

$$\int_\Omega R(u) W \, d\Omega + \int_{\Gamma_1} R_1(u) W_1 \, d\Gamma + \int_{\Gamma_2} R_2(u) W_2 \, d\Gamma = 0 \quad (9.2.7)$$

where W, W_1, W_2 are weighting functions. Let $W_1 = \partial W/\partial n$ (otherwise the equation will not have the correct dimension) and $W_2 = -W$; Eq. (9.2.7) is simplified to

$$\int_\Omega R(u) W \, d\Omega = -\int_{\Gamma_1} R_1(u) \frac{\partial W}{\partial n} \, d\Gamma + \int_{\Gamma_2} R_2(u) W \, d\Gamma. \quad (9.2.8)$$

Consider first that the governing equation is Laplace's equation, then Eq. (9.2.8) is changed to

$$\int_\Omega (\nabla^2 u) W \, d\Omega = -\int_{\Gamma_1} (u - \bar{u}) \frac{\partial W}{\partial n} \, d\Gamma + \int_{\Gamma_2} (q - \bar{q}) W \, d\Gamma \quad (9.2.9)$$

where u is the approximate solution. Integrating Eq. (9.2.9) by parts, yields

$$\int_\Omega \frac{\partial u}{\partial x_k} \frac{\partial W}{\partial x_k} d\Omega = \int_{\Gamma_2} \bar{q} W \, d\Gamma + \int_{\Gamma_1} q W \, d\Gamma - \int_{\Gamma_2} \bar{u} \frac{\partial W}{\partial n} \, d\Gamma + \int_{\Gamma_1} u \frac{\partial W}{\partial n} \, d\Gamma. \quad (9.2.10)$$

This is a weak formulation of Eq. (9.2.9), as it reduces the order of the derivative of the unknown function. Hence the requirement of continuity of the approximate function of u is reduced.

Integrating Eq. (9.2.10) by parts once again, the dual equation of Eq. (9.2.9) is obtained:

$$\int_\Omega (\nabla^2 W) u \, d\Omega = -\int_{\Gamma_2} \bar{q} W \, d\Gamma - \int_{\Gamma_1} \frac{\partial u}{\partial n} W \, d\Gamma + \int_{\Gamma_1} \bar{u} \frac{\partial W}{\partial n} \, d\Gamma + \int_{\Gamma_2} u \frac{\partial W}{\partial n} \, d\Gamma. \quad (9.2.11)$$

For Poisson's equation, Eq. (9.2.11) becomes

$$\int_\Omega (u \nabla^2 W - f W) \, d\Omega = -\int_{\Gamma_2} \bar{q} W \, d\Gamma - \int_{\Gamma_1} \frac{\partial u}{\partial n} W \, d\Gamma$$
$$+ \int_{\Gamma_1} \bar{u} \frac{\partial W}{\partial n} \, d\Gamma + \int_{\Gamma_2} u \frac{\partial W}{\partial n} \, \partial \Gamma. \quad (9.2.12)$$

Equations (9.2.11) and (9.2.12) require that the second order derivatives of the weighting function are continuous and only require the continuity of the

9.2 Boundary element equations

function u. Equation (9.2.12) can be written in a compact form as below:

$$\int_\Omega (\nabla^2 W - f) u \, d\Omega + \int_\Gamma u \frac{\partial W}{\partial n} d\Gamma = \int_\Gamma W \frac{\partial u}{\partial n} d\Gamma \qquad (9.2.13)$$

where $\Gamma = \Gamma_1 + \Gamma_2$.

Equation (9.2.13) is the fundamental equation of the boundary element method.

9.2.2 Green's theorem

Equation (9.2.13) can also be derived by Green's theorem. Using the second identity of Green's theorem,

$$\int_\Omega (W \nabla^2 u - u \nabla^2 W) \, d\Omega = \int_\Gamma \left(W \frac{\partial u}{\partial n_1} - u \frac{\partial W}{\partial n} \right) d\Gamma \qquad (9.2.14)$$

Eq. (9.2.9) can be written as

$$\int_\Omega (u \nabla^2 W - fW) \, d\Omega + \int_\Gamma \left(W \frac{\partial u}{\partial n} - u \frac{\partial W}{\partial n} \right) d\Gamma = -\int_{\Gamma_1} (u - \bar{u}) \frac{\partial W}{\partial n} d\Gamma$$

$$+ \int_{\Gamma_2} (q - \bar{q}) W \, d\Gamma . \qquad (9.2.15)$$

Here the governing equation is assumed as Poisson's equation. Eliminating the terms present on both sides of the above equation, one obtains

$$\int_\Omega (u \nabla^2 W - fW) \, d\Omega = -\int_{\Gamma_2} \bar{q} W \, d\Gamma - \int_{\Gamma_1} \frac{\partial u}{\partial n} W \, d\Gamma$$

$$+ \int_{\Gamma_1} \bar{u} \frac{\partial W}{\partial n} d\Gamma + \int_{\Gamma_2} u \frac{\partial W}{\partial n} d\Gamma . \qquad (9.2.16)$$

This equation is exactly the same as the one derived using integration by parts.

9.2.3 Variational principle

Recall now the corresponding functional of Laplace's equation is

$$I(u) = \tfrac{1}{2} \int_\Omega (\operatorname{grad} u)^2 \, d\Omega - \int_{\Gamma_2} \bar{q} u \, d\Gamma . \qquad (9.2.17)$$

Taking the variation of the functional I, and let $\delta I = 0$, yields

$$\int_\Omega \nabla u \cdot \nabla \delta u \, d\Omega - \int_{\Gamma_2} \bar{q} \delta u \, d\Gamma = 0 \, . \tag{9.2.18}$$

Using the vector identity,

$$\nabla \cdot (\delta u \nabla u) = \nabla \delta u \cdot \nabla u + \delta u \nabla^2 u$$

and the divergence theorem, Eq. (9.2.18) is transformed to

$$\int_\Omega (\nabla^2 u) \delta u \, d\Omega + \int_\Gamma \delta u \nabla u \, d\Gamma - \int_{\Gamma_2} \bar{q} \delta u \, d\Gamma = 0 \tag{9.2.19}$$

where $\Gamma = \Gamma_1 + \Gamma_2$.

Let $\delta u = W$, the integral equation

$$\int_\Omega (\nabla^2 u) W \, d\Omega = - \int_{\Gamma_2} \bar{q} W \, d\Gamma - \int_{\Gamma_1} \frac{\partial u}{\partial n} W \, d\Gamma$$

$$+ \int_{\Gamma_1} \bar{u} \frac{\partial W}{\partial n} d\Gamma + \int_{\Gamma_2} u \frac{\partial W}{\partial n} d\Gamma \tag{9.2.20}$$

is obtained.

The integral equation derived from the functional is identical to the one derived using the method of weighted residuals. This is true only if δu is chosen as the weighting function.

9.2.4 Boundary integral equation

Choose the fundamental solution F, which satisfies Eq. (9.2.21), as the weighting function, i.e.

$$\nabla^2 F = - \delta_i(\mathbf{r} - \mathbf{r}') \tag{9.2.21}$$

where δ_i is a Dirac's delta function, this function has the property that

$$\int_\Omega u \nabla^2 F \, d\Omega = - \int_\Omega u \delta_i(\mathbf{r} - \mathbf{r}') \, d\Omega = - u_i \tag{9.2.22}$$

while point 'i' is in the domain Ω, then Eq. (9.2.16) becomes

$$u_i + \int_\Omega f F \, d\Omega = \int_{\Gamma_2} \bar{q} \, F \, d\Gamma + \int_{\Gamma_1} \frac{\partial u}{\partial n} F \, d\Gamma - \int_{\Gamma_1} \bar{u} \frac{\partial F}{\partial n} d\Gamma - \int_{\Gamma_2} u \frac{\partial F}{\partial n} d\Gamma \, . \tag{9.2.23}$$

The compact form of Eq. (9.2.23) is:

$$u_i = \int_\Gamma \left(F \frac{\partial u}{\partial n} - u \frac{\partial F}{\partial n} \right) d\Gamma - f F \, d\Omega \, . \tag{9.2.24}$$

9.2 Boundary element equations

Due to the property of the δ function, if the point 'i' is outside the domain Ω, the integral of Eq. (9.2.22) is zero.

Substitution of the fundamental solution by the 3-D Poisson's equation, $F = 1/4\pi r$, into Eq. (9.2.24) yields

$$u_i = \frac{1}{4\pi}\int_\Gamma \left[\frac{1}{r}\frac{\partial u}{\partial n} - u\frac{\partial}{\partial n}\left(\frac{1}{r}\right)\right]d\Gamma - \frac{1}{4\pi}\int_\Omega f\frac{1}{r}d\Omega. \qquad (9.2.25)$$

This result is identical to the one derived in Chap. 1. *Thus the potentials in the domain are determined by the boundary values of the potential and its normal derivative and the source density within the domain.*

Equation (9.2.25) is valid for points inside the domain. If the point 'i' is on the boundary, the singularity must be considered.

Suppose the boundary is smooth, draw a small sphere centred on the point 'i' with a radius of $\varepsilon(\varepsilon \to 0)$, as shown in Fig. 9.2.2.

First the point of singularity is considered on the boundary Γ_2. The boundary Γ_2 is divided into two parts, i.e. $\Gamma_2 = \Gamma_\varepsilon + \Gamma_{2-\varepsilon}$. The last term of the RHS of Eq. (9.2.23) becomes

$$\int_{\Gamma_2} u\frac{\partial F}{\partial n}d\Gamma = \int_{\Gamma_2-\varepsilon} u\frac{\partial F}{\partial n}d\Gamma + \int_{\Gamma_\varepsilon} u\frac{\partial F}{\partial n}d\Gamma. \qquad (9.2.26)$$

Substituting the fundamental solution into the second integral on the RHS of Eq. (9.2.26) and take the limit, the result is

$$\lim_{\varepsilon \to 0}\left\{\int_{\Gamma_\varepsilon} u\frac{\partial F}{\partial n}d\Gamma\right\} = \lim_{\varepsilon \to 0}\left\{\int_{\Gamma_\varepsilon} u\frac{1}{4\pi\varepsilon^2}d\Gamma\right\}$$

$$= \lim_{\varepsilon \to 0}\left\{-\frac{2\pi\varepsilon^2}{4\pi\varepsilon^2}u_i\right\} = -\frac{1}{2}u_i. \qquad (9.2.27)$$

On the other hand, note that as $\varepsilon \to 0$, the boundary $\Gamma_{2-\varepsilon}$ is almost identical to Γ_2. Substitution of Eq. (9.2.27) into Eq. (9.2.23), obtains the boundary integral

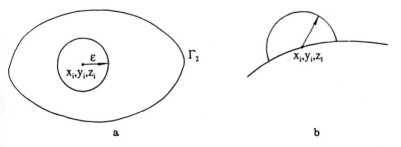

Fig. 9.2.2a, b. Integration around singularity of the boundary

equation for the points on the boundary, i.e.

$$\frac{1}{2}u_i + \int_\Omega fF\,d\Omega = \int_{\Gamma_2} \bar{q}F\,d\Gamma + \int_{\Gamma_1} \frac{\partial u}{\partial n}F\,d\Gamma - \int_{\Gamma_1} \bar{u}\frac{\partial F}{\partial n}\,d\Gamma - \int_{\Gamma_2} u\frac{\partial F}{\partial n}\,d\Gamma. \quad (9.2.28)$$

This equation is still usable for 2-D problems. For a 2-dimensional Laplace's equation, the fundamental solution and its derivatives are:

$$F = \frac{1}{2\pi}\ln\left(\frac{1}{r}\right) \quad (9.2.29)$$

and

$$\frac{\partial F}{\partial n} = \frac{\partial F}{\partial r} = -\frac{1}{2\pi r} \quad (9.2.30)$$

then

$$\lim_{\varepsilon\to 0}\left\{\int_{\Gamma_\varepsilon} u\frac{1}{2\pi\varepsilon}\,d\Gamma\right\} = \lim_{\varepsilon\to 0}\left(-\frac{u\pi\varepsilon}{2\pi\varepsilon}\right) = -\frac{u_i}{2}. \quad (9.2.31)$$

If the point 'i' is on the Γ_1, the same result is obtained.

Both for 3-D and 2-D problems, the first integral of the RHS of Eq. (9.2.23) is regular because

$$\lim_{\varepsilon\to 0}\left\{\int_{\Gamma_\varepsilon} \bar{q}F\,d\Gamma\right\} = \lim_{\varepsilon\to 0}\left(\bar{q}\frac{2\pi\varepsilon^2}{4\pi\varepsilon}\right) = 0 \quad (9.2.32)$$

$$\lim_{\varepsilon\to 0}\left\{\int_{\Gamma_\varepsilon} \bar{q}F\,d\Gamma\right\} = \lim_{\varepsilon\to 0}\left\{-\int_{\Gamma_\varepsilon} \bar{q}\frac{\ln\varepsilon}{2\pi}\,d\Gamma\right\} = \lim_{\varepsilon\to 0}\left\{-\bar{q}\frac{\ln\varepsilon}{2\pi}\pi\varepsilon\right\}$$

$$= \frac{\bar{q}}{2}\lim_{\varepsilon\to 0}\frac{\ln\varepsilon}{1/\varepsilon} = -\frac{\bar{q}}{2}\lim_{\varepsilon\to 0}\varepsilon = 0. \quad (9.2.33)$$

Combining Eqs. (9.2.31)–(9.2.33), the boundary integral equation is obtained

$$\frac{1}{2}u_i = \int_\Gamma \left(F\frac{\partial u}{\partial n} - u\frac{\partial F}{\partial n}\right)d\Gamma - \int_\Omega fF\,d\Omega. \quad (9.2.34)$$

Summarizing the above cases, Eqs. (9.2.28) and (9.2.34) are written as

$$C_i u_i = \int_\Gamma \left(F\frac{\partial u}{\partial n} - u\frac{\partial F}{\partial n}\right)d\Gamma - \int_\Omega fF\,d\Omega, \quad (9.2.35)$$

where

$$C_i = \begin{cases} 1 & \text{in domain } \Omega \\ 1/2 & \text{on smooth boundary} \\ 0 & \text{outside domain } \Omega. \end{cases} \quad (9.2.36)$$

Equation (9.2.35) is the typical form of the boundary element method.

9.3 Matrix formulations of the boundary integral equation

By using the same method as to derive Eq. (9.2.34), the boundary integral equation of the Helmholtz equation has the same form as the integral equation of the Laplace equation, only the fundamental solution is different.

9.2.5 Indirect boundary integral equation

Recall Eq. (9.2.25)

$$u_i = \frac{1}{4\pi}\int_\Gamma \left[\frac{1}{r}\frac{\partial u}{\partial n} - u\frac{\partial}{\partial n}\left(\frac{1}{r}\right)\right]d\Gamma - \frac{1}{4\pi}\int_\Omega f\frac{1}{r}d\Omega.$$

Note that if the boundary values of u and $\partial u/\partial n$ and the source function are known, the value of u within Ω can be calculated.

Let us define that Ω' is the exterior region of the boundary Γ and u' is the solution of Laplace's equation in the exterior region. Then, in Ω', Eq. (9.2.35) reduces to

$$\int_\Gamma \left(F\frac{\partial u'}{\partial n} - u'\frac{\partial F}{\partial n}\right)d\Gamma = 0. \tag{9.2.37}$$

Subtraction of Eq. (9.2.37) from Eq. (9.2.35) leads to

$$u_i = -\int_\Omega fF\,d\Omega + \int_\Gamma F\left(\frac{\partial u}{\partial n} - \frac{\partial u'}{\partial n}\right)d\Gamma - \int_\Gamma (u-u')\frac{\partial F}{\partial n}d\Gamma. \tag{9.2.38}$$

Considering the boundary conditions (Eqs. (1.3.5) and (1.3.18))

$$\frac{\partial u}{\partial n} - \frac{\partial u'}{\partial n} = \frac{\sigma}{\varepsilon} \qquad u' - u = \frac{\tau}{\varepsilon} \tag{9.2.39}$$

and supposing $f = -\rho/\varepsilon$, the following equation is obtained

$$u_i = \frac{1}{4\pi\varepsilon}\int_\Gamma \frac{\sigma}{r}d\Gamma + \frac{1}{4\pi\varepsilon}\int_\Gamma \tau\frac{\partial}{\partial n}\left(\frac{1}{r}\right)d\Gamma + \frac{1}{4\pi\varepsilon}\int_\Omega \rho\frac{1}{r}d\Omega. \tag{9.2.40}$$

The first term of the RHS of Eq. (9.2.40) represents the single layer source while the second term represents the double layer source. Here σ is the surface charge density and τ is the dipole charge density. Equation (9.2.40) indicates that both the SSM and the magnetic surface charge method are special cases of BEM.

9.3 Matrix formulations of the boundary integral equation

In this section, the discretization form of the boundary integral equation will be derived by using the constant and linear elements both in homogeneous and piece-wise homogeneous media in 2-D and 3-D cases.

9.3.1 Discretization and shape functions

The most commonly used discretization elements are constant, linear, quadratic or the combinations of the constant and linear elements. In the case of combined elements [3], the unknown function varies linearly but its normal derivative is constant. For 2-dimensional problems, the boundary is a contour. The three types of the elements are shown in Fig. 9.3.1.

The constant element is defined as one where the function and its normal derivative are constants along each element. The centre point of each element is the representation of the element as shown in Fig. 9.3.1(a). In the case of the linear or the quadratic element, both the potential function and its normal derivative vary linearly or quadratically within each element. In local coordinates the unknown function and its normal derivative of linear and quadratic elements are expressed as

$$\begin{cases} u(\xi) = a + b\xi = \psi_1 u_1 + \psi_2 u_2 \\ \dfrac{\partial u}{\partial n}(\xi) = \psi_1 \left(\dfrac{\partial u}{\partial n}\right)_1 + \psi_2 \left(\dfrac{\partial u}{\partial n}\right)_2 \end{cases} \quad (9.3.1)$$

and

$$\begin{cases} u(\xi) = a + b\xi + c\xi^2 = \psi_1 u_1 + \psi_2 u_2 + \psi_3 u_3 \\ \dfrac{\partial u}{\partial n}(\xi) = \psi_1 \left(\dfrac{\partial u}{\partial n}\right)_1 + \psi_2 \left(\dfrac{\partial u}{\partial n}\right)_2 + \psi_3 \left(\dfrac{\partial u}{\partial n}\right)_3 \end{cases} \quad (9.3.2)$$

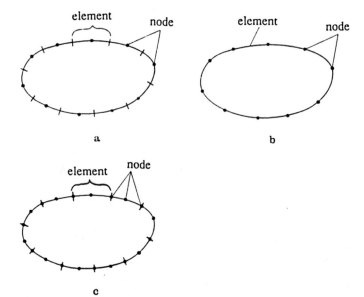

Fig. 9.3.1a–c. Constant, linear and quadratic element

9.3 Matrix formulations of the boundary integral equation

where the u_1, u_2 and u_3 are nodal values of the function and ψ_1, ψ_2 and ψ_3 are interpolation functions or shape functions. In the case of linear elements, the terminal points of the elements are nodes as shown in Fig. 9.3.1(b). The shape functions are

$$\psi_1 = \tfrac{1}{2}(1 - \xi) \qquad \psi_2 = \tfrac{1}{2}(1 + \xi) \tag{9.3.3}$$

at terminal 1, $\psi_1 = 1$, $\psi_2 = 0$; at terminal 2, $\psi_1 = 0$, $\psi_2 = 1$. In a quadratic element, the terminal points and the centre point of the element are nodes as shown in Fig. 9.3.2(c). The shape functions are

$$\psi_1 = \tfrac{1}{2}\xi(\xi - 1) \qquad \psi_2 = \tfrac{1}{2}\xi(\xi + 1) \qquad \psi_3 = (1 - \xi)(1 + \xi) \tag{9.3.4}$$

at node 1, $\psi_1 = 1$, $\psi_2 = 0$, $\psi_3 = 0$; at node 2, $\psi_1 = 0$, $\psi_2 = 1$, $\psi_3 = 0$ and at node 3 $\psi_1 = 0$, $\psi_2 = 0$, $\psi_3 = 1$.

Higher order elements can be obtained by using the same principle. For instance, a cubic shape function of the approximate function

$$u(x) = \alpha_1 + \alpha_2 x + \alpha_3 x^2 + \alpha_4 x^3$$

is obtained by taking four nodes over each element, as shown in Fig. 9.3.3

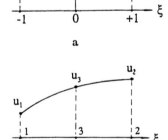

Fig. 9.3.2a–c. Shape functions of constants, linear and quadratic elements

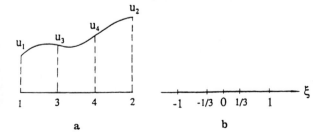

Fig. 9.3.3a, b. A cubic order element

Node	ξ	ψ_1	ψ_2	ψ_3	ψ_4
1	-1	1	0	0	0
2	1	0	1	0	0
3	$-1/3$	0	0	1	0
4	$1/3$	0	0	0	1

(9.3.5)

here

$$u(\xi) = \psi_1 u_1 + \psi_2 u_2 + \psi_3 u_3 + \psi_4 u_4 . \tag{9.3.6}$$

Then the shape functions are

$$\begin{cases} \psi_1 = \dfrac{1}{16}(1 - \xi)[-10 + 9(\xi^2 + 1)] \\ \psi_2 = \dfrac{1}{16}(1 + \xi)[-10 + 9(\xi^2 + 1)] \\ \psi_3 = \dfrac{9}{16}(1 - \xi^2)(1 - 3\xi) \\ \psi_4 = \dfrac{9}{16}(1 - \xi^2)(1 + 3\xi) . \end{cases} \tag{9.3.7}$$

Other kinds of higher order elements are given in reference [3]. For instance one could take the function to be the unknowns at the two interior points of the element and consider that the derivatives are unknowns at the two end points. Then the continuity of the derivative of the function is guaranteed at the intersection of the elements.

For 3-dimensional problems, the boundary elements of linear or quadratic triangular and quadrilateral elements are all the same as defined for the 2-D case of the finite element method and as discussed in Chap. 6.

9.3.2 Matrix equation of a 2-dimensional constant element

Consider that the boundary has been divided into N elements. The integral equation of Eq. (9.2.34) for the case of $f = 0$ is approximated as

$$\frac{1}{2} u_i + \sum_{j=1}^{N} \int_{\Gamma_j} u \frac{\partial F}{\partial n} d\Gamma = \sum_{j=1}^{N} \int_{\Gamma_j} \frac{\partial u}{\partial n} F \, d\Gamma \tag{9.3.8}$$

where Γ_j is the contour of the element j. As u, $\partial u/\partial n$ are constants, Eq. (9.3.8) is simplified to

$$\frac{1}{2} u_i + \sum_{j=1}^{N} u_j \left(\int_{\Gamma_j} \frac{\partial F}{\partial n} d\Gamma \right) = \sum_{j=1}^{N} \left(\frac{\partial u}{\partial n} \right)_j \left(\int_{\Gamma_j} F \, d\Gamma \right) . \tag{9.3.9}$$

9.3 Matrix formulations of the boundary integral equation

Applying Eq. (9.3.9) to each element under consideration, and let

$$\hat{H}_{ij} = \int_{\Gamma_j} \frac{\partial F}{\partial n} d\Gamma \qquad (9.3.10)$$

$$G_{ij} = \int_{\Gamma_j} F \, d\Gamma \qquad (9.3.11)$$

Eq. (9.3.9) becomes

$$\frac{1}{2} u_i + \sum_{j=1}^{N} \hat{H}_{ij} u_j = \sum_{j=1}^{N} G_{ij} \left(\frac{\partial u}{\partial n} \right)_j. \qquad (9.3.12)$$

In the above equations, F is the fundamental solution of the governing equation, for 2-D Laplace's equation

$$F = \frac{1}{2\pi} \ln \frac{1}{|r - r'|}. \qquad (9.3.13)$$

As a simplification, it is defined that

$$H_{ij} = \begin{cases} \hat{H}_{ij} & i \neq j \\ \hat{H}_{ij} + \frac{1}{2} & i = j \end{cases} \qquad (9.3.14)$$

then Eq. (9.3.12) is written as

$$\sum_{j=1}^{N} H_{ij} u_j = \sum_{j=1}^{N} G_{ij} q_j. \qquad (9.3.15)$$

The matrix form of Eq. (9.3.15) is

$$[H]^\dagger U = [G] Q \qquad (9.3.16)$$

This is the normal form of the boundary element equation. Here $[H]$, $[G]$ are matrices of the order of $N \times N$, these are full matrices and in general they are asymmetrical. U, Q are two unknown column matrices of the order N. They are potentials and its normal derivative at each nodes. Substitute the known boundary conditions of the first and the second kind into Eq. (9.3.16) and rearrange the knowns and the unknowns on both sides of the equation. One has to solve the following algebraic equation:

$$AX = B. \qquad (9.3.17)$$

It should be noticed that if the domain has a hole, for the outer contour the nodes are numbered counterclockwise, for the inner contour the nodes are

---

† In order to avoid the confusion of the matrix [H] and the magnetic field strength, in this chapter [ ] is used to express the matrix H and G, but the column matrix U and Q are still expressed by bold character only.

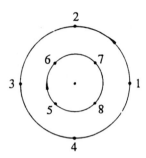

Fig. 9.3.4. Sequence of boundary nodes on outer and inner contours

numbered clockwise as shown in Fig. 9.3.4. Once all the values of u and $\partial u/\partial n$ on the boundary nodes have been solved, the values of u and $\partial u/\partial x$, $\partial u/\partial y$ inside the domain can be calculated by Eq. (9.3.18) and Eq. (9.3.19).

$$u_i = -\int_\Gamma u \frac{\partial F}{\partial n} d\Gamma + \int_\Gamma \frac{\partial u}{\partial n} F d\Gamma = \sum_{j=1}^N G_{ij} q_j - \sum_{j=1}^N H_{ij} u_j \qquad (9.3.18)$$

$$\left(\frac{\partial u}{\partial x}\right)_i = \sum_{j=1}^N \left(\frac{\partial u}{\partial n}\right)_j \frac{\partial F}{\partial x} - \sum_{j=1}^N (u)_j \frac{\partial}{\partial x}\left(\frac{\partial F}{\partial n}\right)$$

$$\left(\frac{\partial u}{\partial x}\right)_i = \sum_{j=1}^N \left(\frac{\partial u}{\partial n}\right)_j \frac{\partial F}{\partial y} - \sum_{j=1}^N (u)_j \frac{\partial}{\partial y}\left(\frac{\partial F}{\partial n}\right). \qquad (9.3.19)$$

In these equations, the partial derivatives are approximated by differences.

9.3.2.1 Evaluation of H_{ij} and G_{ij}

For a 2-dimensional Laplacian problem

$$F = -\frac{1}{2\pi} \ln r \qquad (9.3.20)$$

$$\frac{\partial F}{\partial n} = \text{grad } F \cdot n = -\frac{1}{2\pi}\left(\frac{r \cos \alpha}{r^2}\right) = -\frac{1}{2\pi}\left(\frac{\pm D_{ij}}{r^2}\right) \qquad (9.3.21)$$

where r is the distance between point i and the point on element Γ_j, α is the angle between the vector n and r and $\cos \alpha = n \cdot r$. The $\pm$ sign of D_{ij} is dependent on whether α is acute or obtuse. While $r_1 \cdot r_2' \geq 0$, $\cos \alpha > 0$; $r_1 \cdot r_2' < 0$, $\cos \alpha < 0$. r_1, r_2, r_2' are shown in Fig. 9.3.5 where n is the normal direction of the element. r_2' is orthogonal to r_2.

From Fig. 9.3.5, let

$$T = \frac{y_{j+1} - y_j}{x_{j+1} - x_j},$$

the distance between point i and line $j, j+1$, is:

$$D_{ij} = \frac{|T(x_i - x_j) - y_i + y_j|}{(1+T^2)^{1/2}} \qquad (9.3.22)$$

9.3 Matrix formulations of the boundary integral equation

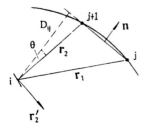

Fig. 9.3.5. Relationship between variables

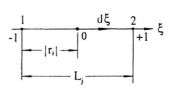

Fig. 9.3.6. Integration of element

Then

$$\hat{H}_{ij} = -\frac{1}{2\pi} \int_{\Gamma_j} \frac{\pm D_{ij}}{r^2} d\Gamma = -\frac{1}{2\pi} \int_{-1}^{1} \frac{\pm D_{ij}}{r^2} \frac{L_j}{2} d\xi \qquad (9.3.23)$$

$$G_{ij} = \frac{1}{2\pi} \int_{-1}^{1} \ln\left(\frac{1}{r}\right) \frac{L_j}{2} d\xi \qquad (9.3.24)$$

where $d\Gamma = \frac{1}{2} L_j d\xi$, L_j is the length of the jth element as shown in Fig. 9.3.6.

9.3.2.2 Evaluation of H_{ii} and G_{ii}

For constant elements the diagonal element G_{ii}

$$G_{ii} = \frac{1}{2\pi} \int_{(1)}^{(0)} \ln\left(\frac{1}{r}\right) d\Gamma + \frac{1}{2\pi} \int_{(0)}^{(2)} \ln\left(\frac{1}{r}\right) d\Gamma$$

$$= \frac{1}{\pi} |r_1| \left[\ln\left(\frac{1}{|r_1|}\right) + \int_0^1 \ln\left(\frac{1}{\xi}\right) d\xi \right]$$

$$= \frac{1}{\pi} |r_1| \left[\ln\left(\frac{1}{|r_1|}\right) + 1 \right]. \qquad (9.3.25)$$

For element H_{ii}, as the radius vector r (from point i to element j) is orthogonal to the normal direction n of the contour, it follows that

$$\hat{H}_{ii} = 0 .$$

Due to discretization, the boundary is not smooth, $C_i \neq \frac{1}{2}$ and H_{ii} is evaluated by:

$$H_{ii} = -\sum_{i \neq j} H_{ij} . \qquad (9.3.26)$$

This is because if the boundary condition statisfies the homogeneous boundary condition of the second kind, then

$$[H]U = 0. \tag{9.3.27}$$

Consequently, Eq. (9.3.26) exsists.

One remaining problem is that the interior solution near the boundary exhibits large numerical inaccuracies. Paulsem [14] regards it as stemming from the inability of the quadrature to account for Green's function near singularity. A simple scheme is offered to remove the numerical inaccuracy by increasing the quadrature in the nearby boundary element.

9.3.3 Matrix equation of 2-D linear elements

For linear elements

$$u(\xi) = \psi_1 u_j + \psi_2 u_{j+1} = [\psi_1 \ \psi_2] \begin{Bmatrix} u_j \\ u_{j+1} \end{Bmatrix} \tag{9.3.28}$$

$$q(\xi) = \psi_1 q_j + \psi_2 q_{j+1} = [\psi_1 \ \psi_2] \begin{Bmatrix} q_j \\ q_{j+1} \end{Bmatrix} \tag{9.3.29}$$

substitution of the above equations into the two integral terms of Eq. (9.3.8) yields

$$\int_{\Gamma_j} u \frac{\partial F}{\partial n} d\Gamma = \int_{\Gamma_j} [\psi_1 \ \psi_2] \frac{\partial F}{\partial n} d\Gamma \begin{Bmatrix} u_j \\ u_{j+1} \end{Bmatrix} = [H_{ij}^{(1)} \ H_{ij}^{(2)}] \begin{Bmatrix} u_j \\ u_{j+1} \end{Bmatrix} \tag{9.3.30}$$

$$\int_{\Gamma_j} F \frac{\partial u}{\partial n} d\Gamma = \int_{\Gamma_j} [\psi_1 \ \psi_2] F d\Gamma \begin{Bmatrix} q_j \\ q_{j+1} \end{Bmatrix} = [G_{ij}^{(1)} \ G_{ij}^{(2)}] \begin{Bmatrix} q_j \\ q_{j+1} \end{Bmatrix}. \tag{9.3.31}$$

hence

$$H_{ij}^{(1)} = \int_{\Gamma_j} \psi_1 \frac{\partial F}{\partial n} d\Gamma \qquad H_{ij}^{(2)} = \int_{\Gamma_j} \psi_2 \frac{\partial F}{\partial n} d\Gamma \tag{9.3.32}$$

$$G_{ij}^{(1)} = \int_{\Gamma_j} \psi_1 F d\Gamma \qquad G_{ij}^{(2)} = \int_{\Gamma_j} \psi_2 F d\Gamma. \tag{9.3.33}$$

In above equations, $H_{ij}^{(1)}$, $G_{ij}^{(1)}$ are contributions of the first node of elements j, $H_{ij}^{(2)}$ and $G_{ij}^{(2)}$ are contributions of the second node of element j, hence

$$\begin{cases} \hat{H}_{ij} = H_{ij}^{(1)} + H_{i,j-1}^{(2)} \\ G_{ij} = G_{ij}^{(1)} + G_{i,j-1}^{(2)} \end{cases} \tag{9.3.34}$$

$$G_{ij}^{(1)} = \int_{\Gamma_j} \psi_1 F d\Gamma = \int_{-1}^{1} 0.5(1-\xi) \frac{1}{2\pi} \ln\left(\frac{1}{r}\right) |J| d\xi \tag{9.3.35}$$

9.3 Matrix formulations of the boundary integral equation

$$G_{ij}^{(2)} = \int_{\Gamma_j} \psi_1 F \, d\Gamma = \int_{-1}^{1} 0.5(1+\xi)\frac{1}{2\pi}\ln\left(\frac{1}{r}\right)|\mathbf{J}|\,d\xi . \tag{9.3.36}$$

Figure 9.3.6 indicates that

$$\xi = -1, \quad \Gamma = -\tfrac{1}{2}\Gamma_j; \qquad \xi = 1, \quad \Gamma = \tfrac{1}{2}\Gamma_j$$

i.e.

$$\Gamma = \frac{\Gamma_j}{2}\xi, \qquad d\Gamma = \frac{\Gamma_j}{2}d\xi ,$$

hence

$$\begin{cases} x = \psi_1 x_j + \psi_2 x_{j+1} \\ y = \psi_1 y_j + \psi_2 y_{j+1} \end{cases} \tag{9.3.37}$$

The Jacobian is

$$|\mathbf{J}| = \frac{\partial \Gamma}{\partial \xi} = \left[\left(\frac{\partial x}{\partial \xi}\right)^2 + \left(\frac{\partial y}{\partial \xi}\right)^2\right]^{1/2}$$

$$= \tfrac{1}{2}[(x_{j+1} - x_j)^2 + (y_{j+1} - y_j)^2]^{1/2} = \tfrac{1}{2}L_j \tag{9.3.38}$$

therefore

$$H_{ij}^{(1)} = \int_{\Gamma_j} \psi_1 \frac{\partial F}{\partial n} d\Gamma = \int_{-1}^{1} 0.5(1-\xi)\frac{1}{2\pi}\frac{\partial}{\partial n}\ln\left(\frac{1}{r}\right)|\mathbf{J}|\,d\xi \tag{9.3.39}$$

$$H_{ij}^{(2)} = \int_{\Gamma_j} \psi_2 \frac{\partial F}{\partial n} d\Gamma = \int_{-1}^{1} 0.5(1+\xi)\frac{1}{2\pi}\frac{\partial}{\partial n}\ln\left(\frac{1}{r}\right)|\mathbf{J}|\,d\xi . \tag{9.3.40}$$

The term $\partial/\partial n[\ln(1/r)]$ has been discussed in Sect. 9.3.2.1.

Note that while $j = i$ and $j = i-1$, due to $\mathbf{r}\cdot\mathbf{n} = 0$, $H_{i,i-1}^{(1)} = H_{i,i-1}^{(2)} = H_{i,i}^{(1)} = H_{i,i}^{(2)} = 0$; $G_{ij}^{(1)}$, $G_{ij}^{(2)}$ are singular integrals, they can be integrated analytically. For $j = i$,

$$G_{ij}^{(1)} = \int_{\Gamma_j} \psi_1 F \, d\Gamma = \int_{-1}^{1} 0.5(1-\xi)\frac{1}{2\pi}\ln\left(\frac{1}{r}\right)|\mathbf{J}|\,d\xi$$

$$= \frac{1}{4\pi}L_j(1.5 - \ln L_j) \tag{9.3.41}$$

$$G_{ij}^{(2)} = \int_{\Gamma_j} \psi_2 F \, d\Gamma = \int_{-1}^{1} 0.5(1+\xi)\frac{1}{2\pi}\ln\left(\frac{1}{r}\right)|\mathbf{J}|\,d\xi$$

$$= \frac{1}{4\pi}L_j(0.5 - \ln L_j) . \tag{9.3.42}$$

For $j = i - 1$,

$$G_{ij}^{(1)} = \frac{1}{4\pi} L_j (0.5 - \ln L_j) \tag{9.3.43}$$

$$G_{ij}^{(2)} = \frac{1}{4\pi} L_j (1.5 - \ln L_j) . \tag{9.3.44}$$

If the analytical integration is difficult, the Gaussian quadrature [15] may be used.

9.3.4 Matrix form of Poisson's equation

Recall Eq. (9.2.35)

$$C_i u_i + \int_\Omega u \frac{\partial F}{\partial n} d\Gamma = \int_\Gamma \frac{\partial u}{\partial n} F d\Gamma - \int_\Omega fF d\Omega .$$

After discretization (including the source area), the above equation is transformed to:

$$\mathbf{B} + [\mathbf{H}]\mathbf{U} = [\mathbf{G}]\mathbf{Q} . \tag{9.3.45}$$

B is a column matrix of the order N, each element of **B** is

$$B_i = \int_\Omega fF d\Omega = \sum_{i=1}^{M} \left(\int_{\Omega_i} fF d\Omega \right) = \sum_{i=1}^{M} \left[\sum_{r=1}^{K} W_r(fF) \right] A_i \tag{9.3.46}$$

where M is the number of segments of the source area, K is the number of abscissas of the Gauss integration, W_r are the weighting coefficients and A_i is the area of subelement. It should be noticed that there is no increase of unknowns due to the discretization of the source area.

Once the values of u and q are known over the whole boundary, the values of u and q at any interior points are calculated:

$$u_i = \sum_{i=1}^{N} G_{ij} q_j - \sum_{i=1}^{N} H_{ij} u_j - B_i . \tag{9.3.47}$$

If the source is a constant, the volume integral of Eq. (9.3.46) may be transformed to a boundary integral. Define a function satisfying $F = \nabla^2 v$, using Green's second identity:

$$\int_\Omega (f\nabla^2 v - v\nabla^2 f) d\Omega = \int_\Gamma \left[f \frac{\partial v}{\partial n} - v \frac{\partial f}{\partial n} \right] d\Gamma . \tag{9.3.48}$$

Due to f being a constant, Eq. (9.3.46) becomes

$$\int_\Omega fF d\Omega = \int_\Gamma f \frac{\partial v}{\partial n} d\Gamma . \tag{9.2.49}$$

9.3 Matrix formulations of the boundary integral equation

9.3.5 Matrix equation of a piecewise homogeneous domain

Consider a general case as shown in Fig. 9.3.7, the domain is composed of different materials. Each sub-region $\Omega_1, \Omega_2, \Omega_3$, is homogeneous. In each area, the following equations exist

$$[G_1 \quad G_{12}^1 \quad G_{31}^1] \begin{Bmatrix} q_1 \\ q_{12}^1 \\ q_{31}^1 \end{Bmatrix} = [H_1 \quad H_{12}^1 \quad H_{31}^1] \begin{Bmatrix} u_1 \\ u_{12}^1 \\ u_{31}^1 \end{Bmatrix} \qquad (9.3.50)$$

$$[G_2 \quad G_{23}^2 \quad G_{12}^2] \begin{Bmatrix} q_2 \\ q_{23}^2 \\ q_{12}^2 \end{Bmatrix} = [H_2 \quad H_{23}^2 \quad H_{12}^2] \begin{Bmatrix} u_2 \\ u_{23}^2 \\ u_{12}^2 \end{Bmatrix} \qquad (9.3.51)$$

$$[G_3 \quad G_{31}^3 \quad G_{23}^3] \begin{Bmatrix} q_3 \\ q_{31}^3 \\ q_{23}^3 \end{Bmatrix} = [H_1 \quad H_{31}^3 \quad H_{23}^3] \begin{Bmatrix} u_3 \\ u_{31}^3 \\ u_{23}^3 \end{Bmatrix}. \qquad (9.3.52)$$

The single number of the sub- and the superscripts: 1, 2, 3 represent the external boundary and the domain of each subarea respectively. The combined numbers: 12, 23, 31 represent the interface boundary. Based on the continuity of the interface boundary, we have

$$\begin{cases} u_{12}^1 = u_{12}^2 = u_{12} & q_{12}^1 = -\dfrac{1}{\beta_1} q_{12}^2 = q_{12} \\ u_{23}^2 = u_{23}^3 = u_{23} & q_{23}^2 = -\dfrac{1}{\beta_2} q_{23}^3 = q_{23} \\ u_{31}^3 = u_{31}^1 = u_{31} & q_{31}^3 = -\dfrac{1}{\beta_3} q_{31}^1 = q_{31} \end{cases} \qquad (9.3.53)$$

The sign $-$ve before $1/\beta_i$ is because the normal direction of the interface boundary is opposite the neighbouring region. Inserting Eq. (9.3.53) into Eq. (9.3.50) to (9.3.52) and combining them, the final matrix equation is

$$\begin{bmatrix} G_1 & 0 & 0 & G_{12}^1 & -H_{12}^1 & 0 & 0 & -\beta_3 G_{31}^1 & -H_{31}^1 \\ 0 & G_2 & 0 & -\beta_1 G_{12} & -H_{12}^2 & G_{23}^2 & -H_{23}^2 & 0 & 0 \\ 0 & 0 & G_3 & 0 & 0 & -\beta_2 G_{23}^3 & -H_{23}^3 & G_{31}^3 & -H_{31}^3 \end{bmatrix}$$

$$\begin{Bmatrix} q_1 \\ q_2 \\ q_3 \\ q_{12} \\ u_{12} \\ q_{23} \\ u_{23} \\ q_{31} \\ u_{31} \end{Bmatrix} = \begin{bmatrix} H_1 & 0 & 0 \\ 0 & H_2 & 0 \\ 0 & 0 & H_3 \end{bmatrix} \begin{Bmatrix} u_1 \\ u_2 \\ u_3 \end{Bmatrix} \qquad (9.3.54)$$

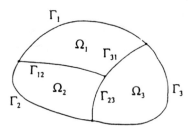

Fig. 9.3.7. Piecewise homogeneous regions

This equation may be solved once the boundary conditions on $\Gamma_1, \Gamma_2, \Gamma_3$ are prescribed. The total number of unknowns is equal to the number of nodal degrees of freedom over the external boundaries plus twice the number of nodal degrees of freedom over the whole internal boundaries.

Subdivision of the region into several zones may be used in homogeneous media as a way of avoiding numerical problems or improving computational efficiency. For instance, if the problem includes cracks or notches, then the region can be divided into two zones to avoid any numerical difficulties due to the nodes that are very close to each other.

9.3.6 Matrix equation of axisymmetric problems

Assuming that all boundaries and consequently all domain values are axisymmetric, the boundary integral equation in cylindrical coordinates is

$$C_i u_i + \int_{\Gamma'} u \frac{\partial F}{\partial n} r \, d\Gamma' = \int_{\Gamma'} \frac{\partial u}{\partial n} F r \, d\Gamma' \qquad (9.3.55)$$

where

$$d\Gamma(x, y, z) = r \, d\theta \, d\Gamma'(r, z) \qquad (9.3.56)$$

and Γ' is the intersection of the problem boundary Γ with the $r - z$ half plane, shown in Fig. 9.3.8.

The fundamental solution of Laplace's equation in axisymmetric domains is

$$F(r, r') = \frac{K(k)}{2\pi^2 (a + b)^{\frac{1}{2}}} . \qquad (9.3.57)$$

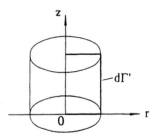

Fig. 9.3.8. Boundary contour of an axisymmetric field

9.3 Matrix formulations of the boundary integral equation

This is obtained by assuming a filament ring source. In Eq. (9.3.57)

$$k = \frac{2b}{a+b} \qquad a = r'^2 + r^2 + (z'-z)^2 \qquad b = 2rr' \qquad (9.3.58)$$

where r', z' are coordinates of the source point, while r, z are coordinates of the field point.

The normal derivative of the fundamental solution along the contour Γ' is

$$\frac{\partial F}{\partial n} = \frac{1}{2\pi^2(a+b)^{\frac{1}{2}}} \left\{ \frac{1}{2r} \left[\frac{r'^2 - r^2 + (z'-z)^2}{a-b} E(k) - K(k) \right] n_r \right.$$

$$\left. + \frac{z'-z}{a-b} E(k) n_z \right\} \qquad (9.3.59)$$

Equation (9.3.57) to (9.3.59) indicate that if $r' = 0$, then $k = 0$, thus $K(k) = E(k) = \pi/2$. The ring source contracted to a point source on the axis of symmetry.

Approximating Eq. (9.3.55) and summing the contributions from all boundary elements, a system matrix equation similar to Eq. (9.3.45) is obtained. The terms $H_{ij}, G_{ij} (i \neq j)$ are evaluated numerically using Gauss quadratures. The diagonal terms H_{ii}, G_{ii} are the results of evaluating singular integrals. In order to facilitate the evaluation of these integrals, the fundamental solution and its normal derivative are written in terms of Legendre functions of the second kind, detailed formulations for which can be found in reference [3].

9.3.7 Discretization of 3-dimensional problems

In the case of 3-dimensions, the boundary Γ is a 2-dimensional surface. It is discretized by flat or curved triangles or quadrilaterals and the potentials and their normal derivatives over an elementary surface are assumed to be piecewise constant, linear or quadratic. These have been employed in 2-dimensional finite element analysis.

Consider the isoparametric elements, the following equations are valid

$$u = \sum_{k=1}^{N} \psi_k(\xi_1, \xi_2, \eta) u_k \qquad (9.3.60)$$

$$\begin{cases} x = \sum_{k=1}^{N} \psi_k(\xi_1, \xi_2, \eta) x_k \\ y = \sum_{k=1}^{N} \psi_k(\xi_1, \xi_2, \eta) y_k \\ z = \sum_{k=1}^{N} \psi_k(\xi_1, \xi_2, \eta) z_k \end{cases} \qquad (9.3.61)$$

The shape functions ψ_k are the same as listed in App. 6.1 denoted by N_k. The differential surface area and the volume element are expressed as

$$ds = \left| \frac{\partial \mathbf{r}}{\partial \xi_1} \times \frac{\partial \mathbf{r}}{\partial \xi_2} \right| d\xi_1 d\xi_2 = |\mathbf{J}| d\xi_1 d\xi_2 \tag{9.3.62}$$

$$d\Omega = \left| \left(\frac{\partial \mathbf{r}}{\partial \xi_1} \times \frac{\partial \mathbf{r}}{\partial \xi_2} \right) \frac{\partial \mathbf{r}}{\partial \eta} \right| d\xi_1 d\xi_2 d\eta = |\mathbf{G}| d\xi_1 d\xi_2 d\eta \tag{9.3.63}$$

where

$$|\mathbf{J}| = \left| \frac{\partial \mathbf{r}}{\partial \xi_1} \times \frac{\partial \mathbf{r}}{\partial \xi_2} \right| = \begin{vmatrix} \mathbf{i} & \mathbf{j} & \mathbf{k} \\ \frac{\partial x}{\partial \xi_1} & \frac{\partial y}{\partial \xi_1} & \frac{\partial z}{\partial \xi_1} \\ \frac{\partial x}{\partial \xi_2} & \frac{\partial y}{\partial \xi_2} & \frac{\partial z}{\partial \xi_2} \end{vmatrix} \tag{9.3.64}$$

$$|\mathbf{G}| = \left| \left(\frac{\partial \mathbf{r}}{\partial \xi_1} \times \frac{\partial \mathbf{r}}{\partial \xi_2} \right) \frac{\partial \mathbf{r}}{\partial \eta} \right| = \frac{\partial x}{\partial \eta} \begin{vmatrix} \frac{\partial y}{\partial \xi_1} & \frac{\partial z}{\partial \xi_1} \\ \frac{\partial y}{\partial \xi_2} & \frac{\partial z}{\partial \xi_2} \end{vmatrix} - \frac{\partial y}{\partial \eta} \begin{vmatrix} \frac{\partial x}{\partial \xi_1} & \frac{\partial z}{\partial \xi_1} \\ \frac{\partial x}{\partial \xi_2} & \frac{\partial z}{\partial \xi_2} \end{vmatrix}$$

$$+ \frac{\partial z}{\partial \eta} \begin{vmatrix} \frac{\partial x}{\partial \xi_1} & \frac{\partial y}{\partial \xi_1} \\ \frac{\partial x}{\partial \xi_2} & \frac{\partial y}{\partial \xi_2} \end{vmatrix} . \tag{9.3.65}$$

Then the boundary integral equation is

$$C_i u_i = \int_\Gamma \left(F \frac{\partial u}{\partial n} - u \frac{\partial F}{\partial n} \right) |\mathbf{J}| d\xi_1 d\xi_2 - \int_\Omega fF |\mathbf{G}| d\xi_1 d\xi_2 d\eta . \tag{9.3.66}$$

Suppose the boundary is divided into 3-node triangles where u and $\partial u/\partial n$ are linear functions within the element. The discretized form of Eq. (9.3.66) is

$$C_i u_i + \sum_{e=1}^{N} [h_1 \ h_2 \ h_3]_e \begin{Bmatrix} u_1 \\ u_2 \\ u_3 \end{Bmatrix}_e = \sum_{e=1}^{N} [g_1 \ g_2 \ g_3]_e \begin{Bmatrix} q_1 \\ q_2 \\ q_3 \end{Bmatrix}_e \tag{9.3.67}$$

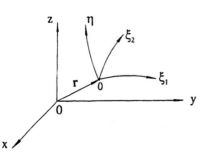

Fig. 9.3.9. Calculation of an infinitesimal surface and volume

9.3 Matrix formulations of the boundary integral equation

The system matrix equation is

$$[H]U = [G]Q \qquad (9.3.68)$$

Reference [16] gives a good example for solving a 3-D magnetic problem.

9.3.8 Use of symmetry

The application of symmetry in integral equation method is different from the domain method. For instance, the central line is the line of symmetry of the square slot as shown in Fig. 9.3.10. If FEM is used to calculate the field distribution, only the half area is the domain for calculation. If BEM is used, the influence of the whole contour must be added in the whole integral equation.

Dividing the boundary into eight elements, the LHS of Eq. (9.3.68) is

$$\begin{bmatrix} h_{11} & h_{12} & h_{13} & h_{14} & h_{18} & h_{17} & h_{16} & h_{15} \\ h_{21} & & & & h_{28} & & & h_{25} \\ h_{31} & & & & h_{38} & & & h_{35} \\ h_{41} & & & & h_{48} & & & h_{45} \\ \hdashline h_{18} & h_{17} & h_{15} & h_{11} & & & h_{14} & \\ h_{28} & & & h_{21} & & & h_{24} & \\ h_{38} & & & h_{31} & & & h_{34} & \\ h_{48} & & h_{45} & h_{41} & & & h_{44} & \end{bmatrix} \begin{bmatrix} u_1 \\ u_2 \\ u_3 \\ u_4 \\ \text{---} \\ u_8 \\ u_7 \\ u_6 \\ u_5 \end{bmatrix} \qquad (9.3.69)$$

Due to symmetry, $u_1 = u_8$, $u_2 = u_7$ and so on, only u_1 to u_4 are the unknowns, thus the unknowns of Eq. (9.3.69) are reduced to half of (9.3.69). The final matrix of **H** is

$$\begin{bmatrix} h_{11} + h_{18} & h_{12} + h_{17} & h_{13} + h_{16} & h_{14} + h_{15} \\ h_{21} + h_{28} & h_{22} + h_{27} & h_{23} + h_{26} & h_{24} + h_{25} \\ h_{31} + h_{38} & h_{32} + h_{37} & h_{33} + h_{36} & h_{34} + h_{35} \\ h_{41} + h_{48} & h_{42} + h_{47} & h_{43} + h_{46} & h_{44} + h_{45} \end{bmatrix} \qquad (9.3.70)$$

Hence the order of the matrix is reduced to half of the original one.

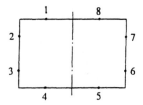

Fig. 9.3.10. A symmetric problem

9.4 Eddy current problems

Eddy current problems are significant in engineering practice. It was studied from the beginning of the ninteenth century [17]. The physical properties of the eddy current and the analytical solution methods are given in references [18, 19]. Due to the complexities of the eddy current problems, there are many different formulations to solve them [20]. There are still many unsolved problems in 3-dimensional cases.

9.4.1 Eddy current equations

Consider a steady-state case with skin effects, the displacement currents are neglected. Maxwell's equations and the constitutive equations are

$$\nabla \times \mathbf{H} = \mathbf{J} \tag{9.4.1}$$

$$\nabla \times \mathbf{E} - \frac{\partial \mathbf{B}}{\partial t} \tag{9.4.2}$$

$$\nabla \cdot \mathbf{B} = 0 \tag{9.4.3}$$

$$\nabla \cdot \mathbf{J} = 0 \tag{9.4.4}$$

$$\mathbf{B} = \mu \mathbf{H} \tag{9.4.5}$$

$$\mathbf{J} = \gamma \mathbf{E} . \tag{9.4.6}$$

Take the curl of Eq. (9.4.1) then combine it with Eq. (9.4.2) and use the vector identity, $\nabla \times \nabla \times \mathbf{A} = -\nabla^2 \mathbf{A} + \nabla(\nabla \cdot \mathbf{A})$, to obtain

$$\begin{aligned}\nabla^2 \mathbf{H} &= \nabla \cdot \mathbf{H} - \gamma \nabla \times \mathbf{E} - (\nabla \gamma) \times \mathbf{E} \\ &= -\nabla \cdot \left(\mathbf{H} \cdot \frac{1}{\mu} \nabla \mu \right) + \gamma \mu \frac{\partial \mathbf{H}}{\partial t} - \frac{1}{\gamma}(\nabla \gamma) \times \nabla \times \mathbf{H} .\end{aligned} \tag{9.4.7}$$

For linear magnetic or non-magnetic materials with constant conductivity and permeability, Eq. (9.4.7) reduces to

$$\nabla^2 \mathbf{H} = \mu \gamma \frac{\partial \mathbf{H}}{\partial t} . \tag{9.4.8}$$

This is the diffusion equation in linear materials in terms of the field density $\mathbf{H}$. By introducing the potential function one can obtain several different kinds of formulations.

9.4.1.1 A-φ formulations

Based on Eqs. (9.4.2) and (9.4.3), a magnetic vector potential **A** and a scalar potential φ are introduced. Then

$$\mathbf{E} = -\frac{\partial \mathbf{A}}{\partial t} - \nabla\varphi = \mathbf{E}_e - \mathbf{E}_s \tag{9.4.9}$$

where the subscripts e and s correspond to the induced and the impressed components. By taking the curl of **B** and using Eqs. (9.4.1), (9.4.5) and (9.4.9), one obtains

$$\nabla^2 \mathbf{A} - \mu\gamma\left(\frac{\partial \mathbf{A}}{\partial t} + \nabla\varphi\right) + \frac{1}{\mu}(\nabla\mu) \times \nabla \times \mathbf{A} = 0 \tag{9.4.10}$$

$$\nabla^2 \varphi + \frac{1}{\gamma}\nabla\gamma\left(\frac{\partial \mathbf{A}}{\partial t} + \nabla\varphi\right) = 0 . \tag{9.4.11}$$

In deriving Eq. (9.4.10), the Coulomb's gauge $\nabla \cdot \mathbf{A} = 0$ is considered. When the constitutive parameters are constants, the above equations reduce to

$$\nabla^2 \mathbf{A} = -\mu \mathbf{J} = -\mu(\mathbf{J}_s + \mathbf{J}_e) \tag{9.4.12}$$

$$\nabla^2 \varphi = 0 \tag{9.4.13}$$

where **J** is the total measurable current density including the impressed current density $\mathbf{J}_s$ and the induced current density $\mathbf{J}_e$

$$\mathbf{J}_s = \gamma \mathbf{E}_s = \gamma(-\nabla\varphi) \tag{9.4.14}$$

$$\mathbf{J}_e = \gamma \mathbf{E}_e = -j\omega\gamma \mathbf{A} . \tag{9.4.15}$$

In a 2-D case, $\nabla\varphi$ is a constant. Substitute Eq. (9.4.15) into Eq. (9.4.13), one obtains

$$(\nabla^2 + \beta^2)\mathbf{A} = -\mu \mathbf{J}_s \tag{9.4.16}$$

$$\beta^2 = -j\omega\mu\gamma . \tag{9.4.17}$$

Equation (9.4.16) is the nonhomogeneous Helmholtz equation in terms of the vector potential **A**. Solving the eddy current problem based on the Eq. (9.4.10) and (9.4.11) is called the $\mathbf{A} - \varphi$ method.

9.4.1.2 T-Ω formulations

Similar to the magnetic vector potential A, an electric vector potential $\mathbf{T}$ ($\mathbf{J} = \nabla \times \mathbf{T}$) and a scalar potential Ω ($\mathbf{H} = \mathbf{T} - \nabla\Omega$), as defined in Sect. 1.1.1, can be used to analyse eddy current problems. The divergence of **T** is defined as

$$\nabla \cdot \mathbf{T} = -\beta^2 \Omega \tag{9.4.18}$$

then the following equation is obtained

$$\nabla^2 \mathbf{T} + \beta^2 \mathbf{T} = -\nabla \times \mathbf{J}_s = \mathbf{S} . \qquad (9.4.19)$$

This is because

$$\nabla \times \nabla \times \mathbf{T} = \nabla \times \mathbf{J}$$
$$-\nabla^2 \mathbf{T} + \nabla(\nabla \cdot \mathbf{T}) = \nabla \times \mathbf{J}_s - j\omega\mu\gamma \mathbf{H}$$

then

$$(\nabla^2 + \beta^2)\mathbf{T} = -\nabla \times \mathbf{J}_s + \nabla(\nabla \cdot \mathbf{T} + \beta^2 \Omega) . \qquad (9.4.20)$$

Based on the definition of fundamental solution and Eq. (9.4.20), the following integral equation yields

$$CT_i = \int_\Omega S_i F d\Omega - \int_\Gamma \left(T_i \frac{\partial F}{\partial n} - F \frac{\partial T_i}{\partial n} \right) d\Gamma \qquad (9.4.21)$$

where T_i is one component of the $\mathbf{T}$, S_i is the correspondent component of the source.

As $\mathbf{B}$, $\mathbf{A}$, $\mathbf{T}$ are vectors, the solution of Eqs. (9.4.8), (9.4.12) and (9.4.21) is not easy. In a 2-D case, $A = A_z \mathbf{k}$ the vector partial differential equation is simplified to a scalar equation, yielding an easier solution. For a 2-D case, the $A - \varphi$ formulation consists of 3 equations, and the boundary conditions have several components. Therefore the choice of the method of formulations of eddy current problems is important [21]. A comparison of the CPU time required when using the $A - \varphi$ method and $\mathbf{T} - \Omega$ method for solving a 3-D eddy current problem is given in reference [22]. It shows when solving a given problem the $A - \varphi$ method is much more time consuming than the $\mathbf{T} - \Omega$ method. The other formulations such as the $\mathbf{R} - \psi$ formulation [23], the reduced and the total magnetic scalar potential formulations [24] are well known.

9.4.2 One-dimensional solution of an eddy current problem

Assume a long conductor with circular cross section as shown in Fig. 9.4.1(a). In consideration of the sinusoidal excitation and the circular symmetry, Eq. (9.4.8) reduces to

$$\frac{d^2 H}{dr^2} + \frac{1}{r} \frac{dH}{dr} - \left(\beta^2 + \frac{1}{r^2} \right) H = 0 \qquad (9.4.22)$$

where $\beta^2 = +j\omega\mu\gamma$, $H = H_\theta$. The solution of Eq. (9.4.22) is

$$H = AI_1(\beta r) + BK_1(\beta r) \qquad (9.4.23)$$

where $I_1(\beta r)$ and $K_1(\beta r)$ are the first order modified Bessel functions [25] of the first and the second kind, respectively. A and B are constants determined by the boundary conditions as follows.

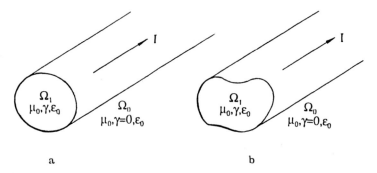

Fig. 9.4.1a, b. Current carrying conductor. **a** Circular cross section; **b** arbitrary cross-section

At $r = 0$, H must be finite, but the function $K_1(\beta r)$ tends to infinity as r approaches zero as shown in Fig. A.9.2, hence $B = 0$.
At $r = a$, $H = I/2\pi a$, hence

$$H = \frac{I}{2\pi a} \frac{I_1(\beta r)}{I_1(\beta a)} \tag{9.4.24}$$

consequently,

$$J = J_z = \frac{dH}{dr} + \frac{H}{r} = \frac{\beta I}{2\pi a} \frac{I_0(\beta r)}{I_1(\beta a)} \tag{9.4.25}$$

$$A = A_z = -\frac{\mu I}{2\pi a} \frac{I_0(\beta r)}{I_1(\beta a)} \tag{9.4.26}$$

where $I_0(\beta r)$ is the modified Bessel function of the first kind of zero order. $I_1(\beta r)$ is the modified Bessel function of the first kind of the first order. Equations (9.4.24) and (9.4.25) indicate that both the amplitude and the phase of H and J are changing along the radius.

9.4.3 BEM for solving eddy current problems

To solve eddy current problems with FEM, the mesh generation is complex as it depends on frequency. Transient cases are even more difficult since the eddy current distribution depends on the steepness of the transient. In this case, the mesh should be regenerated at each time step. However, for BEM, the mesh discretization of the boundary does not need to consider the influence of the eddy current distribution within the domain. Hence the thickness of the penetration depth can be arbitrary if BEM is used to solve eddy current problems. A general description for using the BEM to solve the eddy current problem is given in reference [26]. In this section a 2-D problem is used as an example.

A 2-dimensional current carrying conductor with an arbitrary cross-section as shown in Fig. 9.4.1(b) is considered. The vector potential **A** is chosen as the unknown variable, then

$$\nabla^2 A_{z1} + \beta^2 A_{z1} = -\mu J_s \quad \text{in } \Omega_1 . \tag{9.4.27}$$

$$\nabla^2 A_{z0} = 0 \quad \text{in } \Omega_0 . \tag{9.4.28}$$

Actually, the measurable current I is the known condition, and

$$I = \int_s J_s \, ds + \int_s J_e \, ds \tag{9.4.29}$$

where s is the area of the cross-section of the conductor. The current density J_s is uniformly distributed. Thus

$$J_s = \frac{1}{S}\left(I + j\omega\gamma \int_s A_{z1} \, ds\right). \tag{9.4.30}$$

Substitution of Eq. (9.4.30) into Eq. (9.4.16), an nonhomogeneous integral differential equation yields

$$\nabla^2 A_{z1} + \beta^2 A_{z1} + \frac{1}{S} j\omega\mu\gamma \int_s A_{z1} \, ds = -\frac{\mu I}{S} . \tag{9.4.31}$$

The constraint condition I is included in Eq. (9.4.31). In reference [27] this integral differential equation was solved by FEM. In this section it will be solved by BEM, which is especially powerful with high frequency.

To avoid discretization of the source area, Eq. (9.4.16) is transformed into Eq. (9.4.35) by the following method.

For simplification the subscript 'z' is ommitted let

$$A = A_e - A_s \tag{9.4.32}$$

$$J_s = -j\omega\gamma A_s . \tag{9.4.33}$$

As J_s is a constant it follows that A_s is a constant. Assuming the Helmholtz operator is a linear operator Eq. (9.4.27) is expanded to

$$(\nabla^2 - \beta^2) A_s + (\nabla^2 + \beta^2) A_e = -\mu J_s . \tag{9.4.34}$$

By substituting Eq. (9.4.33) into Eq. (9.4.34) and considering that A_s is constant, $\nabla^2 A_s = 0$, then in domain Ω_1 one obtains

$$-(\nabla^2 + \beta^2) A_e = 0 . \tag{9.4.35}$$

This is a homogeneous Helmholtz equation. By using the standard boundary element formulation, Eq. (9.4.35) yields

$$[\mathbf{H}_1]\{A_e\} - [\mathbf{G}_1]\left\{\frac{\partial A_e}{\partial n}\right\} = 0 \quad \text{in } \Omega_1 . \tag{9.4.36}$$

In the free space region, a similar equation is

$$-[\mathbf{H}_0]\{A_0\} + [\mathbf{G}_0]\left\{\frac{\partial A_0}{\partial n}\right\} = 0 \quad \text{in } \Omega_0 \tag{9.4.37}$$

9.4 Eddy current problems

By combining Eq. (9.4.36) and Eq. (9.4.37) by the interfacial boundary conditions

$$\begin{cases} A_1 = A_0 = A \qquad A = A_e - A_s \\ \dfrac{1}{\mu_1}\dfrac{\partial A}{\partial n} = \dfrac{1}{\mu_2}\dfrac{\partial A_0}{\partial n} \end{cases} \qquad (9.4.38)$$

and in addition to the known constrained boundary condition I is expressed by Ampere's law, i.e.

$$\oint H\,dl = I \qquad (9.4.39)$$

then the resulting system equation is

$$\begin{bmatrix} [H_1] & -[G_1] & hh \\ -[H_0] & [G_0] & 0 \\ 0 & L & 0 \end{bmatrix} \begin{Bmatrix} \{A\} \\ \{\partial A/\partial n\} \\ \{A_s\} \end{Bmatrix} = \begin{Bmatrix} 0 \\ 0 \\ -\mu I \end{Bmatrix} \qquad (9.4.40)$$

where the subscripts 1 and 0 correspond to the different areas Ω_1 and Ω_0, the column vectors $\{A\}, \{\partial A/\partial n\}, \{A_s\}$ are unknowns along the boundary and **hh** is a column vector. The element of **hh** is the summation of the element of $[H_1]$ in each row. The component $\{A_s\}$ is proportional to the impressed source voltage. The elements of the matrices $[H], [G]$ are

$$H_{ij}^{(k)} = \int_{\Gamma_j} \psi_k \frac{\partial F}{\partial n}\,d\Gamma \qquad (9.4.41)$$

$$G_{ij}^{(k)} = \int_{\Gamma_j} \psi_k F\,d\Gamma \qquad (9.4.42)$$

where ψ_k are shape functions of each element. For the constant element $\psi_k = 1$. F is the fundamental solution corresponding to a different area, as listed in Table 1.3.1. For a 2-D Helmholtz equation,

$$F(\mathbf{r}\cdot\mathbf{r}') = \frac{1}{4j} H_0^{(2)}(\beta|\mathbf{r}-\mathbf{r}'|) \qquad (9.4.43)$$

$H_0^{(2)}(\beta|\mathbf{r}-\mathbf{r}'|)$ is the Hankel function of the second kind of zero order (see Appendix A.9.1 and reference [28]).

The elements of the column matrix **L** in Eq. (9.4.40) is

$$l_j^{(k)} = \int_{\Gamma_j} \psi_k\,d\Gamma \qquad (9.4.44)$$

To solve Eq. (9.4.40), the boundary values of A and $\partial A/\partial n$ are obtained. In a 2-D case, $E = -j\omega A$ and $H = 1/\mu\,\partial A/\partial n$, then the power loss within the conductor may be easily calculated by using the Poynting theorem:

$$\mathbf{P} + j\mathbf{Q} = \oint_s (\mathbf{E}\times\mathbf{H})^* \cdot d\mathbf{S} \qquad (94.45)$$

where $\mathbf{H}^*$ is the conjugate of the vector $\mathbf{H}$.

Table 9.4.1. A comparison of the BEM to the analytical solution

r (mm)	0.5	1.0	1.5	2.0	2.5	3.0	3.5	4.0	4.5
$\Delta_{|A_z|}$ (%)	1.186	1.185	1.186	1.186	1.186	1.186	1.186	1.187	1.077
Δ_{ph} (%)	2.94	2.34	1.76	1.31	1.01	0.80	0.66	0.56	0.45

Example 9.4.1. Assume the radius of an infinitely long conducor is $R = 5$ mm, the conductivity of the material is 5.6×10^7 S/m, the frequency of the current is 1000 Hz ($R/\delta = 2.35$, δ is the penetration depth). While the circle of the conductor is divided into 24 constant elements, the relative errors of the magnitude and phase of the vector potential along a radius are given in Table 9.4.1. If the number of elements are increased to 36, then the maximum errors of the potential value and phase are decreased to 0.52% and 1.2%. This shows that if the number of the elements are enough the accuracy of the BEM is sufficient. This method can be used to a very high frequency. For example, if $R/\delta = 40.7$, the error of the power loss is 0.722%. However, it should be noted, that the size of the discretized elements must be related to the wave length of the electromagnetic field. Otherwise the results may be inaccurate.

The formulations and computer program for calculating the modified Bessel functions are given in [28, 29]. This method is also useful for solving eddy current problems in multiple conductors [30].

9.4.4 Surface impedance boundary conditions

If the penetration depth is sufficiently small and the radii of the conductors are much larger than the wave length, then the wave impedance condition can be used to reduce the size of the problem by 50%.

Consider a plane wave, the electric and magnetic field strength satisfies the impedance boundary condition which can be expressed

$$\mathbf{n} \times \mathbf{E} = Z_s \mathbf{n} \times (\mathbf{n} \times \mathbf{H}) \qquad (9.4.46)$$

$$Z_s = \left(\frac{\mu}{\varepsilon - (j\gamma/\mu)}\right)^{1/2} \approx (1+j)\omega\mu\delta \qquad (9.4.47)$$

$$\delta = \left(\frac{2}{\omega\mu\gamma}\right)^{1/2} \qquad (9.4.48)$$

where Z_s is the boundary impedance. Consider the same problem given in Sect. 9.4.3, by substitution of Eq. (9.4.46) into Eq. (9.4.40), the resulting equation is

$$\begin{bmatrix} [\mathbf{H}] - \frac{j\omega\mu}{Z_s}[\mathbf{G}] & \mathbf{hh} \\ \frac{j\omega\mu}{Z_s}\mathbf{L} & 0 \end{bmatrix} \begin{Bmatrix} A \\ A_s \end{Bmatrix} = \begin{Bmatrix} 0 \\ -\mu I \end{Bmatrix} \qquad (9.4.49)$$

where
$$[H] = [H_1] - [H_0] \tag{9.4.50}$$

In Eq. (9.4.49) only the component of the electric field is chosen as unknowns. Where the matrices $[H_1]$ and $[H_0]$ are the same as in Eq. (9.4.40). More detailed applications and discussions were shown in references [31] and [32]. In a high frequency range this condition is very useful in field computation.

9.5 Non-linear and time-dependent problems

9.5.1 BEM for non-linear problems

As the superposition principle is implicit when using BEM it cannot be generally used to solve non-linear problems. In electromagnetic fields, most of the non-linear problems are due to the non-linearity of the materials. If the non-linearity is not too strong, iterative procedure [33–35] may be successfully used.

The operator equation of a non-linear problem may be written as

$$\mathscr{L}u + \mathscr{N}u = f \tag{9.5.1}$$

where $\mathscr{L}$ and $\mathscr{N}$ represent linear and non-linear operators, respectively. If the non-linear term $\mathscr{N}u$ can be treated as a known function, then Eq. (9.5.1) is rearranged to

$$\mathscr{L}u = f - \mathscr{N}u. \tag{9.5.2}$$

Thus the LHS of Eq. (9.5.2) is a linear term. All the influence of the non-linear component is included in the RHS as a source term.

Assume $u = u^{(0)}$, then

$$\mathscr{L}u = f - \mathscr{N}u^{(0)}. \tag{9.5.3}$$

This equation is solved by conventional procedures and the first approximate solution $u^{(1)}$ is obtained. Repeat the procedure, by using the recurrent method until a certain criterion is satisfied. Finally, the solution of Eq. (9.5.1) is obtained. The main advantage of this method is that during each iteration, the matrix of H and G (the coefficient matrix of the potential and its normal derivative) of the LHS of Eq. (9.5.3) are unaltered and only the term $\mathscr{N}u^{(k)}$ has to be calculated.

Example 9.5.1. A practical cable shown in Fig. 9.5.1 is chosen as an example where the shield pipe is made out of ferromagnetic material. Assume the magnetic strength in the pipe is less than the value of saturation, calculate the magnetic flux density in the ferromagnetic pipe.

For non-linear permeability, Eq. (9.4.1) is written as

$$\nabla \times \left(\frac{1}{\mu} \nabla \times A \right) = J \tag{9.5.4}$$

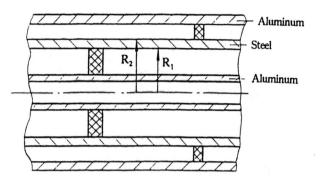

Fig. 9.5.1. The cross section of a power cable

or

$$\frac{1}{\mu}(\nabla \times \nabla \times A) - \nabla \times A \times \nabla\left(\frac{1}{\mu}\right) = J. \tag{9.5.5}$$

By using the same methods as in the Sect. 9.4.3, one obtains

$$(\nabla^2 + \beta^2)A = -\nabla \times A \times \nabla\left(\frac{1}{\mu}\right). \tag{9.5.6}$$

If μ is non-linear, β is the function of A. Let

$$\beta^2 = \beta_0^2 + \tilde{\beta}^2 \tag{9.5.7}$$

Assuming all the variables are sinusoidal function, then

$$\beta_0^2 = -j\omega\mu_0\gamma \tag{9.5.8}$$

$$\tilde{\beta}^2 = -j\omega\tilde{\mu}\gamma = -j\omega\gamma(\mu - \mu_0) \tag{9.5.9}$$

where μ_0 is assumed as an initial value of the permeability. By substituting Eq. (9.5.7) into Eq. (9.5.6) and expand the RHS of Eq. (9.5.6) in cylindrical coordinates, the result is

$$\nabla^2 A + \beta_0^2 A = -j\omega\gamma(\mu - \mu_0)A - \frac{1}{\mu}\left(\frac{\partial \mu}{\partial r}\frac{\partial A}{\partial r} + \frac{1}{r^2}\frac{\partial \mu}{\partial \alpha}\frac{\partial A}{\partial \alpha}\right). \tag{9.5.10}$$

The RHS of Eq. (9.5.10) is considered to be a source term. It varies with the changing of the vector potential A. The corresponding boundary integral equation of Eq. (9.5.10) is

$$\frac{1}{2}A + \int_\Gamma A \frac{\partial F}{\partial n} d\Gamma - \int_\Gamma F \frac{\partial A}{\partial n} d\Gamma + \int_\Omega PF d\Omega = 0 \tag{9.5.11}$$

where

$$P = j\omega\gamma(\mu - \mu_0)A + \frac{1}{\mu}\left(\frac{\partial \mu}{\partial r}\frac{\partial A}{\partial r} + \frac{1}{r^2}\frac{\partial \mu}{\partial \alpha}\frac{\partial A}{\partial \alpha}\right). \tag{9.5.12}$$

9.5 Non-linear and time-dependent problems

In Eq. (9.5.11), F is the fundamental solution of a linear Helmholtz equation. By using boundary conditions of Eq. (9.4.41) and writing Eq. (9.5.11) in an iterative form, one obtains

$$\frac{1}{2}(A^{K+1}) + \sum_{j=1}^{N} \int_{\Gamma_j} A^{K+1} \frac{\partial F}{\partial n} d\Gamma - \sum_{j=1}^{N} \int_{\Gamma_j} \mu_r^K \left(\frac{\partial A}{\partial n}\right)^{K+1} F d\Gamma$$

$$= - \sum_{i=1}^{N_e} \left(\sum_{q=1}^{m} W_q (P_i^K F)_q \right) S_{ne} = b^k \qquad (9.5.13)$$

where N is the number of the boundary elements, N_e is the number of the discretized elements of the domain, m is the number of the integrate points of each element, w_q are weighting coefficients of integration and S_{ne} is the area of element. The superscript k is the time of iterations. The iterative steps of Eq. (9.5.13) are

(1) Start with a given μ_0.
(2) Calculate β_0, P^0, b^k.
(3) Solve Eq. (9.5.13); obtain A^k, $\left(\dfrac{\partial A}{\partial n}\right)^k$.
(4) Calculate the value of A_i^k, B_{xi}^k, B_{yi}^k inside the conductor, and the variation rate $\left(\dfrac{\partial A}{\partial r}\right)^k$, $\left(\dfrac{\partial A}{\partial \alpha}\right)^k$.
(5) Determine the value of $\bar{\mu}$ from the $B - \mu_r$ curve, then calculate $\dfrac{\partial \mu}{\partial \alpha}$, $\dfrac{\partial \mu}{\partial r}$.
(6) If $\max|\mu^{k+1} - \mu^k| < \varepsilon$, then stop the iteration, otherwise

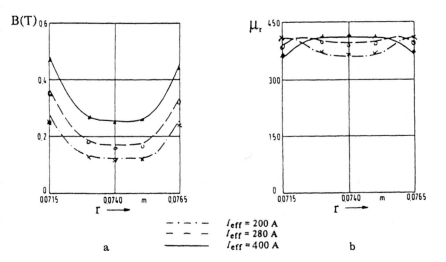

Fig. 9.5.2a, b. The distribution of B and μ_r in the pipe

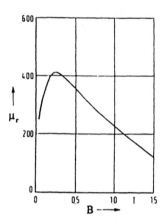

Fig. 9.5.3. The μ_r – B curve of the steel

(7) Calculate $\dfrac{\partial \mu}{\partial r}$, $\dfrac{\partial \mu}{\partial \alpha}$, P^k, b^k.

(8) Return to step (3) until the criterion, $\max|\mu_i^{k+1} - \mu_i^k| < \varepsilon$, is satisfied. During iteration, in order to accelerate the convergence, the permeability is approximated by:

$$\mu^{k+1} = \mu^k + \alpha(\bar{\mu} - \mu^k) \quad 0 < \alpha < 1 \tag{9.5.14}$$

where $\bar{\mu}$ is the value obtained from the $B - \mu_r$ curve at the kth iteration.

The distribution of magnetic flux density and the $\mu_r - r$ curve in the pipe are shown in Fig. 9.5.2(a) and (b), respectively. The smooth curves is obtained in reference [36] by using FEM, the discretized points are calculated using this method. The $\mu_r - B$ curve of the steel is given in Fig. 9.5.3.

9.5.2 Time-dependent problems

For time-dependent problems, incremental sequences are used. This means that the problem is solved in subsequent time intervals by conventional BEM. For example, consider a diffusion equation

$$\nabla^2 u = \frac{\partial u}{\partial t}. \tag{9.5.15}$$

At any time, this equation can be approximated as

$$\nabla^2(u_t) = \frac{u_t - u_0}{\Delta t}$$

or

$$\nabla^2(u_t) - \frac{u_t}{\Delta t} = -\frac{u_0}{\Delta t}. \tag{9.5.16}$$

The boundary integral equation of Eq. (9.5.16) is

$$C_i u_t = \int_\Gamma \left(\frac{\partial F}{\partial n}\right)_t u_t \, d\Gamma - \int_\Gamma F_t \left(\frac{\partial u}{\partial n}\right)_t d\Gamma + \frac{1}{\Delta t}\int_\Omega u_0 F_t \, d\Omega \,. \qquad (9.5.17)$$

Assuming an initial value u_0 within the domain and the boundary value

$$u = \bar{u}_t \quad \text{on } \Gamma_1$$

and

$$\frac{\partial u}{\partial n} = \bar{q}_t \quad \text{on } \Gamma_2$$

then the solution of Eq. (9.5.17) is the result of u at the time t. Let this be a new initial value of Eq. (9.5.17), and repeat the same procedure until a stable result is obtained. In Eq. (9.5.17), the fundamental solution is obtained from the following equation, i.e.

$$\nabla^2 F_t - \frac{1}{\Delta t} F_t + \delta_i = 0 \,. \qquad (9.5.18)$$

9.6 Summary

BEM is based on the boundary integral equation and the fundamental solution of the governing equation of the problem. The boundary integral equation is developed from the principle of weighted residuals, Green's theorem and the variational principle. It is useful for solving open boundary and three-dimensional problems.

A typical discretized equation of BEM is in the form of

$$[H]U = [G]Q$$

where the components of U are the nodal values of the potential function on the boundary. The components of Q are the normal derivatives of the potential function of the boundary nodes. The coefficients of matrices H and G are integrated from the fundamental solution of the governing equation and the shape function of the discretization, i.e.

$$H_{ij}^{(k)} = \int_{\Gamma_j} \psi_k \frac{\partial F}{\partial n} d\Gamma \qquad G_{ij}^{(k)} = \int_{\Gamma_j} \psi_k F d\Gamma$$

Because the discretization is carried out only on the boundary, the size of the matrix is much smaller than the one obtained using differential methods. The pre- and post-data processing is simpler than with FEM. It is useful for solving 3-D problems and eddy current problems in high frequency and transient cases.

The corresponding matrix of the discretization equation is usually unsymmetrical, and full, Gauss's elimination or Cholesky's decomposition methods are used for solving the matrix equation.

During the procedure of approximation, the principle of superposition is implied, so this method is usually not suitable for non-linear problems. If the material contained in the problem domain is non-linear, the iterative method is more successful for such problems.

References

1. C.A. Brebbia, The Boundary Element Method in Engineering Practice, *Engineering Analysis*, 1(1), 3–12, 1984
2. C.A. Brebbia, *Boundary Element Method*. Pentech Press, 1978
3. C.A. Brebbia, J. Telles, L. Wrobel, *Boundary Element Methods – Theory and Applications in Engineering*. Springer-Verlag, 1984
4. K.R. Shao, K.D. Zhou, Transient Response in Slot-Embedded Conductor for Voltage Source Solved by Boundary Element Method, *IEEE Trans. on Mag.* 21(6), 2257–2259, 1985
5. I.D. Mayergozy, M.V.K. Chari, A. Konrad, Boundary Galerkin's Method for Three-Dimensional Finite Element Electromagnetic Field Computation, *IEEE Trans, Mag.* 19(6), 2333–2336, 1983
6. J.P. Adriaens, et al., Computation of Electric Field in HV Devices Using Boundary Element Method, *6th ISH*, 33.12, 1987
7. T.H. Fawzi, Two-Dimensional and Quasi-Two-Dimensional Induction Problems, Ph.D. Thesis, University of Toronto, 1973
8. M.H. Lean, Electromagnetic Field Solution with the Boundary Element Method, Ph.D. Thesis, Winnipeg, Manitoba, Canada, 1981
9. M.T. Ahmed, Application of the Boundary Element Method and the Impedance Boundary Condition in T-W and 2-D Eddy Current Poblems, Ph.D. Thesis, University of Toronto, 1986
10. P.K. Banerjee, R. Butterfeld, *Boundary Element Method in Engineering Science*, McGraw-Hill, 1981
11. C.A. Brebbia, J. Dominguej, *Boundary Elements, An Introduction Course*, McGraw-Hill, 1989
12. C.A. Brebbia (Ed), Topics in Boundary Element Research, Vol. 6, *Electromagnetic Applications*, Springer-Verlag, 1989
13. M.H. Lean and, A. Wexler, Accurate Field Computation with Boundary Element Method, *IEEE Trans. on Mag.* 18(2), 331–335, 1982
14. Kieth D. Paulsen, Calculation of Interior Values by the BEM, *Communication of Applied Numerical Methods*, 5(1), 7–14, 1989
15. H. Tim & Tullberg, More on Boundary Elements for 3-Dimensional Potential Problems. In *Boundary Element Methods, Proc. of 7th Int. Conf.* 1985. C.A. Brebbia (Ed.), Springer-Verlag, 2-13-2-14
16. L. Li, J. Luomi, Three-Dimensional Magnetostatic Field Analysis Using Vector Variables. In *Topics in Boundary Element Research*, Vol. 6. C.A. Brebbia (Ed.), Springer-Verlag, 1989
17. H.B. Dwight, Skin Effect in Tublar and Flat Conductors. *Trans. AIEE*, 37, 1379–1403, 1918
18. R.L. Stoll, *The Analysis of Eddy Currents*. Clarendon Press, Oxford, 1974
19. J.A. Tegopoulos, E.E. Kriezis, *Eddy Current in Linear Conduction Media*. Elsevier, 1985
20. A. Krawzyk, J. Turowski, Recent Development in Eddy Current Analysis, *IEEE Trans. on Mag.*, 23(5), 3032–3037, 1987
21. C.J. Carpenter, Comparison of Alternative Formulations of Three-Dimensional Magnetic Field and Eddy Current Problem at Power Frequency, *Proc. IEE*, 124(11), 1026–1034, 1977
22. T. Nakata, N. Takahashi, K. Fujwara, Y. Shiraki, Comparison of Different Finite Element for 3-D Eddy Current Analysis, *IEEE Trans. on Mag.*, 1989
23. C.S. Biddlecombe, E.A. Heighway, J. Simkin, C.W. Trowbridge, Methods for Eddy Current Computation in Three Dimensions, *IEEE Trans. on Mag.*, 18(2), 492–497, 1982

24. J. Simkin, C.W. Trowbridge, On the Use of the Total Scalar Potential in the Numerical Solution of Field Problem in Electromagnetics, *Int. J. Num. Meth. of Eng.*, **14**, 23–44, 1979
25. N.W. McLachlan, *Bessel Functions for Engineers*, Clarendon Press, Oxford, 1955
26. S. Kalaichelvan, J.D. Lavers, Boundary Element Methods for Eddy Current Problems. In *Topics of Boundary Element Research*, Vol. 6, Electromagnetic Applications, C.A. Brebbia (Ed.), Springer-Verlag, 1989
27. A. Konrad, Integro-differential Finite Element Formulation of Two-Dimensional Steady-State Skin Effect, *IEEE Trans. on Mag.*, **18**(1), 284–292, 1981
28. M. Abramowitz, I.A. Stegun (Ed.), *Handbook of Mathematical Functions*. Dover Publications, New York, 1970
29. Keith H. Burrell, Evaluation of the Modified Bessel Functions $K_0(z)$ and $K_1(z)$ for Complex Argument [417], *Communications of the ACM*, **17**(9), 524–526, 1974
30. Zhou Peibai, J.D. Lavers, High Frequency Losses in Multi-Turns Coils Using a Boundary Element Method, *IEEE Trans. on Mag.*, **22**(5), 1060–1062, 1986
31. T.H. Fawzi, M. Taher Ahmed, P.E. Burke, On the Use of the Impedance Boundary Conditions in Eddy Current Problem, *IEEE Trans. on Mag.*, **21**, 1835–1840, 1985
32. Zhou Peibai, J.D. Lavers, On the Use of the Surface Impedence Boundary Condition in BEM for Calculating High Frequency Power Losses, Electromagnetic Fields in Electrical Engineering, *Proc. of Beijing International Symposium BISEF 88*, pp 469–472, 1988
33. I. Mayergoyz, Iteration Methods for the Calculation of Steady Magnetic Fields in Nonlinear Ferromagnetic Media, *COMPEL* **1**(2), pp. 89–110, 1982
34. M.H. Lean, D.S. Bloomberg, Nonlinear Boundary Element Method for Two-Dimensional Magnetostatics, *J. Appl. Phys.* **55**(6), 2195–2197, 1984
35. K.R. Shao, K.D. Zhou, The Iterative Boundary Element Method for Nonlinear Electromagnetic Field Calculations, *IEEE Trans. on Mag.*, **24**(1), 150–152, 1988
36. Detmar Arlt, Berechnung der Wirbelstrom-und Hystereseverluste in beliebig geformten einphasigen Kapselungen, *ETZ Archiv Bd.* 9, H.9, pp 275–281, 1987

Appendix 9.1   Bessel functions

Bessel functions, $J_\nu(z)$, $J_{-\nu}(z)$, $Y_\nu(z)$ are the solutions of a particular differential equation (called Bessel equation) in the form of

$$z^2 \frac{d^2 f}{dz^2} + z \frac{df}{dz} + (\lambda^2 z^2 - \nu^2) f = 0 \qquad (A.9.1)$$

which was published by the German astronomer F.W. Bessel in 1826.

$J_\nu(z)$, $J_{-\nu}(z)$ are Bessel functions of the first kind, $Y_\nu(z)$ is a Bessel function of the second kind. They are expressed by the infinite series

$$J_\nu(\lambda z) = \sum_{k=0}^{\infty} (-1)^k \frac{1}{k!\,\Gamma(\nu + k + 1)} \left(\frac{\lambda z}{2}\right)^{\nu + 2k} \qquad (A.9.2)$$

$$J_{-\nu}(\lambda z) = \sum_{k=0}^{\infty} (-1)^k \frac{1}{k!\,\Gamma(-\nu + k + 1)} \left(\frac{\lambda z}{2}\right)^{-\nu + 2k} \qquad (A.9.3)$$

$$Y_\nu(\lambda z) = \frac{J_\nu(\lambda z) \cos(\nu\pi) - J_{-\nu}(\lambda z)}{\sin(\nu\pi)} \qquad (A.9.4)$$

where Γ is the Gamma function, the subscript ν may be integer or noninteger. $J_\nu(z)$, $J_{-\nu}(z)$ are linearly independent except when ν is an integer. If $\lambda z = x$ is

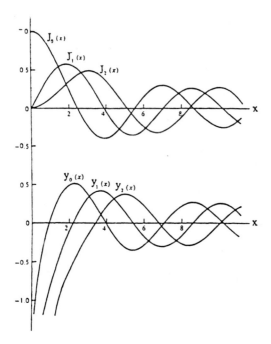

Fig. A.9.1. The Bessel functions of $J_n(x)$, $Y_n(x)$, $n = 0, 1, 2$

a real variable, and $v = n$ ($n = 0, 1, 2, \ldots$), then $J_{-n} = (-1)^n J_n$. $J_n(x)$ and $Y_n(x)$ are damped oscillation functions, as shown in Fig. A.9.1.

The linear combinations of $J_v(z)$ and $Y_v(z)$ are Hankel functions of the first and second kind respectively, i.e.

$$H_v^{(1)}(z) = J_v(z) + iY_v(z) \tag{A.9.5}$$

$$H_v^{(2)}(z) = J_v(z) - iY_v(z). \tag{A.9.6}$$

Hankel functions are Bessel functions of the third kind. If $v = n = 0$, then

$$H_0^{(1)}(z) = J_0(z) = iY_0(z) \tag{A.9.7}$$

$$H_0^{(2)}(z) = J_0(z) - iY_0(z) \tag{A.9.8}$$

where $H_0^{(1)}(z)$ and $H_0^{(2)}(z)$ are Hankel functions of the first and second kind of zero order, respectively.

Modified Bessel function

It is often desirable in applications to give the solution in real terms instead of in a complex form. To do so, the Bessel function must be modified. If $x = -iz$, then

Appendix 9.1 Bessel functions

Eq. (A.9.1) becomes

$$x^2 \frac{d^2 f}{dx^2} + x \frac{df}{dx} - (\lambda^2 x^2 + v^2) f = 0 . \tag{A.9.9}$$

This is a modified Bessel equation. The solutions of a modified Bessel equation are modified Bessel functions i.e.

$$I_v(x) = i^{-v} J_v(ix) = \sum_{k=0}^{\infty} \frac{1}{k! \Gamma(v+k+1)} \left(\frac{x}{2}\right)^{v+2k} \tag{A.9.10}$$

$$I_{-v}(x) = i^v J_{-v}(ix) = \sum_{k=0}^{\infty} \frac{1}{k! \Gamma(-v+k+1)} \left(\frac{x}{2}\right)^{-v+2k} \tag{A.9.11}$$

and

$$K_v(x) = \frac{\pi}{2} \left(\frac{I_{-v}(x) - I_v(x)}{\sin v\pi} \right) . \tag{A.9.12}$$

$I_v(x)$ and $K_v(x)$ are linearly independent, hence the complete solution of Eq. (A.9.9) is

$$f = A I_v(x) + B K_v(x) . \tag{A.9.13}$$

$I_{\pm v}(x)$ and $K_v(x)$ are modified Bessel functions of the first and second kind, respectively. If $v = n = 0, 1$, the functions of $I_0(x)$, $I_1(x)$, $K_0(x)$ and $K_1(x)$ are given in Fig. A.9.2.

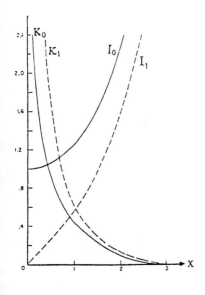

Fig. A.9.2. The modified Bessel functions of $I_0(x)$, $I_1(x)$, $K_0(x)$, $K_1(x)$

ber, bei, ker and kei functions

In solving some electrical problems, ber(x), bei(x) have been introduced by Thomson. These functions are the real and the image part of Bessel functions i.e.,

$$\text{ber}_\nu(x) + i\text{bei}_\nu(x) = I_\nu(i^{1/2}x) = J_\nu(i^{1/2}xi) \tag{A.9.14}$$

$$\begin{aligned}\text{ker}_\nu(x) \pm i\text{kei}_\nu(x) &= e^{\mp \nu i(\pi/2)} K_\nu(xe^{\pm i\pi/4}) \\ &= \pm \tfrac{1}{2}\pi i H_\nu^{(1)}(xe^{\pm 3\pi i/4}) \\ &= \mp \tfrac{1}{2}\pi i e^{-\nu\pi i} H_\nu^{(2)}(xe^{\mp \pi i/4}) .\end{aligned} \tag{A.9.15}$$

The fundamental solution of a Helmholtz equation is:

$$F(\mathbf{r} \cdot \mathbf{r}') = \frac{1}{4i} H_0^{(2)}(\beta|\mathbf{r} - \mathbf{r}'|) = \frac{1}{2\pi} K_0(\beta|\mathbf{r} - \mathbf{r}'|) \tag{A.9.16}$$

where

$$\beta = \sqrt{\omega\mu\varepsilon}\, e^{\pi i/4} \tag{A.9.17}$$

hence

$$\frac{1}{2\pi} K_0(xe^{-\pi i/4}) = \frac{1}{4j} H_0^{(2)}(xe^{-\pi i/4}) = \frac{1}{2\pi}(\text{ker}_0 x + i\text{kei}_0 x) . \tag{A.9.18}$$

The computation programs for calculating ber, bei, ker, kei are given in reference [29].

Chapter 10

Moment Methods

10.1 Introduction

As outlined in Chapter 2, the method of moments is a generalized method based on the principle of weighted residuals. It covers the many specific methods discussed such as the charge simulation method, the surface charge simulation method, boundary element method and even the finite element method which is regarded as one of the special cases of the method of moments. The name 'moment' is understood here as the product of an appropriate weighting function with an approximate solution. Any method whereby an operator equation is reduced to a matrix equation can be interpreted as a method of moments. It is also considered as the unified treatment of a matrix method (R.F. Harrington [1]).

Following Harrington, the basic principle of the moment method is to assume an approximate function

$$u = \sum_n \alpha_n \psi_n \qquad (10.1.1)$$

to replace the unknown function of the operator equation

$$\mathscr{L}u = f. \qquad (10.1.2)$$

In Eq. (10.1.1), α_n are unknown constants and ψ_n are basis functions or expansion functions. Because ψ_n is assumed to be a complete sequence of linearly independent functions, thus $\sum_n \alpha_n \psi_n$ approximates to the actual solution only if $n \to \infty$.

Substituting Eq. (10.1.1) into (10.1.2), and using the *linearity* of $\mathscr{L}$, yields:

$$\sum_n \alpha_n \mathscr{L}\psi_n = f. \qquad (10.1.3)$$

Taking the inner product of Eq. (10.1.3) with a weighting function W_m, one obtains:

$$\sum_n \alpha_n \langle \mathscr{L}\psi_n, W_m \rangle = \langle f, W_m \rangle \quad m = 1, \ldots, N. \qquad (10.1.4)$$

It can be written in a matrix form, i.e.

$$\mathbf{A}\{\alpha\} = \mathbf{B} \qquad (10.1.5)$$

where

$$A = \begin{bmatrix} \langle \mathscr{L}\psi_1, W_1\rangle \langle \mathscr{L}\psi_2, W_1\rangle \ldots \langle \mathscr{L}\psi_n, W_1\rangle \\ \vdots \qquad \vdots \qquad \qquad \vdots \\ \langle \mathscr{L}\psi_1, W_N\rangle \langle \mathscr{L}\psi_2, W_N\rangle \ldots \langle \mathscr{L}\psi_n, W_N\rangle \end{bmatrix} \quad (10.1.6)$$

$$B = \begin{Bmatrix} \langle f, W_1\rangle \\ \vdots \\ \langle f, W_N\rangle \end{Bmatrix} \quad (10.1.7)$$

$\{\mathbf{x}\}$ is a column vector which consists of the unknown parameters of the approximate solution.

The operator of Eq. (10.1.2) may be a differential or integral operator. W_m are linearly independent functions. In choosing the type of basis function, if the properties of the basis functions coincide with the properties of the real solution, then the approximate solution will quickly converge. In other words, a few terms of the basis function ψ_n is sufficient to approximate the real function. Other aspects to be considered in choosing the basis and weighting functions are:

(1) the accuracy of solution desired,
(2) the ease of evaluating the matrix elements,
(3) the size of the matrix, and
(4) the realization of a well-conditioned matrix **A**.

The procedures of the moment method are

(1) Assume an approximate function to replace the unknown function in the operator equation.
(2) Select a suitable function as a weighting function and construct the inner product to the operator equation.
(3) Evaluate the integrals of the inner product and form the matrix equation.
(4) Solve the matrix equation to obtain an approximate solution.

Example 10.1. Solve the problem as shown in Eq. (10.1.8)

$$\begin{cases} \dfrac{d^2 u}{dx^2} = 1 + 2x^2 & 0 \leqslant x \leqslant 1 \\ u(0) = u(1) = 0 \end{cases} \quad (10.1.8)$$

Solution. Let $u = \sum_n \alpha_n \psi_n$, because the exact solution of Eq. (10.1.8) is the linear combination of power functions, hence

$$\psi_n(x - x^{n+1}) \quad (10.1.9)$$

is chosen as the basis function. If the weighting function W_m is the same as the basis function, i.e.

$$W_m = (x - x^{m+1}) \quad (10.1.10)$$

10.1 Introduction

therefore

$$a_{mn} = \langle \mathscr{L}\psi_n, W_m \rangle = \int_0^1 (x - x^{m+1})\left[\frac{d^2}{dx^2}(x - x^{n+1})\right]dx$$

$$= \int_0^1 n(n+1)[x^n - x^{m+n}]dx = n(n+1)\left[\frac{x^{n+1}}{n+1}\bigg|_0^1 - \frac{x^{m+n+1}}{m+n+1}\bigg|_0^1\right]$$

$$= \frac{mn}{m+n+1} \tag{10.1.11}$$

$$\mathbf{B} = \langle f, W_m \rangle = \int_0^1 (1 + 2x^2)(x - x^{m+1})dx = \frac{m(m+3)}{(m+2)(m+4)}$$

$$(m = n).$$

If $n = 3$, the matrix equation and the solution are

$$\begin{bmatrix} 1/3 & 1/2 & 3/5 \\ 1/2 & 4/5 & 1 \\ 3/5 & 1 & 9/7 \end{bmatrix} \begin{Bmatrix} \alpha_1 \\ \alpha_2 \\ \alpha_3 \end{Bmatrix} = \begin{Bmatrix} 4/15 \\ 5/12 \\ 18/35 \end{Bmatrix} \tag{10.1.12}$$

and

$$\alpha = \begin{Bmatrix} 1/2 \\ 0 \\ 1/6 \end{Bmatrix} \tag{10.1.13}$$

Then

$$u = \tfrac{1}{2}(x - x^2) + \tfrac{1}{6}(x - x^4) = \tfrac{2}{3}x - \tfrac{1}{2}x^2 - \tfrac{1}{6}x^4. \tag{10.1.14}$$

This solution is exactly the same as the closed form solution.

In Example 10.1, the term 'moment' can be understood as a product of the approximate function and of the moment x, x^2, x^3 where

$$W = \beta_1\psi_1 + \beta_2\psi_2 + \ldots \tag{10.1.15}$$

when $n = 1$, $W = x - x^2$, then $\psi_1 = x$, $\psi_2 = x^2$, and so on.

Recall the basic idea of the method using weighted residuals introduced in Sect. 2.3.1. If the boundary conditions are exactly satisfied, then the average error principle remains

$$\int \varepsilon W d\Omega = 0 \tag{10.1.16}$$

In this case, the error is distributed in proportion to the weighting function. Consequently, the method of moments consists of taking moments of the weighting function and the error function. *It displays them while the weighting function equals the approximate function, moment methods are equivalent to the Rayleigh Ritz method* (see Sect. 5.4.1).

The method of moments is also useful for solving integral equations. For instance, a Fredholm integral equation of the first kind is

$$\int_a^b K(x, x') f(x) \, dx = g(x') \tag{10.1.17}$$

where $K(x, x')$ is the kernel of the integral equation and $g(x')$ is a known function. Assume $f(x) = \sum_n \alpha_n \psi_n(x)$ and substitute it into Eqs. (10.1.17) and (10.1.6), the element coefficients of matrix **A** are evaluated by double integrations, i.e.

$$a_{mn} = \int_a^b \int_a^b K(x, x') \psi_n(x) W_m(x) \, dx \, dx' \tag{10.1.18}$$

It is obvious more computation time is needed.

A different selection of weighting functions, basis functions and applications of the moment method to solve different problems are introduced in following sections.

10.2 Basis functions and weighting functions

The choice of the basis and weighting function are important to obtain solutions using the method of moments. They influence the accuracy of the solution and the computation time, even the success of the method. As shown in Example 10.1, the choice of the basis function is dependent on the property of the solution sought. In this section a different choice of using weighting function is illustrated by the same example. Usually the basis functions are divided into a global and subregion basis function. *The global basis function is defined in the whole region of the operator space where it cannot be zero. The sub-region basis function is defined in the whole region and is nonzero only over a small sub-region of the solution domain.*

The most commonly used global basis functions are

Fourier series: $\quad u(x) = \alpha_1 \cos(\pi x/2) + \alpha_2 \cos(3\pi x/2) \quad$ (10.2.1)
$\qquad\qquad\qquad\quad + \alpha_3 \cos(5\pi x/2) + \cdots$

Power series: $\quad u(x) = \alpha_1 + \alpha_2 x^2 + \alpha_3 x^4 + \cdots \quad$ (10.2.2)

Chebyshev polynomial: $u(x) = \alpha_1 T_0(x) + \alpha_2 T_2(x) + \alpha_3 T_4(x) + \cdots \quad$ (10.2.3)

Legendre polynomial $\quad u(x) = \alpha_1 P_0(x) + \alpha_2 P_2(x) + \alpha_3 P_4(x) + \cdots \quad$ (10.2.4)

The commonly used sub-region basis functions are

Pulse function: $\quad P(x - x_j) = \begin{cases} 1 & |x - x_j| < \dfrac{1}{2(N+1)} \\ 0 & |x - x_j| > \dfrac{1}{2(N+1)} \end{cases} \quad$ (10.2.5)

10.2 Basis functions and weighting functions

Triangular function:
$$T(x) = \begin{cases} 1 - |x|(N+1) & |x| < \dfrac{1}{N+1} \\ 0 & |x| > \dfrac{1}{N+1} \end{cases} \quad (10.2.6)$$

Piecewise sinusoidal function:
$$S(x) = \begin{cases} \dfrac{x_j \sin K(x_{j+1} - x) + x_{j+1} \sin K(x - x_j)}{\sin K \Delta x_j} & x_j < x < x_{j+1} \\ 0 \end{cases} \quad (10.2.7)$$

Quadratic interpolation function:
$$Q(x) = \begin{cases} a_j + b_j(x - x_j) + c_j(x - x_j)^2 & x_j < x < x_{j+1}, \\ 0 \end{cases} \quad (10.2.8)$$

Spline interpolation function:
$$\Omega_2(t) = \begin{cases} \tfrac{1}{2}(t + 3/2)^2 & -3/2 < t < -1/2 \\ -t^2 + 3/4 & -1/2 < t < 1/2 \\ \tfrac{1}{2}(t - 3/2)^2 & 1/2 < t < 3/2 \end{cases} \quad (10.2.9)$$

In Eqs. (10.2.5) and (10.2.6), N is the number of subdivisions in the domain. In Eq. (10.2.9), $t = x/\Delta$, Δ is the length of each sub-element.

The above functions are shown in Fig. 10.2.1. A linear combination of the triangular functions of the form
$$\psi = \sum_n \alpha_n T(x - x_n) \quad (10.2.10)$$
gives a piecewise linear approximation function, as shown in Fig. 10.2.1(c).

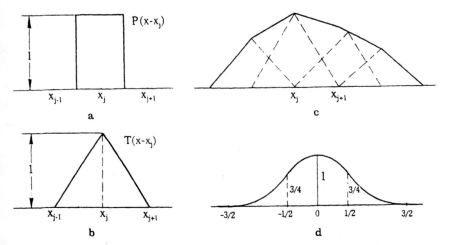

Fig. 10.2.1a–d. Subregion basis functions. **a** pulse function; **b** triangular function; **c** piecewise linear function; **d** spline function

The usual choice of basis and weighting functions is listed in Table 10.2.1.

Table 10.2.1. Collocation between the basis and weighting functions

Basis functions	Weighting functions
Quadratic or spline function	Dirac delta function
Triangular function	Rectangular pulse function
Rectangular pulse function	Triangular function
Pulse function	Pulse function
Dirac delta function	Quadratic or spline function

10.2.1 Galerkin's methods

If the weighting function is identical to the basis function (e.g. Example 10.1) then the resultant moment method is called Galerkin's method. It is similar to the Rayleigh-Ritz method which has been introduced in Sect. 5.4.1. *In Galerkin's method, if the operator is self-adjoint, the matrix* **A** *is symmetrical* (Eq. (10.1.12)). Consequently the matrix equation is may be easily solved. If the order of the matrix equation is not high, quick convergence and good accuracy may be expected. The only disadvantage is that it takes a longer time to calculate the elements of the matrix, especially for the integral operator.

10.2.2 Point matching method

In order to simplify the evaluation of the integration in calculating the coefficients of matrix **A**, *the domain is presented by a set of discrete points and the approximate solution is forced to satisfy the operator equation only at these discrete points.* Hence, it is called a point matching method. In this method, the Dirac delta function is chosen as the weighting function. The procedures of point matching are illustrated in Example 10.2.1.

Example 10.2.1. Use the point matching method to solve Eq. (10.1.8)

Let
$$u = \sum_{n=1}^{N} \alpha_n(x - x^{n+1}) \tag{10.2.11}$$
and
$$W_m = \delta(x - x_m) \tag{10.2.12}$$
then
$$\frac{d^2 u}{dx^2} = \sum_{n=1}^{N} \alpha_n \left[\frac{d^2}{dx^2}(x - x^{n+1}) \right] = 1 + 2x^2.$$

10.2 Basis functions and weighting functions

Divide the area by N points and take the matching points $x_m = m/(N + 1)$ ($m = 1, 2, \ldots, N$). They are equi-distance within the interval of $0 \leqslant x \leqslant 1$, as shown in Fig. 10.2.2. Calculate the coefficients

$$a_{mn} = \int_0^1 \delta(x - x_m) \frac{d^2}{dx^2}(x - x^{n+1}) dx = n(n + 1)\left(\frac{m}{N + 1}\right)^{n-1}$$

$$b_m = 1 + 2\left(\frac{m}{N + 1}\right)^2$$

If $N = 3$, then

$$A = \begin{bmatrix} 2 & 2 & 3/4 \\ 2 & 4 & 3 \\ 2 & 9/2 & 27/4 \end{bmatrix}, \quad B = \begin{Bmatrix} 9/8 \\ 3/2 \\ 17/8 \end{Bmatrix}, \quad \alpha = \begin{Bmatrix} 1/2 \\ 0 \\ 1/4 \end{Bmatrix}$$

The result is

$$u = \tfrac{1}{2}(x - x^2) + \tfrac{1}{6}(x - x^4) = \tfrac{2}{3}x - \tfrac{1}{2}x^2 - \tfrac{1}{6}x^4.$$

It is identical to the solution obtained in Example 10.1.

It is concluded that because the delta function is chosen as the weighting function, the calculation of the coefficients of the matrix in using the point matching method is simpler than Galerkin's method. However, the matrix A is no longer symmetrical as it was when using Galerkin's method. Even though the matrix A of the point matching method and Galerkin's method are different in value they yield the same solution. The only difference is that they have different speeds of convergence and different computing time requirements in evaluating the matrices. Using the point matching method, the accuracy and the convergence are dependent upon the number and the position of the matching points. The charge simulation method described in Chap. 7 is a special case of the point matching method.

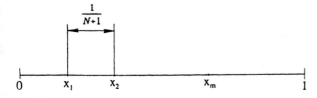

Fig. 10.2.2. Subdivision of a one-dimensional area

10.2.3 Sub-regions and sub-sectional basis

With point discretization, the domain is subdivided into a number of sub-sections, Ω_m, and let

$$\psi_j = \begin{cases} 1 & \text{inside } \Omega_m \\ 0 & \text{outside } \Omega_m \end{cases} \qquad (10.2.13)$$

This equation shows that the basis function exists only in the sub-region of the domain. The following example is an illustration using this method.

Example 10.2.2. Use the sub-region method to solve Eq. (10.1.8).

Subdivide the region $0 \leqslant x \leqslant 1$ into N equal sub-sections. The triangular function is chosen as the basis function; it is a sectional linear function as shown in Fig. 10.2.1(c), i.e. the unknown function is approximated by

$$u(x) = \sum \alpha_n T(x - x_n). \qquad (10.2.14)$$

Referring to Table 10.2.1, the rectangular impulse function is chosen as the weighting function, i.e.,

$$W_m = P(x - x_m) = \begin{cases} 1 & |x - x_m| < \dfrac{1}{2(N+1)} \\ 0 & |x - x_m| > \dfrac{1}{2(N+1)} \end{cases}.$$

Thus

$$a_{mn} = \langle \mathscr{L} T(x - x_n), P(x - x_m) \rangle$$

$$= \int_0^1 (N+1)[-\delta(x - x_{n-1}) + 2\delta(x - x_n)$$

$$\quad - \delta(x - x_{n+1})] P(x - x_m) \, dx$$

if

$$m = n \qquad a_{mn} = 2(N+1)$$

if

$$\begin{cases} m > n+1 \\ m < n+1 \end{cases} \quad a_{mn} = 0 \qquad (10.2.15)$$

and if

$$m = \begin{cases} n+1 \\ n-1 \end{cases} \quad a_{mn} = -(N+1).$$

Based on Eq. (10.1.7),

$$b_m = \langle 1 + 2x^2, P(x - x_m) \rangle = \langle 1, P(x - x_m) \rangle + \langle 2x^2, P(x - x_m) \rangle$$

$$= \int_0^1 P(x - x_m)dx + \int_{x_m - \frac{1}{2(N+1)}}^{x_m + \frac{1}{2(N+1)}} 2x^2 P(x - x_m)dx$$

$$= \frac{1}{N+1} + \frac{2}{3}x^3 \Big|_{x_m - \frac{1}{2(N+1)}}^{x_m + \frac{1}{2(N+1)}}$$

$$= \frac{1}{N+1} + \left[1 + \frac{2m^2 + 1/2}{(N+1)^2}\right]. \qquad (10.2.16)$$

It is obvious that if the pulse function is used as a weighting function, then the evaluation of matrix **A** is simpler than in any other methods used in this section.

For the operator equation given in this example, the rectangular impulse function is not suitable to be used directly as the basis function. This is due to the fact that the second derivative of $P(x - x_m)$ is not definite in the range of the operator. The domain of the operator must be extended by redefining the operator as a new function, and the extended operator does not change the original operation in its domain. More detailed analysis is shown in references [2, 3].

10.3 Interpretation using variations

The method of moments can also be interpreted by variations as discussed in reference [2]. In this section a special example is used to illustrate that the method of moments is identical to the variational principle.

For an electrostatic problem

$$-\varepsilon \nabla^2 \varphi = \rho \qquad (10.3.1)$$

or

$$\varphi(\mathbf{r}) = \int_{\Omega'} \frac{\rho(\mathbf{r}')}{4\pi\varepsilon R} d\Omega' \qquad (10.3.2)$$

where $R = |\mathbf{r} - \mathbf{r}'|$ is the distance from a source point to a field point, here

$$\mathscr{L} = -\varepsilon \nabla^2 \quad \text{and} \quad \mathscr{L}^{-1} = \int_{\Omega'} \frac{1}{4\pi\varepsilon R} d\Omega'. \qquad (10.3.3)$$

The two operators in Eq. (10.3.3) are reciprocal only if the boundary condition $r\varphi \to \text{constant}|_{r \to \infty}$ is satisfied.

Take a suitable inner product

$$\langle \mathcal{L}\varphi, \psi \rangle = \int_{\Omega'} (-\varepsilon \nabla^2 \varphi) \psi \, d\Omega' \tag{10.3.4}$$

and use Green's identiy

$$\int_{\Omega} (\psi \nabla^2 \varphi - \varphi \nabla^2 \psi) d\Omega = \oint_{\Gamma} \left(\psi \frac{\partial \varphi}{\partial n} - \varphi \frac{\partial \psi}{\partial n} \right) d\Gamma . \tag{10.3.5}$$

Let Γ be a spherical surface of radius r, φ and ψ are constants in the limit of $r \to \infty$, thus the RHS of Eq. (10.3.5) vanishes. This equation then reduces to

$$\int_{\Omega} \psi \nabla^2 \varphi \, d\Omega = \int_{\Omega} \varphi \nabla^2 \psi \, d\Omega . \tag{10.3.6}$$

Considering the vector identity $\psi \nabla^2 \varphi = \nabla(\psi \nabla \varphi) - \nabla \psi \cdot \nabla \varphi$ and the divergence theorem, Eq. (10.3.4) becomes

$$\langle \mathcal{L}\varphi, \psi \rangle = \int_{\Omega} \varepsilon \nabla \psi \cdot \nabla \varphi \, d\Omega - \oint_{\Gamma} \varepsilon \psi \nabla \varphi \, d\Gamma \tag{10.3.7}$$

The last term of Eq. (10.3.7) vanishes as $r \to \infty$ for the same reasons as in Eq. (10.3.5). Let $\psi = \varphi$, then

$$\langle \mathcal{L}\varphi, \psi \rangle = \int_{\Omega} \varepsilon |\nabla \varphi|^2 \, d\Omega . \tag{10.3.8}$$

Equation (10.3.8) shows that the inner product of $\langle \mathcal{L}\varphi, \psi \rangle$ equals the equivalent functional of the Laplacian operator. This means that the method of moments is identical to the variational principle.

10.4 Moment methods for solving static field problems

To illustrate the use of the moment methods for solving static electromagnetic field problems, two examples are shown.

10.4.1 Charge distribution of an isolated plate

A charged plate is shown in Fig. 10.4.1. The potential at any observation point is expressed by a Fredholm integral equation of the first kind

$$\varphi(\mathbf{r}) = \frac{1}{4\pi\varepsilon} \int_{-a}^{a} dx' \int_{-b}^{b} dy' \frac{\sigma(x', y')}{[(x - x')^2 + (y - y')^2 + (z - z')^2]^{1/2}} . \tag{10.4.1}$$

The corresponding operator equation of Eq. (10.4.1) is

$$\mathcal{L}\sigma = f . \tag{10.4.2}$$

If the RHS of Eq. (10.4.2) is known, then the charge distribution $\sigma(x', y')$ can be determined.

10.4 Moment methods for solving static field problems

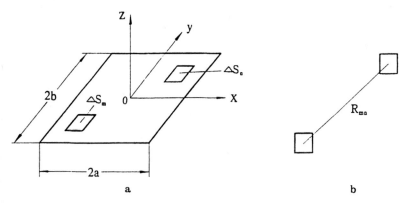

Fig. 10.4.1a, b. A charged plate

Let

$$\sigma(\mathbf{r}') = \sum_{n=1}^{N} \alpha_n \psi_n(\mathbf{r}') . \tag{10.4.3}$$

If the pulse function is chosen as the subsectional basis function and Dirac's delta function is chosen as the weighting function, respectively, i.e.

$$\psi_n = \begin{cases} 1 & \text{within } \Delta S_n \\ 0 & \text{outside } \Delta S_n \end{cases} \tag{10.4.4}$$

and

$$W_m = \delta(x - x_m)\delta(y - y_m) \tag{10.4.5}$$

Equation (10.4.4) shows that the charge density is a constant within a small area ΔS_n. As the potential of the plate is known (U_0), let the observation points be located on the plate, using Eqs. (10.1.6) and (10.1.7), the coefficients of the matrices **A** and **B** are:

$$a_{mn} = \langle W_m, \mathscr{L}\psi \rangle = \delta(x - x_m)\delta(y - y_m) \frac{1}{4\pi\varepsilon} \int_{-a}^{a} dx'$$

$$\times \int_{-b}^{b} \frac{\Psi_n(x', y')}{[(x - x')^2 + (y - y')^2]^{1/2}} dy'$$

$$= \frac{\Delta s_n}{4\pi\varepsilon_0 R_{mn}} \quad m \neq n \tag{10.4.6}$$

$$a_{mn} = \frac{2e}{\pi\varepsilon_0} \ln(1 + 2^{1/2}) \quad m = n \tag{10.4.7}$$

$$b_m = \langle W_m, f \rangle = \langle W_m, U_0 \rangle$$

$$= \int_{-a}^{a} dx' \int_{-b}^{b} \delta(x - x_m)\delta(y - y_m) U_0 \, dy' = U_0 \tag{10.4.8}$$

where
$$R_{mn} = [(x - x_m)^2 + (y - y_m)^2]^{1/2} \tag{10.4.9}$$

where Δs_m is the area of the sub-region. In Eq. (10.4.7) e is the equivalent side length of the sub-region. The results obtained by solving the matrix $\mathbf{A}\{\alpha\} = \mathbf{B}$ are identical to the ones obtained in Sect. 8.1 by using SSM. In other words, the constant element of SSM is equivalent to the moment method while the pulse and Dirac delta function are chosen as the basis and weighting functions, respectively. Thus, SSM is one of the special cases of the moment method.

10.4.2 Charge distribution of a charged cylinder

A charged cylinder is shown in Fig. 10.4.2. The potential at any point $P(r, \alpha, z)$ is

$$\varphi(\mathbf{r}) = \frac{1}{4\pi\varepsilon} \int_{-h}^{h} \sigma(z') \int_{0}^{2\pi} \frac{1}{R} r d\alpha' dz' \tag{10.4.10}$$

where
$$R = |\mathbf{r} - \mathbf{r}'| = [r^2 - r'^2 - 2rr'\cos(\alpha - \alpha') + (z - z')^2]^{1/2}. \tag{10.4.11}$$

If both the source point and the field point are located on the surface of the cylinder and in the plane of $\alpha = 0$, then

$$R = [2a^2 - 2a^2 \cos\alpha' + (z - z')^2]^{1/2}$$
$$= \left[4a^2 \sin^2\left(\frac{\alpha'}{2}\right) + (z - z')^2\right]^{1/2} \tag{10.4.12}$$

$$U_0 = \frac{a}{4\pi\varepsilon_0} \int_{-h}^{h} \sigma(z') \int_{0}^{2\pi} \left[4a^2 \sin\left(\frac{\alpha'}{2}\right) + (z - z')^2\right]^{-1/2} d\alpha' dz'$$

$$= \frac{a}{2\varepsilon_0} \int_{-h}^{h} \frac{\sigma(z')}{[a^2 + (z - z')^2]^{1/2}} dz' \tag{10.4.13}$$

U_0 is the potential of the charged cylinder and $\sigma(z')$ is the unknown surface

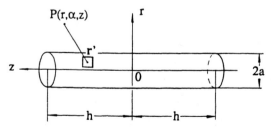

Fig. 10.4.2. A charged cylinder

10.4 Moment methods for solving static field problems

charge distribution. Let

$$\sigma(z') = \sum_{n=1}^{N} \alpha_n \psi_n(z') \tag{10.4.14}$$

and

$$W(z) = \delta(z - z_m) \quad m = 1, \ldots, N. \tag{10.4.15}$$

Substitution of Eq. (10.4.14) into Eq. (10.4.13), leads to

$$\frac{2\varepsilon U_0}{a} = \sum_{n=1}^{N} \alpha_n \int_{-h}^{h} \frac{\psi_n(z')}{[a^2 + (z - z')^2]^{1/2}} \, dz' \tag{10.4.16}$$

where N is the total number of source points. Taking the inner product of Eq. (10.4.16) with weighting function, a set of algebraic equations are then obtained

$$\sum_{n=1}^{N} a_{mn} \alpha_n = b_m \quad m = 1, \ldots, N. \tag{10.4.17}$$

When the matching points (z_m, a) are specified, the coefficients a_{mn} and b_m are evaluated by

$$a_{mn} = \int_{-h}^{h} \frac{\psi_n(z')}{[a^2 + (z_m - z')^2]^{1/2}} \, dz' \tag{10.4.18}$$

$$b_m = U_0. \tag{10.4.19}$$

To evaluate the integrals of Eq. (10.4.18), it is important to note that if the radius of the cylinder, a, is very small, then the integral of Eq. (10.4.18) tends of infinity when $z_m = z$. In this case the resulting singular integral is evaluated by the method discussed in Sect. 8.3.

If $\psi_n(z')$ is a continuous function, Eq. (10.4.18) can be written as

$$a_{mn} = \int_{-h}^{h} \frac{\psi_n(z') - \psi_n(z_m) + \psi_n(z_m)}{[a^2 + (z_m - z')^2]^{1/2}} \, dz'$$

$$= \psi_n(z_m) \int_{-h}^{h} \frac{1}{[a^2 + (z_m - z')^2]^{1/2}} \, dz'$$

$$+ \int_{-h}^{h} \frac{\psi_n(z') - \psi_n(z_m)}{[a^2 + (z_m - z')^2]^{1/2}} \, dz'.$$

$$= \psi_n(z_m) \ln \frac{z_m + h + [a^2 + (z_m + h)^2]^{1/2}}{z_m - h + [a^2 + (z_m - h)^2]^{1/2}}$$

$$+ \int_{-h}^{h} \frac{\psi_n(z') - \psi_n(z_m)}{[a^2 + (z_m - z')^2]^{1/2}} \, dz'. \tag{10.4.20}$$

If $z_m = z'$, then the second term of the RHS of Eq. (10.4.20) equals zero, thus

$$a_{mn} = \psi_n(z_m) \ln \frac{z_m + h + [a^2 + (z_m + h)^2]^{1/2}}{z_m - h + [a^2 + (z_m - h)^2]^{1/2}}. \tag{10.4.21}$$

This equation is used for evaluating the coefficients when the observation point is identical to the source point. In other cases Eq. (10.4.20) is used to evaluate the coefficients where the Gauss quadrature [4] is applied for the integration.

So far the basis function $\psi_n(z)$ has not been specified. If a pulse function is chosen as the basis function and the cylinder is subdivided by N equalized elements, the charge distribution of a cylinder of height $h = 1$ m and radius $a = 10^{-2}$ m is shown in Fig. 10.4.3 where $2\varepsilon_0 U_0/a = 1$ is assumed. The results show that the number of subdivisions strongly influences the accuracy of the approximation.

If a good initial guess such as $\sigma_s(z')$ is chosen, it will be helpful in obtaining an accurate solution by using fewer elements. Consider that near the two ends of the cylinder, $\sigma_s(z')$ is the largest. At the middle of the cylinder, $\sigma_s(z')$ is the minimum and its derivative is zero. Thus a power series is considered as a basis function, i.e. let

$$\psi_n(z') = \left(\frac{z'}{h}\right)^{2n} \tag{10.4.22}$$

and the positions of the matching points are chosen as

$$z_m = \frac{(m - 1/2)}{N} h. \tag{10.4.23}$$

Reference [5] gives the result if Eq. (10.4.22) is used as the basis function, when $N = 6$, the charge distribution is similar to the result obtained by $N = 30$ while the pulse function is used as the basis function. This demonstrates the influence of the basis function. Although the concepts of the moment method are more complicated than SSM, the method of moments allows flexibility in choosing

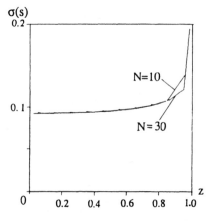

Fig. 10.4.3. Charge distribution along the cylinder surface

the basis functions. Therefore, it is more efficient than SSM in some cases. Moment methods are considered as a generalized form of the methods of approximation.

10.5 Moment methods for solving eddy current problems

10.5.1 Integral equation of a 2-D eddy current problem

Assume that the problem has translational symmetry as shown in Fig. 10.5.1(a). A and J are scalar functions of the coordinates.

First consider the interior problem. Assume that the conductor is composed of filaments with current density $J(\mathbf{r}')$ then

$$A(\mathbf{r}) = \frac{\mu}{2\pi} \int_s J(\mathbf{r}') \ln \frac{1}{|\mathbf{r} - \mathbf{r}'|} \, ds' \tag{10.5.1}$$

$$J(\mathbf{r}) = J_e + J_s = -j\omega\gamma A(\mathbf{r}) + \gamma U_0/L \tag{10.5.2}$$

$$J_s = \gamma U_0/L \tag{10.5.3}$$

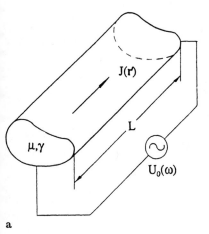

a

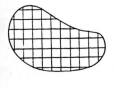

b

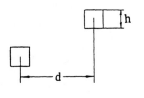

c

Fig. 10.5.1a–c. A 2-D current carrying conductor

where U_0 is the terminal voltage and L and S are the length and the cross-section area of the conductor. By substituting Eq. (10.5.1) into Eq. (10.5.2), a Fredholm integral equation of the second kind is obtained

$$J(\mathbf{r}) = \frac{j\omega\mu\gamma}{2\pi} \int_s J(\mathbf{r}') \ln|\mathbf{r} - \mathbf{r}'| ds' + J_s. \qquad (10.5.4)$$

10.5.2 Sub-sectional basis method

Rewrite Eq. (10.5.4) to

$$J(\mathbf{r}) - \frac{j\omega\mu\gamma}{2\pi} \int_s J(\mathbf{r}') \ln|\mathbf{r} - \mathbf{r}'| ds' = J_s. \qquad (10.5.5)$$

The operator of Eq. (10.5.5) is

$$\mathscr{L} = 1 - \frac{j\omega\mu\gamma}{2\pi} \int_s \ln|\mathbf{r} - \mathbf{r}'| ds'. \qquad (10.5.6)$$

Subdivide the domain into N square elements as shown in Fig. 10.5.1(b). Assume the current density is approximated as

$$J(\mathbf{r}) = \sum_{i=1}^{n} J_i \psi_i \qquad (10.5.7)$$

Then let

$$\psi_i = \begin{cases} 1 & \text{within the sub-element} \\ 0 & \text{outside the sub-element} \end{cases}$$

This means that the pulse function $P(x)$ is chosen as an approximate solution. If $P(x)$ is also chosen as the weighting function, then the coefficients of matrix **A** are

$$\begin{aligned} a_{mn} &= \langle \mathscr{L}\psi_n, W_m \rangle \\ &= \int_s P_m(x, y) ds - \frac{j\omega\mu\gamma}{2\pi} \int_s \int_s P_m(x, y) \\ &\quad \times \ln[(x_m - x_n)^2 + (y_m - y_n)^2]^{1/2} P_n(x, y) ds ds' \\ &= \Delta S_m - \frac{j\omega\mu\gamma}{2\pi} \Delta S_m \Delta S_n \ln[(x_m - x_n)^2 + (y_m - y_n)^2]^{1/2} \end{aligned} \qquad (10.5.8)$$

$$b_m = \langle J_s, W_m \rangle = J_s \Delta S_m. \qquad (10.5.9)$$

Assume that the length of one side of the square elements is h. Eliminate all the terms ΔS_m in the equations of a_{mn} and b_m, and consider the different relative

10.6 Moment methods to solve the current distribution of a line antenna

positions of the source and field points as shown in Fig. 10.5.1(c), the formulations to evaluate the elements of the matrix **A** are [6]

$$a_{mn} = 1 - \frac{j\omega\mu\gamma}{4\pi} h^2 \ln[(x_m - x_n)^2 + (y_m - y_n)^2] \quad d/h > 2 \quad (10.5.10)$$

$$a_{mn} = 1 - 1.0065 \frac{j\omega\mu\gamma}{4\pi} h^2 \ln[(x_m - x_n)^2 + (y_m - y_n)^2] \quad d/h = 1$$
(10.5.11)

if $m = n$

$$a_{mn} = 1 - \frac{j\omega\mu\gamma}{2\pi} h^2 \ln(0.44705h) . \quad (10.5.12)$$

Solve the matrix equation

$$\mathbf{A}\{J\} = \{J_s\} \quad (10.5.13)$$

the current distribution is obtained.

For rectangular elements, the more accurate expression for calculating a_{mn} is given in reference [6].

For a 2-D interior eddy current problem, the method described here is the simplest and more efficient one. It is used in reference [7, 8].

For a problem with rotational symmetry, a similar method was used in reference [9].

10.6 Moment methods to solve the current distribution of a line antenna

Allow flexibility in choosing of the basis function and the weighting function. The method of moments is ideally suited to solve electromagnetic radiation and scattering problems. (See the contributions of Harrington [2], Mittra [10] and Moore [11].) A one-dimensional problem is chosen here to illustrate the use of the moment method.

10.6.1 Integral equation of a line antenna

Figure 10.6.1 shows a cylindrical antenna of length $l = 2h$ and with radius a, where a is much smaller than the wavelength λ. Hence the current is assumed to be uniformly distributed along the circle, then the current is dependent only on the variable z'. Thus

$$A_z = \frac{\mu_0}{4\pi} \int_{-h}^{h} I(z') G(z, z') dz' \quad (10.6.1)$$

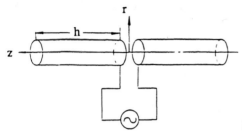

Fig. 10.6.1. A line antenna with small radius

where G is Green's function of free space. In a 3-D case,

$$G(z, z') = \frac{e^{-jkR}}{R}. \tag{10.6.2}$$

Hence Eq. (10.6.1) becomes

$$A_z = \frac{\mu_0}{4\pi} \int_{-h}^{h} \frac{I(z')e^{-jkR}}{R} dz' \tag{10.6.3}$$

where

$$k = \frac{2\pi}{\lambda} = (\omega^2 \mu_0 \varepsilon_0)^{\frac{1}{2}}. \tag{10.6.4}$$

Consider the Lorentz gauge $\nabla \cdot \mathbf{A} = -j\omega\mu_0\varepsilon_0\varphi$ and in the sinusoidal steady state case,

$$\mathbf{E} = -\frac{\partial \mathbf{A}}{\partial t} - \nabla\varphi$$

$$E_z = -j\omega A_z - \frac{\partial \varphi}{\partial z} = \frac{1}{j\omega\mu_0\varepsilon_0}\left(\frac{\partial^2 A_z}{\partial z^2} + k^2 A_z\right) \tag{10.6.5}$$

then

$$j\omega\mu_0\varepsilon_0 E_z = \frac{\partial^2 A_z}{\partial z^2} + k^2 A_z. \tag{10.6.6}$$

Substitution of Eq. (10.6.1) into Eq. (10.6.6) yields

$$4\pi j\omega\varepsilon_0 E_z = \left(\frac{\partial^2}{\partial z^2} + k^2\right) \int_{-h}^{h} I(z')G(z, z') dz'. \tag{10.6.7}$$

This equation is named Pocklington's equation. It was used by Pocklington in 1897 to analyse the current distribution along a line antenna. The result showed that the current distribution is approximately sinusoidal.

10.6 Moment methods to solve the current distribution of a line antenna

In Eq. (10.6.7) E_z is the incident field. Consider the source of the antenna as a δ gap voltage source, $z \cdot E_i(z) = U\delta(z)$, then Eq. (10.6.7) becomes

$$-4\pi j\omega\varepsilon_0 U\delta(z) = \left(\frac{\partial^2}{\partial z^2} + k^2\right) \int_{-h}^{h} I(z')G(z,z')\,dz \ . \quad (10.6.8)$$

Integrate the last equation to obtain

$$\int_{-h}^{h} I(z')G(z,z') = B\cos kz - \frac{jU}{2\eta_0}\text{Sin}k|z| \ . \quad (10.6.9)$$

This is Hallen's equation, where $\eta_0 = (\mu_0/\varepsilon_0)^{1/2}$ is the wave impedance of free space and B is a constant dependent on the imposed condition.

10.6.2 Solution of Hallen's equation

The global basis and point matching method is used here to solve Hallen's equation. As suggested in reference [12] for $kh = 0.5 \to 2.7$ (i.e. $h = 0.0795 \to 0.42993\lambda$), the sinusoidal function is chosen here as the basis function, i.e.,

$$I(z') = \sum_{n=1}^{N} C_n \text{Sin}[nk(h - |z'|)] \quad (10.6.10)$$

when $N = 2$, the matrix equation is

$$\mathbf{A}\begin{Bmatrix} C_1 \\ C_2 \\ -B \end{Bmatrix} = \mathbf{U} \quad (10.6.11)$$

The coefficients of matrix $\mathbf{A}$ are evaluated by

$$\begin{cases} a_{i1} = \int_{-h}^{h} \text{Sin}k(h - |z'|)G(z,z')\,dz' \\ a_{i2} = \int_{-h}^{h} \text{Sin}2k(h - |z'|)G(z,z')\,dz' \quad i = 1, 2, 3 \\ a_{i3} = \text{Cos}kz_i \end{cases} \quad (10.6.12)$$

$$u_i = -\frac{jU}{2\eta_0}\text{Sin}k|z_i| \quad i = 1, 2, 3 \quad (10.6.13)$$

To solve Eq. (10.6.10), the current distribution of the antenna is obtained. While $a/\lambda = 7.022 \times 10^{-3}$, $h/\lambda = 0.0795$ and $h/\lambda = 0.25$, the real and the image components of the current are shown in Fig. 10.6.2(a). If the length of the antenna is

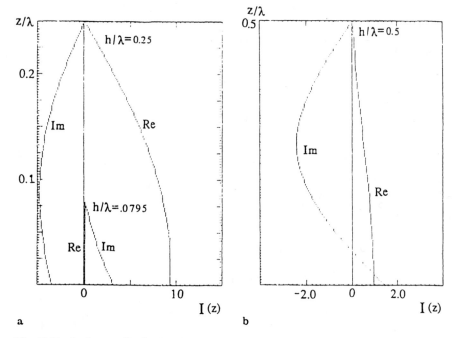

Fig. 10.6.2a, b. Current distribution along the line antenna

longer, e.g. $h = 0.395\lambda - 0.75\lambda$, the basis function

$$I(z') = C_1 \operatorname{Sin} k(h - |z'|) + C_2 (\operatorname{Cos} kz' - \operatorname{Cos} kh)$$
$$+ C_3(\operatorname{Cos} \tfrac{1}{2} kz' - \operatorname{Cos} \tfrac{1}{2} kh) \tag{10.6.14}$$

is a better choice than Eq. (10.6.10). The current distribution of a wave length antenna ($h = 0.5\lambda$) is shown in Fig. 10.6.2(b). In Fig. 10.6.2 the unit of the current $I(z)$ is normalized, i.e. the unit is (A/V).

In solving equation

$$A \begin{Bmatrix} C_1 \\ C_2 \\ C_3 \\ -B \end{Bmatrix} = \begin{Bmatrix} U_1 \\ U_2 \\ U_3 \\ U_4 \end{Bmatrix} \tag{10.6.15}$$

four matching points are chosen at $\tfrac{1}{3}h$, $\tfrac{2}{3}h$ and the two end points. The comparison of the calculated results and the experimental data are given in Table 10.6.1. In Table 10.6.1, the experimental data are taken from reference [13].

The Gaussian quadrature is used to evaluate the coefficients of the matrix, the matrix is solved by Cholesky decomposition. More applications of the moment method to solve scattering and antenna problems are shown in references [2, 14].

10.7 Summary

Table 10.6.1. Current distribution of an antenna ($L = \lambda/2$)

z/λ	Real component		Image component	
	Calculated	Experimental	Calculated	Experimental
0.0	9.4033	9.265	− 3.5736	− 3.825
0.05	9.2096	8.7	− 4.5926	− 4.76
0.10	8.0387	7.65	− 4.8230	− 4.76
0.15	5.9584	5.78	− 4.0324	− 4.08
0.20	3.1723	3.4	− 2.2983	− 2.55
0.25	0.	0.	0.	0.

10.7 Summary

Moment methods are generalized approximate methods based on the principle of weighted residuals. It is also identical to the variational approaches, so is regarded as the universal name for numerical methods.

In numerical solution the operator equation $\mathscr{L}u = f$ is changed to a matrix equation $\mathbf{A}\{\alpha\} = \mathbf{B}$ by using the inner product operation, i.e.

$$\mathbf{A} = \begin{bmatrix} \langle \mathscr{L}\psi_1, W_1 \rangle \langle \mathscr{L}\psi_2, W_1 \rangle \cdots \langle \mathscr{L}\psi_N, W_1 \rangle \\ \vdots \qquad \vdots \qquad \qquad \vdots \\ \langle \mathscr{L}\psi_1, W_N \rangle \langle \mathscr{L}\psi_2, W_N \rangle \cdots \langle \mathscr{L}\psi_N, W_N \rangle \end{bmatrix}$$

$$\mathbf{B} = \begin{Bmatrix} \langle f, W_1 \rangle \\ \vdots \\ \langle f, W_m \rangle \end{Bmatrix}$$

where ψ_n is the basis function of u, i.e., $u = \sum_{n=1}^{N} \alpha_n \psi_n$. W_n are weighting functions.

Due to the flexible choice of the basis function and the weighting function, moment methods cover all of the methods discussed previously. If the best choice of the approximate function and the weighting function is achieved, the efficiency of the computation is high. The problem shown in Fig. 10.4.3 is a good example.

Because of the wide adaptability of this method, it can be used to solve static, quasi-static and dynamic problems expressed by differential or integral equations. The domain to be discretized may be the whole domain or the boundary of the domain. Hence it can be classified as the domain method or the boundary method according to use.

From Sects. 10.4 to 10.6, moment methods have also been used to solve problems expressed by integral equations, so they are known as. integral equation methods. Unlike differential equation methods, integral equation methods as a rule have no need for additional requirements. All the boundary and constrained conditions are part of the kernel function, so a particular kernel is related to particular boundary condition and does not have universal validity. Hence in using the integral equation method, the first step is to derive a specific integral equation for a given problem. During calculation of the coefficients of the matrix, numerical integration is often used.

References

1. R.F. Harrington, Matrix Methods for Field Problems, *Proc. IEEE*, **55**(2), 136–149, 1967
2. R.F. Harrington, *Field Computation by Moment Methods*. Macmillan, 1968
3. Wang Xiznchong, *Theory and Applications of Electromagnetic Field*, Science Press, 1986 (in Chinese)
4. W. Abramowitz, I.A. Stegun (Ed.), *Handbook of Mathematical Functions*, Dover, New York, 1970
5. Herbert P. Neff, *Basic Electromagnetic Fields*, Harper & Row, New York, p. 153, 1981
6. J.T. Higgins, Theory and Applications of Complex Logarithmic and Geometric Mean Distances, *AIEE, Trans.* **66**, 12–16, 1947
7. E.M. Deeley, A Numerical Method of Solving Fredholm Integral Equations Applied to Current Distribution and Inductance in Parallel Conductors, *Int. J. Num. Meth. in Eng.*, **11**, 447–467, 1977
8. M. Cao, P.P. Biringer, Minimizing the Three-Phase Unbalance in an Electric Arc Furnace, *IEEE Trans. on Mag.*, **25**(4), 2849–2851, 1989
9. P. Silvester, S.K. Wong, P.E. Burke, Model Theory of Skin Effect in Single and Multiple Turn Coils, *IEEE Summer Meeting*, Portland, July 1971
10. R. Mittra, *Numerical and Asymptotic Techniques in Electromagnetics*. Springer-Verlag, 1975
11. J. Moore, R. Pijer, *Moment Methods in Electromagnetics*, John Wiley, 1984
12. Li Shi-Zhi, *Moment Methods for Radiating and Scattering Problems of Electromagnetics* (in Chinese), Electronic Publishers, 1985
13. R.E. Collin, F.J. Zucker, *Antenna Theory*, 1969
14. J.R. Mantz, R.F. Harrington, A H-Field Solution for Electromagnetic Scattering by a Conducting Body of Revolution, *COMPEL*, **1**(3), 137–163, 1982

Part Four
Optimization Methods of Electromagnetic Field Problems

In this part the remaining two chapters illustrate the optimization of electromagnetic field problems. Mathematical methods to search the extrema (maximum or minimum) of an objective function are introduced in Chap. 11. These methods are used to solve the problems given in Chap. 12. Strict theoretical problems concerning optimization are beyond the scope of this book. All the material presented in this chapter is for readers who wish to understand the properties of those methods and to use them well.

The optimum design of electromagnetic devices is important and is difficult. This includes field analysis and synthesis. There is no general method to solve these problems. In Chap. 12, several methods are introduced to find the optimum shape of electrodes and the best size of magnets.

Chapter 11

Methods of Applied Optimization

11.1 Introduction

The advanced purpose of the analysis of electromagnetic fields is to determine a better design of a practical problem. Design optimization (sometimes called the inverse problem) deals with the problem of finding the source distribution or the dimensions of devices for a specific purpose. Most inverse problems involve numerical optimization. This chapter provides the mathematical methods used for the optimum design.

The study of optimization methods is a special area of applied mathematics. There are many books available on this topic references [1–5]. The purpose of this chapter is to help readers to understand the general principles of various optimization methods and to choose appropriate algorithms already developed by mathematicians. Most general algorithms of unconstrained and constrained optimization methods are available in references [6, 7], or in the computer libraries of IMSL (International Mathematical Scientific Library) or that of NAG (Numerical Algorithms Group).

11.2 Fundamental concepts

Usually, the optimum design for practical problems has to satisfy some predetermined conditions. Such as

$$\begin{cases} \text{Min}^1 & F(X) \\ \text{Subject to} & h_i(X) = 0 \quad i = 1,\ldots,p \\ & g_j(X) \geq 0 \quad j = 1,\ldots,m \end{cases} \quad (11.2.1)$$

where $F(X)$ is the target or the objective function. X is a vector of the order n in the linear space E_n. The second and third equations in Eq. (11.2.1) are the

---

[1] Max $F(X) = -$ Min $- F(X)$

constrained conditions of equality, inequality, respectively. Hence Eq. (11.2.1) is the mathematical expression of a constrained optimization problem. A typical example used to illustrate the constrained optimum problem is to determine the maximum volume of a rectangular box such that the dimension in any direction is less than or equal to 42 cm and the sum of the girth and the length of one edge is less than 72 cm [8]. The corresponding equations are

$$\begin{cases} \text{Max} \quad F(X) = x_1 x_2 x_3 \\ \text{subject to: } 0 < x_1 \leq 42 \\ \phantom{\text{subject to: }} 0 < x_2 \leq 42 \\ \phantom{\text{subject to: }} 0 < x_3 \leq 42 \\ \phantom{\text{subject to: }} 0 < x_1 + 2x_2 + 2x_3 \leq 72 \end{cases} \quad (11.2.2)$$

This is a constrained non-linear optimization problem, as the target function is non-linear. If one of the constrained conditions is non-linear, then the problem is also non-linear. If the objective function and all of the constrained conditions are linear functions, the problem is one of linear optimization and is a problem of linear programming. If the variables of the objective function have no restrictions, then the problem is one of unconstrained optimization. Actually, unconstrained optimization methods are the foundation of constrained optimization problems. Hence methods to solve unconstrained optimization will be introduced as the main part in this chapter.

11.2.1 Necessary and sufficient conditions for the local minimum

As shown in Fig. 11.2.1, there are two local minima and one global minimum. For a univariate function $y = f(x)$, the necessary and sufficient conditions of local minimum are

$$\begin{cases} f'(x^*) = 0 \\ f''(x^*) > 0 \quad x^* \text{ is minimizer} \\ f''(x^*) < 0 \quad x^* \text{ is maximizer} \end{cases} \quad (11.2.3)$$

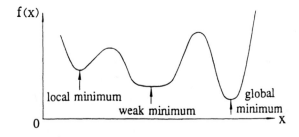

Fig. 11.2.1. Local minima and global minimum

11.2 Fundamental concepts

The minimizer is the minimum point of $f(x)$. In Fig. 11.2.2, a zero slope holds at the minimizer.

For functions with two variables, the necessary and sufficient conditions are

$$\begin{cases} f'_x(x^*, y^*) = f'_y(x^*, y^*) = 0 \\ [f''_{xy}(x^*, y^*)]^2 - f''_{xx}(x^*, y^*) f''_{yy}(x^*, y^*) < 0 \end{cases} \quad (11.2.4)$$

where $f''_{xx}(x^*, y^*) < 0$, (x^*, y^*) is the maximizer, if $f''_{xx}(x^*, y^*) > 0$, (x^*, y^*) is the minimizer.

For the multivariable function $F(X)$, the necessary and sufficient conditions are classified as follows. Expanding $F(X)$ by Taylor's series and neglecting the higher order terms of the series, one obtains

$$F(X + \Delta X) = F(X) + \Delta X^T g(X) + \tfrac{1}{2} \Delta X^T G(X) \Delta X \quad (11.2.5)$$

where

$$g(X) = \nabla F(X) = \begin{bmatrix} \dfrac{\partial F}{\partial x_1} \\ \vdots \\ \dfrac{\partial F}{\partial x_n} \end{bmatrix} \quad (11.2.6)$$

$$G(X) = \nabla g(X) = \nabla^2 F(X) = \begin{bmatrix} \dfrac{\partial^2 F}{\partial x_1^2} & \cdots & \dfrac{\partial^2 F}{\partial x_1 \partial x_n} \\ \vdots & & \vdots \\ \dfrac{\partial^2 F}{\partial x_n \partial x_1} & \cdots & \dfrac{\partial^2 F}{\partial x_n^2} \end{bmatrix} \quad (11.2.7)$$

and $g(X)$ is the first order derivative of the objective function. $G(X)$ is the second order derivative of the objective function, it is known as the Hessian matrix of the objective function. For the linear function, $g(X)$ is constant, for quadratic function, the Hessian matrix is constant.

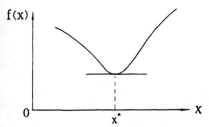

Fig. 11.2.2. Zero slope and non-negative curvature at x^*

From Eq. (11.2.5), it is found that if X is the local minimizer, the following conditions must be satisfied, i.e.

$$g(X^*) = 0 \tag{11.2.8}$$

and

$$F(X + \Delta X) > F(X^*) \quad \text{or} \quad F(X + \Delta X) \geq F(X^*). \tag{11.2.9}$$

In other words, the necessary and sufficient conditions for the optimum of the multivariable functions are

$$\begin{cases} g(X^*) = 0 \\ \langle \Delta X^T G \Delta X \rangle > 0 \quad \text{or} \quad \Delta X^T G \Delta X \geq 0 \end{cases} \tag{11.2.10}$$

$g(X^*) = 0$ means X^* is a saddle point or a stationary point where $F(X^* + \Delta X) - F(X^*)$ is sometimes positive, sometimes negative and sometimes zero, depending upon ΔX. If $\Delta X^T G(X^*) \Delta X > 0$ exists, $G(X^*)$ is a positive definite matrix, X^* is a strong minimizer. It means that the objective function may increase in any direction around the point X^*.

If $\Delta X^T G(X^*) \Delta X \geq 0$, $G(X^*)$ is a positive semi-definite matrix. X^* is a minimizer but not strong. Thus the first equation of Eq. (11.2.10) is the necessary first order condition while the second is the second order sufficient condition. *So the necessary and sufficient conditions for the existence of multivariable functions are $g(X^*) = 0$, and $G(X^*)$ is positive definite or positive semi-definite.*

11.2.2 Geometrical interpretation of the minimizer

For a two-variable function $F = f(x, y)$, let $F = C_1, C_2 \ldots$, then a set of plane curves are obtained as shown in Fig. 11.2.3. These curves are called the contour lines of $f(x, y)$.

It can be seen from Fig. 11.2.3, that in the vicinity of the minimizer the contour lines of the function are approximated to ellipses. The minimizer of the function is the centre of the ellipses. This can be illustrated as follows.

By expanding $f(x, y)$ by Taylor's series at the point of minimizer (where the first order derivative of $f(x, y)$ is zero) and neglecting the third order terms, then

$$f(x, y) \cong f(x^*, y^*) + \tfrac{1}{2}[a_{11}(x - x^*)^2 + 2a_{12}(x - x^*)(y - y^*) + a_{22}(y - y^*)^2] \tag{11.2.11}$$

where

$$a_{11} = f''_{xx} \quad a_{12} = f''_{xy} \quad a_{22} = f''_{yy}. \tag{11.2.12}$$

If $a_{11}a_{22} > a_{12}^2$, the contour lines represented by Eq. (11.2.11) are ellipses and the centre point is x^*, y^*. *The procedures in finding the minimizer of $f(x, y)$ is to find the centre point of the set of ellipses.* Consequently, to find the minimizer of a multivariable function is equivalent to finding the centre of the ellipsoid.

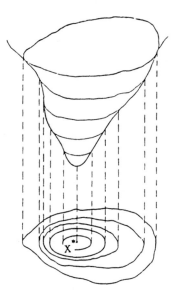

Fig. 11.2.3. Contour lines of the function of two variables

11.2.3 Quadratic functions

The quadratic function is the simplest form of non-linear functions with continuous second derivatives. As illustrated in Sect. 11.2.2. *in the vicinity of the minimizer, the characteristics of any 2-D non-linear function is approximated by a quadratic function.* Hence the quadratic function has specific meaning for discussing the minimization of the non-linear functions. Any quadratic function can be expressed as

$$F(X) = \tfrac{1}{2}X^T AX + \mathbf{b}^T X + C$$
$$= \frac{1}{2}\sum_{i=1}^{n}\sum_{j=1}^{n} a_{ij}x_i x_j + \sum_{i=1}^{n} b_i x_i + C \qquad (11.2.13)$$

where A is a symmetric matrix of the order $n \times n$, and n is the number of independent variables of $F(X)$. Then

$$\nabla F(X) = \mathbf{g}(X) = AX + \mathbf{b} \qquad (11.2.14)$$
$$G(X) = \nabla \mathbf{g}(X) = A \qquad (11.2.15)$$

From Eq. (11.2.14), if $\mathbf{g}(X^*) = 0$, then the minimum point X^* is obtained, i.e.

$$X^* = -A^{-1}\mathbf{b}. \qquad (11.2.16)$$

Thus in the case of a quadratic function, the minimizer can be determined directly by A and $\mathbf{b}$.

Figure 11.2.4 shows the contour lines of function $6x^2 - 5xy + 16y^2 = 1$.

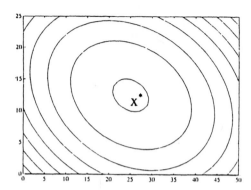

Fig. 11.2.4. The contour lines of a quadratic function

Example 11.2.1. Find the minimizer of the following function
$$F(X) = 4x_1^2 - x_2^2 - 40x_1 - 12x_2 + 136 = \min \quad (11.2.17)$$

Solution. Rewrite the function $F(X)$ in a standard matrix form, i.e.
$$F(X) = 4(x_1 - 5)^2 + (x_2 - 6)^2$$
$$= \frac{1}{2}[x_1 \ x_2]\begin{bmatrix} 8 & 0 \\ 0 & 2 \end{bmatrix}\begin{Bmatrix} x_1 \\ x_2 \end{Bmatrix} - \begin{Bmatrix} 40 \\ 12 \end{Bmatrix}\begin{Bmatrix} x_1 \\ x_2 \end{Bmatrix} + 136$$

thus
$$\mathbf{A} = \begin{bmatrix} 8 & 0 \\ 0 & 2 \end{bmatrix} \quad \mathbf{b} = \begin{Bmatrix} -40 \\ -12 \end{Bmatrix} \quad c = 136$$

solve equation $AX + b = 0$, the result $X^* = \begin{Bmatrix} 5 \\ 8 \end{Bmatrix}$ is obtained. Because **A** is positive definite, hence X^* is the minimizer.

This concept is used in creating a number of optimization methods to solve unconstrained optimization problems.

11.2.4 Basic method for solving unconstrained non-linear optimization problems

In consideration of Eq. 11.2.3, the non-linear unconstrained optimization problems are equivalent to the solution of a set of non-linear equations:
$$\nabla F(X^*) = 0 \quad (11.2.18)$$

Solve the above equation, then determine whether X^* is a minimum or a maximum by using the condition
$$\varDelta X^T G \varDelta X \geq 0 \quad (11.2.19)$$

The solution of Eq. (11.2.18) is a problem of function minimization. However, in practice it is difficult or even impossible to express Eq. (11.2.18) analytically.

11.2 Fundamental concepts

Therefore, the method of solution of unconstrained non-linear optimization involves an iterative process to minimze the objective function approaching a minimum value. The method is composed of the following four steps:

(1) Assign an initial point $X_{(0)}$.
(2) Determine the search direction p_i at the current point X_i.
(3) Find a new point $X_{i+1} = X_i + \lambda p_i$, along the direction p_i to ensure $F(X_{i+1}) \leq F(X_i)$, where λ is the optimum step in the direction p_i.
(4) If $\|\nabla F(X_{i+1})\| < \varepsilon$

or $|F(X_{i+1}) - F(X_i)| < \varepsilon$

or $\|X_{i+1} - X_i\| < \varepsilon$

then $X_{i+1} = X^*$, otherwise return to step (2). Until the convergent criterion in step (4) is satisfied.

The conditions given in (4) are assumed as ideal conditions to obtain the final results. This process is not always workable. It is possible that if the following conditions occur, then the iteration for searching the minimizer could be stopped if

(1) An acceptable estimate of the solution has been found.
(2) The iteration is slow and no progress is observed.
(3) The predetermined number of iterations does not lead to acceptable results.
(4) An acceptable solution does not exist.
(5) The result is oscillatory.

Consequently, a good program for searching the minimizer is needed to deal with the various possibilities.

11.2.5 Stability and convergence

The algorithm is stable or exhibits global convergence if it converges to a minimum (or in some cases, it is a stationary point) no matter the value of the initial point and does not stop at an extraneous point. In non-linear optimization, stability is almost always associated with an iterative scheme which reduces the function value at each iteration until the minimum has been found to any prescribed degree of accuracy. An algorithm is said to exhibit local convergence if the starting point is sufficiently close to the minimum.

The rate of the convergence is defined as:

$$K = \lim_{k \to \infty} \frac{\|X_{k+1} - X^*\|}{\|X_k - X^*\|^r} \quad 0 \leq K \leq 1 \tag{11.2.20}$$

or

$$K = \left[\frac{\lambda_{max} - \lambda_{min}}{\lambda_{max} + \lambda_{min}}\right]^2 = \left[\frac{\gamma - 1}{\gamma + 1}\right]^2. \tag{11.2.21}$$

Equation (11.2.20) measures the closeness of X_{k+1} to X^* compared to that of X_k to X^*. Rapid convergence is associated with large values of r and small values of K. r is the rate of asymptotic convergence. The case $r = 1$ indicates that the sequence is linearly convergent. If $r = 2$, the sequence is said to have quadratic convergence. If $r = 1$, K must be less than 1 in order to obtain convergence. The fastest possible first order rate is that of $K = 0$. This is the rate of superlinear convergence. Generally speaking, the value of r depends upon the algorithm while the value of K depends upon the function being minimized.

In Eq. (11.2.21), the condition number γ of the Hessian matrix is defined as:

$$\gamma = \frac{\lambda_{max}}{\lambda_{min}} \tag{11.2.22}$$

where λ_{max}, λ_{min} are maximum and minimum eigenvalues of the Hessian matrix at the minimum of the function. A function is ill-conditioned if γ is sufficiently large. On the other hand, a function with a relatively small value of γ is a well scaled function. For an ill-conditioned problem, the contours of the objective function are those with a high value of ellipticity. In a well scaled problem, the contour lines of objective function are closed to circles. *It is more difficult to find the minimum if the problem is ill-conditioned than that of a well scaled problem.*

11.3 Linear search and single variable optimization

Optimization methods for a single variable function are not only used to find the extremum value of the univariate function but more important in determining the extrema of multivariable functions. As mentioned in Sect. 11.2.4, to determine the optimum step λ in each iteration is to search the minimum of the function $F(\lambda)$ along a specific direction p_k, e.g.

$$F(\lambda) = F(X_k + \lambda p_k) = \min . \tag{11.3.1}$$

This is a single variable optimization problem with respect to λ. Hence the accuracy of the linear search is extremely important in many optimization methods. It influences the speed of convergence of the methods. If the linear search is not accurate, the solution might diverge.

11.3.1 Golden section method

The golden section method is a simple and effective method for searching the minimum of an unimodal objective function in an interval $[a, b]$. *The strategy of the method is to reduce the interval of the search by comparing the successive values of the objective function iteratively until the minimizer is approached.*

A single evaluation of the function $F(x)$ within the interval $[a, b]$ is not sufficient to reduce the interval by comparing the values $F(a)$, $F(b)$ and $F(c)$ (c is

11.3 Linear search and single variable optimization

any point within $[a, b]$). The reason is that this process cannot distinguish whether x^* is within the interval $[a, c]$ or $[c, b]$ as shown in Fig. 11.3.1. If the values of the objective function are calculated at two points c and d, when $c < d$ and if $F(c) < F(d)$, then a minimum certainly lies either within the interval $[c, d]$ or $[a, c]$, as shown in Fig. 11.3.1(b) and (c). Thus the interval of uncertainty of the minimizer is reduced to $[c, b]$ or $[a, d]$. When $c < d$ and if $F(c) > F(d)$, the interval of uncertainty is reduced to $[a, d]$ or $[c, b]$ as shown in Fig. 11.3.1(a), (d). If $F(c) = F(d)$, then the interval of uncertainty is $[c, d]$. Repeating these procedures, the initial range $[a, b]$ is reduced continuously to $[a_k, b_k]$ (a_k and b_k represent the interval in each iteration) until the optimizer x^* is obtained.

In the process of iteration, e.g. at the kth iteration, the position of the points c_k and d_k within the interval $[a_k, b_k]$ may be determined by several methods. For example, c_k and d_k may be determined by the trisection of $[a_k, b_k]$ or determined by multiplying a changeable coefficient, the Fibonacci number (Fibonacci method). *If a constant such as the so-called optimum coefficient (0.618), is used to reduce the interval in each iteration then the method is called 'golden section search'*. The number 0.618 is obtained by the following derivation.

Let us assume

$$c_k = b_k - \alpha(b_k - a_k) \tag{11.3.2}$$

$$d_k = a_k + \alpha(b_k - a_k) \tag{11.3.3}$$

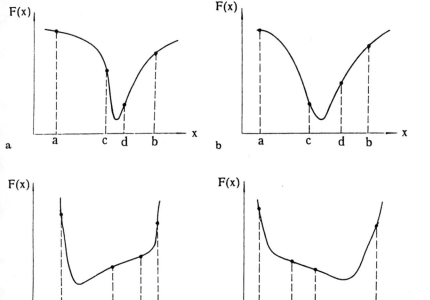

Fig. 11.3.1a–d. The optimum process of an unimodal function

when $k = 1$,
$$c_1 = b - \alpha(b - a) \quad (11.3.4)$$
$$d_1 = a + \alpha(b - a). \quad (11.3.5)$$

If $f(c_1) < f(d_1)$, the next interval is $[a, a + \alpha(b - a)]$, then
$$c_2 = b_1 - \alpha(b_1 - a_1) = a_1 + (1 - \alpha)(b_1 - a_1)$$
$$= a + (1 - \alpha)[\alpha(b - a)] \quad (11.3.6)$$
$$d_2 = a_1 + \alpha(b_1 - a_1) = a + \alpha^2(b - a). \quad (11.3.7)$$

Compare the above equations, if $c_2 = d_1$, then $\alpha = 0$. This is a trivial case of no importance. If $d_2 = c_1$, yields
$$a + \alpha^2(b - a) = b - \alpha(b - a) \quad (11.3.8)$$
then
$$\alpha = \frac{-1 \pm \sqrt{5}}{2} = 0.618033988. \quad (11.3.9)$$

This α is the ratio of the lengths of the subsequent intervals. It is always positive. By using the factor of 0.618 during each iteration the interval is reduced by 38.2% and only one function value is to be calculated in each step. The ratio of the lengths of the final interval to the first interval after n iterations is $\alpha^{n-1} (\alpha = 0.618)$.

The computer program of the golden section method is given in reference [8].

11.3.2 Methods of polynomial interpolation

In the method of polynomial interpolation, *the minimum of an objective function in a given domain is replaced by the minimum of an equivalent polynomial in the same domain*. This method may have better rates of convergence compared with the method of comparison the function value only.

Powell's quadratic interpolation

Assume that the objective function $F(x)$ is approximated by a quadratic interpolation function of the form:
$$\phi(x) = a_0 + a_1 x + a_2 x^2. \quad (11.3.10)$$
It is shown by the curved line of dashes in Fig. 11.3.2. The three coefficients a_0, a_1, a_2 are determined by 3 values of the original objective function $F(x)$.

11.3 Linear search and single variable optimization

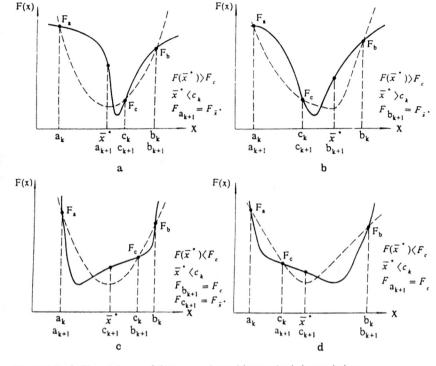

Fig. 11.3.2a–d. The minimum of $F(x)$ approximated by quadratic interpolation

Since the minimizer of Eq. (11.3.10) must satisfy the following equation

$$x^* = -\frac{a_1}{2a_2} \tag{11.3.11}$$

only the ratio a_1/a_2 is of interest. Assume that the interval of interest of the optimized function is $[a, b]$ and that c is an internal point. Let F_a, F_b, F_c indicate the function value at points a, b, c. Hence the minimizer of Eq. (11.3.11) is determined by:

$$\bar{x}^* = \frac{1}{2}\frac{(b^2 - c^2)F_a + (c^2 - a^2)F_b + (a^2 - b^2)F_c}{(b - c)F_a + (c - a)F_b + (a - b)F_c}. \tag{11.3.12}$$

After $\bar{x}^*$ and $F(\bar{x}^*)$ is calculated, according to the four cases shown in Fig. 11.3.2, the interval $[a_k, b_k]$ is reduced sequentially. If the criterion

$$b_k - a_k < \varepsilon_1 \tag{11.3.13}$$

or

$$\frac{F(c_k) - F(c_{k+1})}{F(c_k)} < \varepsilon_2 \tag{11.3.14}$$

is satisfied, then the approximate minimizer is obtained.

In the minimization of multivariable functions, the minimizer is searching in a specific directions, p_k in each step. In the direction p_k an optimum step λ_k is determined by a linear search, i.e.

$$F(\lambda) = F\langle X_k + \lambda p_k) = \min . \tag{11.3.15}$$

In this case, the three points are on the line of $X_k + \lambda p_k$. The optimum of an univariate function is one step of the minimization of the multivariable function. Detailed procedures of this method are given in reference [2].

Davidson's cubic interpolation

The cubic interpolation method uses the minimizer of a cubic function

$$\varphi(x) = a_0 + a_1 x + a_2 x^2 + a_3 x^3$$

to approximate the minimizer of the function. Two function values together with their directional derivatives within the range $[a, b]$ are used to determine the four constants of a cubic interpolation. If the directional derivatives of $F(x)$ can be evaluated, then the cubic interpolation method has higher accuracy than the method of quadratic interpolation. Detailed formulations are derived in reference [2].

11.4 Analytic methods of unconstrained optimization problems

Methods for solving unconstrained minimization of an arbitrary multivariable function $F(X)$ are divided into two kinds: the analytic methods and the direct methods. The analytic method is also called the gradient method, it is based on the derivatives of the objective function. Direct methods are based only on the value of the objective function itself, hence it is a function comparison method. These two kind of methods are described in Sects. 11.4 and 11.5.

Analytical methods are dependent on the analytical properties of the objective function. At the starting point of the kth iteration, the variables are denoted as X_k, the search direction p_k and the optimum step λ_k are used to estimate the new point, i.e.

$$X_{k+1} = X_k + \lambda_k p_k \tag{11.4.1}$$

where the optimum step λ_k is determined by the univariate minimization as discussed in Sect. 11.3, i.e.

$$F(\lambda) = F(X_k + \lambda_k p_k) = \min . \tag{11.4.2}$$

If λ_k is determined by Eq. (11.4.2), the method is called an exact linear search. Opposite to the exact linear search, the simple method is to assume that $\lambda_k = 1$.

11.4 Analytic methods of unconstrained optimization problems

A number of analytical optimization methods are dependent on the different choices of the search direction p_k. These are discussed as follows.

11.4.1 Steepest descent method

The direction opposite to that of the gradient of the objective function is an obvious choice as the steepest descent direction for minimizing the function, i.e.

$$p_k = -\nabla F(X) = -g_k(X). \tag{11.4.3}$$

The exact linear search is then taken to determine the optimum step λ_k, i.e.

$$F(\lambda) = F(X_k + \lambda_k p_k) = \min \tag{11.4.4}$$

then the following equation is obtained.

$$\frac{dF}{d\lambda} = [\nabla F(X_k + \lambda_k p_k)]^T p_k = 0 \tag{11.4.5}$$

i.e.

$$[g_{k+1}(X)]^T p_k = 0. \tag{11.4.6}$$

Equation (11.4.5) indicates that the sequential directions p_{k+1} and p_k of the search are orthogonal to each other.

The steps of the steepest descent method are as follows:

(1) Assume X_0, ε
(2) Calculate g_k, let $p_k = -g_k$
(3) Find the optimum step λ_k along the direction p_k
(4) Evaluate $X_{k+1} = X_k + \lambda_k p_k$
(5) Calculate g_{k+1}, if $\|g_{k+1}\| < \varepsilon$ then output the results otherwise $k = k + 1$ and return back to step 2.

The rate of convergence of the steepest descent method is slow when the search is close to the minimum. The zig-zag progress of the search direction for a typical quadratic function is shown in Fig. 11.4.1. *If the contours are oblate ellipsoidal, in other words the Hessian matrix of the objective function is ill-conditioned, the convergence is very slow in the vicinity of the minimizer.* If the contours of the objective function are hyperspherical, the direction of steepest descent points directly to the minimum. The convergent rate is quadratic.

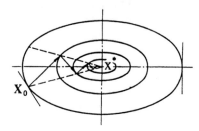

Fig. 11.4.1. The convergent process of the steepest descent method

Consequently, the disadvanatage of the method of steepest descent is that the speed of the convergence depends significantly on the property of the objective function. For example, if $F(X) = x_1^2 + 25x_2^2$, let $X_0 = (2, 2)^T$, after 4th iteration the minimum is obtained. If the variable x_2 is changed by $y = 5x_2$, then $F(x, y) = x^2 + y^2$, the minimum is obtained in a single computation. Hence the method of steepest descent is not recommended directly as a very effective method.

11.4.2 Conjugate gradient method

The aim of this method is to obtain an effective process with quadratic convergence and with the requirement of a lower order of derivatives of the objective function in the process of searching. The definition of conjugate direction, quadratic termination and the procedures of the conjugate gradient method are introduced in this section.

11.4.2.1 Conjugate direction

Let **G** be a symmetric positive definite matrix of the order $n \times n$. p_k ($k = 0, 1, \ldots, n - 1$) are non-zero vectors of the order n. If for any value of i the following condition is satisfied

$$p_i^T G p_j = 0 \quad (i \neq j, i, j = 0, 1, \ldots, n - 1). \tag{11.4.7}$$

Then p_i is said to be mutually conjugate with respect to a positive definite matrix **G**.

11.4.2.2 Quadratic convergence

If the design variables of the objective function are denoted by n and the minimizer is obtained during the nth iteration, i.e.

$$X_n = X^* \tag{11.4.8}$$

then the method has the property of quadratic convergence. This result is obtained if *a quadratic function with a positive definite Hessian matrix* **G** *and the conjugate directions p_i are chosen as the directions of search* and *the exact linear search is used for determining λ_k.* This characteristic is analysed as below.

For a quadratic function

$$F(X) = \tfrac{1}{2} X^T A X + bX + c \tag{11.4.9}$$

the first order derivative of $F(X)$ is

$$g(X) = \nabla F(X) = AX + b \tag{11.4.10}$$

11.4 Analytic methods of unconstrained optimization problems

then

$$g_{k+1} = AX_{k+1} + b = A(X_k + \lambda_k p_k) + b = g_k + \lambda_k A p_k . \quad (11.4.11)$$

Substitution of Eq. (11.4.11) into Eq. (11.4.5) leads to

$$g_k^T p_k + \lambda_k p_k^T A p_k = 0 \quad (11.4.12)$$

or

$$\lambda_k = -\frac{g_k^T p_k}{p_k^T A p_k} . \quad (11.4.13)$$

Rewriting Eq. (11.4.11) in a sequential form yields

$$g_k = g_{k-1} + \lambda_{k-1} A p_{k-1} = g_{k-2} + \lambda_{k-2} A p_{k-2} + \lambda_{k-1} A p_{k-1} . \quad (11.4.14)$$

Considering that

$$\Delta g_k = g_{k+1} - g_k = A \Delta X_k = \lambda_k A p_k . \quad (11.4.15)$$

then

$$g_k = g_{j+1} + \sum_{i=j+1}^{k-1} \Delta g_i \quad (11.4.16)$$

whereby multiplying p_j^T on both sides of Eq. (11.4.16), leads to

$$(p_j)^T g_k = (p_j)^T g_{j+1} + \sum_{i=j+1}^{k-1} (p_j)^T \Delta g_i \quad j = 0, 1, \ldots, k-1 . \quad (11.4.17)$$

Due to the first term of the RHS of Eq. (11.4.17) being zero and combining this equation with Eq. (11.4.15), the following result is obtained

$$(p_j)^T g_k = \sum_{i=j+1}^{k-1} (p_j)^T A \Delta X_i = \sum_{i=j+1}^{k-1} \lambda_k (p_j)^T A p_i$$

$$(j = 0, \ldots, k-1) . \quad (11.4.18)$$

If $k = n$, then

$$(p_j)^T g_n = 0 \quad (11.4.19)$$

i.e. $\nabla F(X_n) = 0$, thus $X_n = X^*$. Therefore, the minimizer is obtained during the nth iteration.

11.4.2.3 Selection of conjugate directions

As demonstrated in the former section, if the conjugate directions were taken as the search directions of search, the solution is obtained during nth iterations. However the directions of conjugates were not specified. There are a number of possibilities to determine the direction p_k. p_0 can be chosen arbitrarily. The sequential direction p_k is any vector orthogonal to Δg_j. The conjugate gradient methods are based on the choice of the conjugate gradient directions as p_k. It starts by finding the direction of gradient, i.e. $p_0 = -g_0$. Then the continuously searching directions are determined by the following formulation

$$p_{k+1} = -g_{k+1} + \beta_k p_k \tag{11.4.20}$$

where

$$\beta_k = g_{k+1}^T g_{k+1} / g_k^T g_k . \tag{11.4.21}$$

These formulations were developed by Fletcher and Reeves [9]. Thus it is called the F-R Algorithm. Only three vectors are used. In order to ensure the second order convergence and to avoid the round-off errors which could ruin the conjugacy of the search direction, one uses the exact linear search in each iteration, and the technique of restarting (after several iterations, the routine is restarted at p_0) is used.

The conjugate gradient method is not only used to find the minimum of functions. It is also used to find the solution of the large sparse matrix equation composed by linear and non-linear algebraic equations. For example, this method is used to solve the simultaneous equations derived from FDM and FEM as introduced in Section 4.3.2.2. The steps of this method are shown in Example 11.4.1.

Example 11.4.1

$$\text{Min } F(X) = \tfrac{3}{2}x_1^2 + \tfrac{1}{2}x_2^2 - x_1 x_2 - 2x_1$$

Solution. Rewrite the above equation as a matrix form shown in Eq. (11.4.9), i.e.

$$A = \begin{bmatrix} 3 & -1 \\ -1 & 1 \end{bmatrix} \qquad b = [-2 \quad 0]^T .$$

The derivative of the objective function is

$$g(X) = \nabla F(X) = [3x_1 - x_2 - 2 \quad x_2 - x_1]^T .$$

Let
$$X_0 = [-2 \quad 4]^T$$
then
$$\nabla F(X_0) = [-12 \quad 6]^T \qquad p_0 = [12 \quad -6]^T .$$

Based on Eq. (11.4.13),

$$\lambda_0 = -\frac{g_0 p_0}{p_0^T A p_0} = \frac{5}{17}$$

then

$$X_1 = X_0 + \lambda p_0 = \begin{bmatrix} \tfrac{26}{17} & \tfrac{38}{187} \end{bmatrix} \qquad g(X_1) = \begin{bmatrix} \tfrac{6}{17} & \tfrac{12}{17} \end{bmatrix} .$$

The direction p_1 is evaluated by Eq. (11.4.21) and Eq. (11.4.20), i.e.,

$$\beta_0 = \frac{g_1^T g_1}{g_0^T g_0} = \frac{(\tfrac{6}{17})^2 + (\tfrac{12}{17})^2}{(-12)^2 + (6)^2} = \frac{1}{289}$$

then
$$p_1 = -g_1 + \beta_0 p_0 = \left[\frac{90}{289} \quad \frac{210}{289}\right]^T$$

Repeat the above computation to obtain λ_1, X_2, g_2 and so on, the final result is

$$X^* = [1 \quad 1]^T .$$

The computer program of this method is given in reference [7].

11.4.3 Quasi-Newton's methods

The steepest descent method and the conjugate gradient method are based on the computation of the first derivative of the objective function. Quasi-Newton's methods are the alternative of Newton's method (the definition will be seen soon). The main difference between these kind of methods is that quasi Newton's methods are based on the modified second order derivatives of the objective function. Sometimes they are more economical, and converge faster.

Recall that

$$g_k = AX_k + b$$

If

$$X_{k+1} = X_k + p_k \tag{11.4.22}$$

then

$$g_{k+1} = A(X_k + p_k) + b = g_k + A(X)p_k . \tag{11.4.23}$$

If X_{k+1} is the minimizer of the function, then $g_{k+1} = 0$, hence

$$p_k = -A^{-1}g_k . \tag{11.4.24}$$

This equation indicates that if the search direction p_k is determined by the above equation, then the minimizer is immediately obtained. This kind of methods are called Newton's methods. If $A(X^*)$ is positive definite, then the convergence of Newton's method is of second order. This is the fastest rate of convergence obtainable in non-linear optimization. Thus, this kind of method is very important. If the objective function is not a quadratic function, and if p_k is given by Eq. (11.4.24), then $X_k + p_k$ will not be the minimizer directly consequently the process has to be executed iteratively.

In general, it is very difficult to find the second order derivative matrix of the objective function even if it exists. The determination of the positive definite of the Hessian matrix is even more difficult. If A_k is singular, there are two possibilities: either there is no solution of Eq. (11.4.24) or there are an infinite number of solutions. In addition, if X_k is a saddle point and A_k is nonsingular then $g_k = 0$. Equation (11.4.24) is satisfied only if $p_k = 0$. Thus it should not be used as a search vector. Consequently, Newton's method is not a practical method for function minimization. However based on the same idea of

Newton's method, a number of modified Newton's methods, Quasi-Newton's methods, were derived having the same asymptotic rate of convergence.

11.4.3.1 Davidon–Fletcher–Powell (DFP) method

This method was first suggested by W.C. Davidon in 1959 [10]. It was commented on by R. Fletcher and M.J.D. Powell in 1963 [11] and improved by Fletcher in 1970 [12]. The abbreviation of this method is DFP. It is an efficient method for solving the unconstrained minimization of non-linear functions. *It combines the advantage of the method of steepest descent in which the descent of the function is rapid in the beginning of the search and the advantage of the Newton's method, in which convergence is quick in the vicinity of the minimum point. The main idea of the DFP method is to construct an approximate matrix* $\mathbf{H}_k$ *to substitute the second order derivative matrix of the objective function and to force* $\mathbf{H}_n = \mathbf{A}^{-1}$. Hence it is known as a variable matrix method. When the DFP method is used together with exact linear search to minimize a quadratic function having a positive definite Hessian matrix, the vectors p_k in Eq. (11.4.25) satisfy the conjugacy property, then the method has second order convergence.

The procedure of this method is described as follows

$$p_k = -\mathbf{H}_k g_k \tag{11.4.25}$$

$$X_{k+1} = X_k + \lambda_k p_k \tag{11.4.26}$$

$$\mathbf{H}_0 = I. \tag{11.4.27}$$

$\mathbf{H}_0$ is an identity matrix of order $n \times n$, n is the number of design variables of the objective function, $\mathbf{H}_k$ is constructed to approximate the inverse of the Hessian matrix of the objective function $\mathbf{A}_k^{-1}$ and has to satisfy the following conditions:

(1) In order to ensure that the algorithm is stable, the following criterion must be satisfied

$$F(X_{k+1}) < F(X_k). \tag{11.4.28}$$

It is proved [2] that, when $\nabla F(X_k) \neq 0$, only if $\mathbf{H}_k$ is a positive definite matrix, then condition Eq. (11.4.28) is possible.

(2) In order to obtain the quadratic termination, p_k should be conjugated to matrix A, i.e. $p_{k+1}^T \mathbf{A} p_k = 0$.

(3) To simplify the computation, a recurrence formula is used to update the Hessian matrix $\mathbf{H}_{k+1}$, i.e.,

$$\mathbf{H}_{k+1} = \mathbf{H}_k + \Delta \mathbf{H}_k. \tag{11.4.29}$$

Thus $\mathbf{H}_{k+1}$ is

$$\mathbf{H}_{k+1} = \mathbf{H}_k + \mathbf{C}_k + \mathbf{D}_k = \mathbf{H}_k + \frac{\sigma_k \sigma_k^T}{\sigma_k^T \gamma_k} - \frac{\mathbf{H}_k \gamma_k \gamma_k^T \mathbf{H}_k}{\gamma_k^T \mathbf{H}_k \gamma_k} \tag{11.4.30}$$

11.4 Analytic methods of unconstrained optimization problems

where

$$\gamma_k = g_{k+1} - g_k \tag{11.4.31}$$

$$\gamma_k = X_{k+1} - X_k = \lambda_k p_k . \tag{11.4.32}$$

The matrix H_{k+1} defined by Eq. (11.4.30) is positive definite and symmetric only if H_0 is positive definite and symmetric. Consequently $H_n = A^{-1}, X^* = X_n$. This expresses that the DFP method has second order convergence. The procedures of DFP method illustrated by the following example.

Example 11.4.2

$$\text{Min } F(X) = \tfrac{3}{2}x_1^2 + \tfrac{1}{2}x_2^2 - x_1 x_2 - 2x_1 \tag{11.4.33}$$

Solution.

$$g(X) = \nabla F(X) = [3x_1 - x_2 - 2 \quad x_2 - x_1]^T \tag{11.4.34}$$

Let

$$X_0 = [1 \quad 2]^T, \quad H_0 = \begin{bmatrix} 1 & 0 \\ 0 & 1 \end{bmatrix}$$

then

$$g_0 = \nabla F(X_0) = [-1 \quad 1]^T .$$

According to Eq. (11.4.25),

$$p_0 = -H_0 g_0 = -[-1 \quad 1]^T$$

Then

$$F(X_0 + \lambda_0 p_0) = 3\lambda_0^2 - 2\lambda_0 - \tfrac{1}{2} . \tag{11.4.35}$$

The optimum value of λ_0 is obtained by exact linear search, i.e. min $F(X_0 + \lambda_0 p_0)$ the result is $\lambda_0 = 1/3$. Substitution of λ_0 into Eq. (11.4.32) yields

$$\sigma_0 = \lambda_0 p_0 = \tfrac{1}{3}[1 \quad -1]^T$$

then

$$X_1 = X_0 + \sigma_0 = [4/3 \quad 5/3]^T$$

$$g_1 = [1/3 \quad 1/3]^T$$

$$F(X_1) = -5/8 .$$

Using Eqs. (11.4.31), one obtains

$$\gamma_0 = g_1 - g_0 = [4/3 \quad -2/3]^T$$

and then

$$\sigma_0 \sigma_0^T = \frac{1}{9}\begin{bmatrix} 1 & -1 \\ -1 & 1 \end{bmatrix}$$

$$\sigma_0^T \gamma_0 = \tfrac{1}{3}[1 \quad -1]\begin{bmatrix} 4/3 \\ -2/3 \end{bmatrix} = 2/3 .$$

Then the modified matrices $\mathbf{C}_0$, $\mathbf{D}_0$ are obtained

$$\mathbf{C}_0 = \frac{\sigma_0 \sigma_0^T}{\sigma_0^T \gamma_0} = \frac{1}{6}\begin{bmatrix} 1 & -1 \\ -1 & 1 \end{bmatrix}$$

$$\mathbf{D}_0 = -\frac{\mathbf{H}_0 \gamma_0 \gamma_0^T \mathbf{H}_0}{\gamma_0^T \mathbf{H}_0 \gamma_0} = -\frac{\begin{bmatrix} 4/3 \\ -2/3 \end{bmatrix}[4/3 \quad -2/3]}{[4/3 \quad -2/3]\begin{bmatrix} 4/3 \\ -2/3 \end{bmatrix}} = -\frac{1}{5}\begin{bmatrix} 4 & -2 \\ -2 & 1 \end{bmatrix}$$

Substitution of $\mathbf{H}_0$, $\mathbf{C}_0$, $\mathbf{D}_0$ into Eq. (11.4.30) leads to

$$\mathbf{H}_1 = \mathbf{H}_0 + \mathbf{C}_0 + \mathbf{D}_0 = \frac{1}{30}\begin{bmatrix} 11 & 7 \\ 7 & 29 \end{bmatrix}$$

Repeating the same procedures as above, the following vectors are obtained:

$$p_1 = -\mathbf{H}_1 g_1 = -\tfrac{1}{3}[1 \quad 2]^T$$

$$X_2 = X_1 + \lambda_1 p_1 = \begin{bmatrix} 4/3 & -\lambda/5 \\ 5/3 & -2\lambda/5 \end{bmatrix}$$

$$F(\lambda_1) = \frac{3}{50}\lambda - \frac{1}{5}\lambda - \frac{5}{6}$$

$$\text{Min } F(\lambda_1), \quad \lambda_1 = 5/3$$

$$\sigma_1 = \lambda_1 p_1 = -\tfrac{1}{3}[1 \quad 2]^T$$

$$X_2 = X_1 + \sigma_1 = [1 \quad 1]^T$$

$$g_2 = [0 \quad 0]^T$$

where g_2 satisfies the condition to terminate the calculation. Thus $X_2 = [1 \quad 1]^T$ is the minimizer. It is easy to verify that $\mathbf{H}_2 = \mathbf{A}^{-1}$ by using Eq. (11.4.34) to calculate the second order derivative of the objective function and the same procedures to caculate $\mathbf{H}_2$.

This method has been used to calculate the optimum locations of simulated charges in Example 7.6.1. The computer program of DFP algorithm is given in reference [7].

11.4.3.2 BFGS formulation

Sometimes the numerical stability of the DFP method is dependent on the round-off error. Brodgen, Fletcher, Goldfarb and Shanno independently published an improved iterative method

$$\mathbf{H}_{k+1} = \mathbf{H}_k + \frac{\beta_k \sigma_k \sigma_k^T - \mathbf{H}_k \gamma_k \sigma_k^T - \sigma_k \gamma_k^T \mathbf{H}_k}{\sigma_k^T \gamma_k} \qquad (11.4.36)$$

11.4 Analytic methods of unconstrained optimization problems

where

$$\beta_k = 1 + \frac{\gamma_k^T H_k \gamma_k}{\sigma_k^T \gamma_k}. \qquad (11.4.37)$$

If Eq. (11.4.36) is used in the iteration, then the method used is the BFGS formulation.

11.4.3.3 B matrix formulae

It was shown that H_k is the approximate of the inverse of matrix A. If an another matrix B_k is taken to approximate the matrix A itself, a complementary or dual formulation is obtained. Interchange the matrices as follows

$$B_k \leftrightarrow H_k$$

$$B_{k+1} \leftrightarrow H_{k+1}$$

$$\Delta X_k \leftrightarrow \Delta g_k$$

Then Eq. (11.4.30) to Eq. (11.4.32) transform to

$$B_{k+1} = B_k + \frac{\gamma_k \gamma_k^T}{\gamma_k^T \sigma_k} - \frac{B_k \sigma_k \sigma_k^T B_k}{\sigma_k^T B_k \sigma_k} \qquad (11.4.38)$$

$$B_k p_k = -g_k \qquad (11.4.39)$$

$$X_{k+1} = X_k + \lambda_k p_k \qquad (11.4.40)$$

This set of equations represent the B matrix formulation.

11.4.3.4 Cholesky factorization of the Hessian matrix

In view of the points discussed above, during the iteration, matrices B_k and H_k remain positive definite. However in practice, the round-off error may cause matrix B_k or H_k to be singular. One possible way to solve this problem is to restart H_k or B_k as an identity matrix frequently. The drawback of this method is that some useful information is lost at each start. Gill and Murray [3] suggest decomposing the matrix B_k into Cholesky factors and thereby avoiding the singularity and loss of information at each iteration. Let

$$B_k = L_k D_k L_k^T \qquad (11.4.41)$$

D_k is a diagonal matrix and L_k is a lower triangular matrix with diagonal elements of unity. The effect of round-off errors (which causes the updating process to produce a numerically singular matrix) can be detected immediately by monitoring the elements of D_{k+1}. *Whenever any diagonal element of D_{k+1} is less than a small positive quantity δ, then the element is replaced by δ and the rest of the factorization process is modified accordingly. This procedure guarantees that B_{k+1} is numerically positive definite.*

11.4.4 Method of non-linear least squares

The conjugate gradient methods and quasi-Newton methods are efficient for all sufficiently smooth objective functions. If the objective function has the form

$$F(X) = \sum_{i=1}^{m} f_i^2(X) = \|f(X)\|^2 = f^T(X)f(X) . \tag{11.4.42}$$

the gradient and Hessian matrix of the objective function have special structures as below

$$g(X) = \nabla F(X) = 2J^T(X)f(X) \tag{11.4.43}$$

$$G(X) = \nabla g(X) = 2J^T(X)J(X) + 2Q(X) \tag{11.4.44}$$

where

$$J(X) = \begin{bmatrix} \dfrac{\partial f_1}{\partial x_1} & \dfrac{\partial f_1}{\partial x_2} & \cdots & \dfrac{\partial f_1}{\partial x_n} \\ \vdots & & & \\ \dfrac{\partial f_m}{\partial x_1} & \dfrac{\partial f_m}{\partial x_2} & & \dfrac{\partial f_m}{\partial x_n} \end{bmatrix} \tag{11.4.45}$$

$$Q(X) = \sum_{i=1}^{m} f_i(X) K_i(X) \tag{11.4.46}$$

$$K_i(X) = \nabla^2 f_i(X) \tag{11.4.47}$$

$J(X)$ is a Jacobian matrix of the order $m \times n$. Since $F(X)$ is being minimized in the least square sense, the components $\nabla^2 f_i$ are small and $G(X)$ may be approximated by

$$G(X) \cong 2J^T J . \tag{11.4.48}$$

11.4.4.1 Gauss–Newton method

Gauss noted that if $f_i(X)$ are all linear functions of X and $F(X)$ is quadratic then the Jacobian matrix is a constant. He suggested approximating the gradient at a point $X + \Delta X$ as follows

$$\nabla F(X + \Delta X) \cong 2J(X)f(X + \Delta X) . \tag{11.4.49}$$

11.4 Analytic methods of unconstrained optimization problems

The approximate form of $f(X + \Delta X)$ is obtained from the linear terms of Taylor's expansion

$$f(X + \Delta X) \cong f(X) + \mathbf{J}^T(X)\Delta X . \tag{11.4.50}$$

Combination of Eqs. (11.4.49) and (11.4.50) gives an estimation of the gradient of $F(X)$ at $X + \Delta X$

$$\nabla F(X + \Delta X) \cong 2\mathbf{J}[f(X) + \mathbf{J}^T(X)(\Delta X)]$$
$$= 2[\mathbf{J}(X)f(X) + \mathbf{J}(X)\mathbf{J}^T(X)\Delta X] . \tag{11.4.51}$$

Since the m rows of $\mathbf{J}$ have been assumed to be linearly independent, the matrix $\mathbf{JJ}^T$ is nonsingular and the inverse matrix $[\mathbf{JJ}^T]^{-1}$ exists. Then the solution of $\nabla F(X + \Delta X)$ is

$$\Delta X = -(\mathbf{JJ}^T)^{-1}\mathbf{J}f(X) \tag{11.4.52}$$

Comparison of Eq. (11.4.52) with Eq. (11.4.1) shows that in the method of least squares

$$p = -(\mathbf{JJ}^T)^{-1}\mathbf{J}f(X) \tag{11.4.53}$$

and

$$\lambda = 1 \tag{11.4.54}$$

Recall that in Newton's method $\mathbf{G}_k p_k = -g_k$, thus

$$(\mathbf{J}_k^T \mathbf{J}_k + \mathbf{Q})p = -\mathbf{J}_k^T f_k(X) \tag{11.4.55}$$

Hence the speed of convergence of the non-linear least square method is the same as quasi-Newton's method and requires only the first order derivative of the objective function. The computer storage required is much less than that using DFP method.

If $\|f(X)\|$ tends to zero as X approaches the minimizer, the matrix $\mathbf{Q}$ also tends to zero. Consequently the least square method is applicable when the first order term $\mathbf{J}^T\mathbf{J}$ of Eq. (11.4.55) dominates compared to the second order term $\mathbf{Q}(X)$. This does not hold when the residuals of the solution are very large. In such a case, one might as well use a general unconstraint method.

11.4.4.2 Levenberg–Marquardt method

The L–M method is an alternative to the Gauss–Newton method, it includes a technique to deal with problems related to the singularity of matrix $\mathbf{J}_k^T \mathbf{J}_k$. In this method, Eq. (11.4.55) is modified to:

$$(\mathbf{J}_k^T \mathbf{J}_k + \mu_k \mathbf{I})p_k = -\mathbf{J}_k^T f_k \tag{11.4.56}$$

where $\mu_k \geq 0$ is a scalar and $\mathbf{I}$ is the unit matrix of order n. For a sufficiently large value of μ_k, the matrix $\mathbf{J}_k^T \mathbf{J}_k + \mu_k \mathbf{I}$ is positive definite and then p_k is the direction of descent. As $X \to X^*$, and $\mu_k \to 0$ the method yields asymptotic convergence in

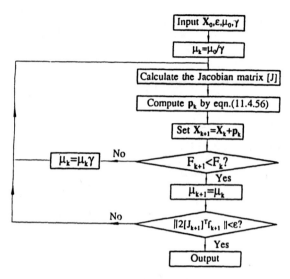

Fig. 11.4.2. The flow-chart of L-M method

the same way as Gauss–Newton's method. The flow-chart of the $L-M$ method is given in Fig. 11.4.2.

The computer program of the least square method is given in reference [7]. In reference [15] the least square algorithm is used to find the positions of the equivalent currents in an exterior eddy current problem.

11.5 Function comparison methods

If the objective function of a physical problem cannot be described analytically or the objective function is discontinuous or if the gradient of the objective function, $g(X)$, is discontinuous, then the analytical method introduced in Sect. 11.4 fails. *Since the function comparison methods only use the value of the objective function in the process of search, they seem to be the simplest. The price paid is reduced speed and reliability.* The method is very sensitive for the accumulation of round off errors. Another disadvantage of function comparison methods is that the convergence cannot be guaranteed. Hence many authors recommend this method only if there is no other suitable alternative method available.

11.5.1 Polytope method

The name 'polytope method' is used to solve unconstrained non-linear optimization problems instead of the 'simplex method' which is defined as linear programming (all of the objective functions and the constrained functions are linear functions). The name 'simplex' or 'polytope' contains the geometrical meaning. A 2-D function in space R_2 describes a plane and the simplest form of the polytope in a plane is a triangle. A 3-D function in space R_3 describes a tetrahedron. Thus a 2-D or 3-D function can be approximated by the function values at the 3 nodes of the triangle or the function values at the 4 nodes of the tetrahedron, respectively. For n-D problem, the function can be approximated by the values of the $n + 1$ vertices of a polytope in n-dimensional space, e.g. the function values are $F_{n+1} \geq F_n \geq \ldots F_2 \geq F_1$ at points $X_{n+1}, X_n, \ldots, X_1$. In the optimization process, compare the given function values on the vertices and replace the 'worse' point (the definition of the worse point is the point where the value of the function is the highest) by taking a new point. This new point generates a new polytope. Repeat these procedures until the minimum point is approached. Detailed formulation and the computer program can be found in reference [2, 6].

This method was used in reference [16] to determine the desired contour of an electrode of a gas circuit breaker.

11.5.2 Powell's method of quadratic convergence

Powell's method is a direct search method. It consists of constructing a set of conjugate directions without the explicit use of the gradient vector. The initial directions $p_{0,k}(k = 1, \ldots, n)$ of the search are parallel to the axes of the coordinates. The two subscripts 0 and k of p denote the iterative times and the sequent number of the variables, respectively. Along each direction of search an univariate minimum is required. Here, Powell's method of quadratic interpolation is recommended.

The steps of this methd are

(1) In the kth iteration, let $X_{k0} = X_k$, search $X_{kr}(r = 1, \ldots, n)$ in direction p_{kr}.
(2) Calculate $\Delta = \max\{F(X_{k,r-1}) - F(X_{kr})\} = F(X_{k,q-1}) - F(X_{k,q})$, q is the value of r which maximizes Δ.
(3) Define $F_1 = F_{k0}(X)$, $F_2 = F_{kn}(X)$, evaluate $F_3 = F(2X_{kn} - X_{k0})$.
(4) If either $F_3 \geq F_1$ or $(F_1 - 2F_2 + F_3)(F_1 - F_2 - \Delta)^2 \geq \frac{1}{2}\Delta(F_1 - F_3)^2$ is satisfied, the old direction p_{kj} is used for the $(k + 1)$th iteration and let $X_{k+1} = X_{kn}$. In any other case go to step 5.
(5) In the $(k + 1)$th iteration, the search directions are:

$$p_{k,1}, \ldots p_{k,q}, p_{k,q+1}, \ldots p_{k,n}, \delta_k$$

where $\delta_k = X_{k,n} - X_k$.
If step 2 yields $q = 1$, then the search directions are $p_{k,2}, \ldots p_{k,n}, \delta_k$.

Example 11.5.1

Minimize $F(X) = x_1^2 - x_1 x_2 + 3x_2^2$

Solution.

(1) Let $X_0 = [1\ 2]^T$, then $p_{0,1} = e_1 = [1\ 0]^T$, $p_{0,2} = e_2 = [0\ 1]^T$

In $p_{0,1}$, $X_{01} = X_0 + \lambda p_{0,1}$ Min $F(\lambda)$, yields $X_{01} = [1\ 2]^T$

In $p_{0,2}$, $X_{02} = X_0 + \lambda p_{0,2}$ Min $F(\lambda)$, yields $X_{02} = [1\ 1/6]^T$

(2) Calculate $\Delta = \max\{F(X_0) - F(X_{01}), F(X_{01}) - F(X_{02})\} = \max(11 - 11, 11 - 11/12\} = 121/12$, hence $q = 2$.

(3) Calculate F_1, F_2, F_3: $F_1 = F(X_0) = 11$, $F_2 = F(X_{02}) = 11/12$, $F_3 = F(2X_{02} - X_1) = 11$

(4) As $F_3 = F_1$, hence $p_{1,1} = p_{0,1}$, $p_{1,2} = p_{0,2}$

The next iteration starts with $X_1 = [1\ 1/6]$
Then

$$X_{11} = [1/12\ 1/6], \quad X_{12} = [1/12\ 1/72]$$
$$\Delta = \max\{F(X_1) - F(X_{11}), F(X_{11}) - F(X_{12})\} = 121/144,$$
hence $q = 1$
$$F_1 = F(X_1) = 11/12, \quad F_2(X_{12}) = 33/5184,$$
$$F_3 = F(2X_{12} - X_1) = 275/432$$

Note here, $F_3 < F_1$ and $(F_1 - 2F_2 + F_3)(F_1 - F_2 - \Delta)^2 < \frac{1}{2}\Delta(F_1 - F_3)^2$, then the next search starts with $X_1 = [1/12\ 1/72]^T$ in the direction $p_{1,1}$, δ_1

$$\delta_1 = X_{12} - X_1 = [-11/12\ -11/72]$$

then obtain

$$X^* = [0\ 0]^T.$$

If the problem has n variables, the individual direction of coordinates are chosen as the search direction in the sequence, i.e.,

$$p_k = e_k$$
$$e_k = [0\ 0\ 0\ldots 0\ 1_k\ 0\ 0\ldots 0].$$

The computer program of this method is given in reference [17].

11.6 Constrained optimization methods

For physical problems, there are restrictions on the acceptable range of the variables. These conditions are the constraints of the problem. If the constraints of the variables are simple, the method of variable transformation may be used as the first approach [1]. For example

11.6 Constrained optimization methods

Minimize $f(x)$

subject to $x < c$ (11.6.1)

or $a \leq x \leq b$

The constraints are written as

$$x = c - y^2$$

or

$$x = a + (b - a)\sin^2 y .$$

In this approach, y is chosen as a new variable. This technique has been used in Chap. 7 to find the optimum positions of the simulated charges. In this section, general methods for solving constrained optimization problems are introduced.

In general, the methods for solving constrained optimization problems are two kinds. The first type approaches the constrained conditions directly e.g. the method using feasible direction or the method of gradient-projection and so on see reference [17]. The second type converts the constrained optimization problem into a sequence of unconstrained optimization problems. The second kind of method are more often used as seen in references [18, 19].

11.6.1 Basic concepts of constrained optimization

The problem of constrained optimization is more complicated than the problem of unconstrained optimization. In the case of constrained optimization, the independent variables X_i are restricted in a specified region of interest which is the feasible region. A point X is called feasible if it satisfies the conditions of constraint. The equality constraints define a series of hypersurfaces in the n-dimensional space. The feasible region S is the intersection of all of these hypersurfaces. For instance,

$$\begin{cases} \text{Minimize} & F(x_1, x_2, x_3) \\ \text{Subject to} & h_1(x_1, x_2, x_3) = 0 \\ & h_2(x_1, x_2, x_3) = 0 \end{cases} \quad (11.6.2)$$

The feasible region of this problem is a curve which is the intersection of the two hypersurfaces $h_1 = 0$, and $h_2 = 0$. A point X satisfying $h(X) = 0$ is called a regular point if the gradients ∇h are linearly independent. For inequality constraints, the feasible region lies on one particular side of each hypersurface, for example

$$\begin{cases} \text{Minimize} & F(x) = x_1 \\ \text{Subject to} & g_1(x) = (1 - x_1^2) - x_2 \geq 0 \\ & g_2(x) = x_1 \geq 0 \\ & g_3(x) = x_2 \geq 0 \end{cases} \quad (11.6.3)$$

The feasible region is the shaded area as shown in Fig. 11.6.1.

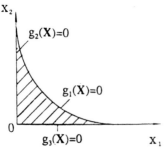

Fig. 11.6.1. Feasible region of Eq. (11.6.3)

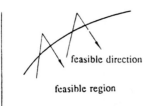

Fig. 11.6.2. Feasible direction of a constrained optimization

An inequality constraint $g(X) \leq 0$ is called active at a feasible point X^* if $g_i(X) = 0$ and it is called passive if $g_i(X) < 0$.

In the process to search the minimum of a function subject to some constraints, the movement of variables in the directions that yield feasible points are called feasible directions. See Fig. 11.6.2.

11.6.2 Kuhn–Tucker conditions

For unconstrained minimization, the necessary condition (Eq. (11.2.8)) indicates that at the minimizer X^*, there is no direction of descent. In constrained optimization, the complication of a feasible region has to be considered. Hence, a local minization must be a feasible point and with no feasible descent directions at X^*.

In 1951, Kuhn and Tucker published some results giving the optimal solution of non-linear constrained optimization [4]. These are necessary conditions for a local optimum and in certain special cases, they are also necessary and sufficient conditions for a global optimum. The conditions are discussed in the following subsections.

11.6.2.1 Lagrange multiplier method

The basic idea to approach the constrained optimization problem was proposed by Lagrange in 1760 [4]. It is to convert the constrained problem

$$\begin{cases} \text{Minimize} & F(X) \\ \text{Subject to} & h_i(X) = 0 \end{cases} \quad (11.6.4)$$

11.6 Constrained optimization methods

into an unconstrained problem

$$\text{Minimize } L(X, \lambda) = F(X) + \sum_{i=1}^{m} \lambda_i h_i(X) . \quad (11.6.5)$$

The new objective function $L(X, \lambda)$ is called Lagrangian. It has $n + m$ unknowns, n is the number of independent variables of $F(X)$ and m is the number of Lagrangian coefficients.

The necessary condition of Eq. (11.6.5) to have a minimum is

$$\begin{cases} \dfrac{\partial L}{\partial x_j} = 0 & j = 1, \ldots, n \\ \dfrac{\partial L}{\partial \lambda_i} = 0 & i = 1, \ldots, m . \end{cases} \quad (11.6.6)$$

The conditions $\partial L/\partial \lambda_i = 0$ guarantee that the constraints are satisfied at the optimal solution. Thus the optimal value of the Lagrangian functions is equal to the optimum of the original problem, i.e.,

$$L(X^*, \lambda^*) = F(X^*) . \quad (11.6.7)$$

The conditions of Eq. (11.6.6) are not sufficient for a constrained minimizer to exist. Sufficient conditions involve second or higher order derivatives of the objective function as indicated in the method of unconstrained optimization.

11.6.2.2 Necessary condition of the first order

For constrained optimization problems

$$\begin{cases} \text{Minimize} & F(X) \\ \text{Subject to} & h_i(X) = 0 \quad i = 1, \ldots, l \\ & g_j(X) \leq 0 \quad j = 1, \ldots, m \end{cases} \quad (11.6.8)$$

the necessary conditions of the first order are

$$\begin{cases} \lambda_0 \nabla F(X^*) + \sum_{i=1}^{l} \lambda_i \nabla h_i(X^*) + \sum_{j=1}^{m} \mu_j \nabla g_j(X^*) = 0 & (11.6.9) \\ \mu_j^* g_j(X^*) = 0 & (11.6.10) \end{cases}$$

where h_i, g_j are functions having continuous derivatives and $\nabla h_i(X^*)$ and $\nabla g_j(X^*)$ are linearly independent and λ_0 and μ_j, λ_i are Lagrange multipliers also known as Kuhn–Tucker multipliers. They are real numbers. For equality constraint, λ_i is unrestricted in sign; for inequality constraints $g_j \geq 0$, $\mu_j \leq 0$ and $g_j \geq 0$, $\mu_j \leq 0$. The following example is given to illustrate the use of Lagrange multipliers and the necessary conditions of the first order.

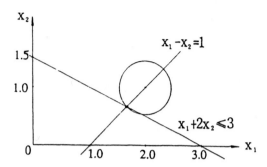

Fig. 11.6.3. The solution of Example 11.6.1

Example 11.6.1

$$\begin{cases} \text{Minimize} & F(X) = (x_1 - 2)^2 + (x_2 - 1)^2 \\ \text{Subject to} & x_1 + 2x_2 \leq 3 \\ & x_1 - x_2 = 1 \\ & x_1, x_2 \geq 0 \end{cases}$$

Solution. The Lagrangian multiplier of the above problem is

$$L(X, \lambda) = (x_1 - 2)^2 + (x_2 - 1)^2 + \lambda(x_1 - x_2 - 1) + \mu(x_1 + 2x_2 - 3).$$

The first order derivatives of $L(X, \lambda)$ regarding X, λ, μ are

$$\frac{\partial L}{\partial x_1} = 2(x_1 - 2) + \lambda + \mu = 0$$

$$\frac{\partial L}{\partial x_2} = 2(x_2 - 1) - \lambda + 2\mu = 0$$

$$\frac{\partial L}{\partial \lambda} = x_1 - x_2 - 1 = 0$$

$$\frac{\partial L}{\partial \mu} = (x_1 + 2x_2 - 3) = 0.$$

After solving these equations, one obtains $X^* = [5/3 \; 2/3]^T$. The figure is shown in Fig. 11.6.3.

11.6.2.3 Necessary and sufficient conditions of the second order

For unconstrained optimization, the positive definite of the Hessian matrix of the objective function has important implications in designing a satisfactory algorithm as it does for constrained optimization. If X is a local minimizer and is

11.6 Constrained optimization methods

the regular point of constraint, then λ_i μ_j, exist such that

$$L(X^*) = \nabla^2 F(X^*) + \sum_{i=1}^{l} \lambda_i \nabla^2 h_i(X^*) + \sum_{j=1}^{m} \mu_j \nabla^2 g_j(X^*) . \tag{11.6.11}$$

Thus the necessary and sufficient conditions of the second order are that Eq. (11.6.11) holds, and

$$y^T L(X^*) y \geq 0 . \tag{11.6.12}$$

It means that $L(X^*)$ is positive definite on the tangent subspace of the active constraints

$$A^{*T} \cdot y = 0 . \tag{11.6.13}$$

A^* denotes the matrix with columns of a_i,

$$a_i = \nabla c_i , \tag{11.6.14}$$

where c_i includes equality and inequality constraints, the Lagrangian function must have non-negative curvature for all feasible directions at X^*. When no constraints are present, Eq. (11.6.12) reduces to the condition that the Hessian matrix is positive definite.

$$y^T L(X^*) y > 0 . \tag{11.6.15}$$

Proof of the necessary and sufficient conditions of the first and the second order are given in reference [5].

11.6.3 Penalty and barrier function methods

These methods are used for problems with equality or inequality constrained optimiation. Consider an equality constrained optimization problem:

$$\begin{cases} \text{Minimize } F(X) \\ \text{Subject to } h_i(X) = 0 \quad i = 1, \ldots, m \end{cases} \tag{11.6.16}$$

A modified objective function $M(X)$ is used to replace the original problem

$$M(X) = F(X) + P(X) \tag{11.6.17}$$

where

$$P(X) = \sum_{i=1}^{m} K_i h_i^2(X) . \tag{11.6.18}$$

In Eq. (11.6.17), $M(X)$ is a generalized objective function and $P(X)$ is a penalty function. K_i are penalty factors, usually these are large values. When $K_i = 0$, *the constraints are ignored, when $K_i = \infty$ the constraints are exactly satisfied*. It can be seen from Eqs. (11.6.17) and (11.6.18), only if $h_i(X) = 0$, the unconstrained minimum of $M(X)$ is the minimum of $F(X)$. Now change the values of K_i until the minimum of $M(X)$ is obtained.

Geometrically, a penalty function replaces a constraint by a steep-step as shown in Fig. 11.6.4 (minimizing problem). It shows that the function value may be evaluated in the non-feasible regions. In each step the 'penalty' for violating a constraint is a high value of the modified objective function.

Example 11.6.2

$$\begin{cases} \text{Minimize} & F(X) = x_1^2 + x_2^2 \\ \text{Subject to} & h(X) = x_1 + x_2 - 1 = 0 \end{cases}$$

Solution. The modified objective function is

$$M(X) = x_1^2 + x_2^2 + K(x_1 + x_2 - 1)^2$$

which according to $\nabla M(X) = 0$, yields

$$x_1 + K(x_1 + x_2 - 1) = 0$$
$$x_2 + K(x_1 + x_2 - 1) = 0$$

then

$$x_1^* = x_2^* = \frac{K}{2K+1} \quad \text{when } K \to \infty, \quad x_1^* = x_2^* = \frac{1}{2}.$$

The above method is the external penalty method. During the search, each extremum solution lies outside the feasible region. The sequence of the extrema converge to the desired solution, as the penalty factor is increased.

For the problem with inequality constraints

$$\begin{cases} \text{Minimize} & F(X) \\ \text{Subject to} & g_j \leq 0 \end{cases} \tag{11.6.19}$$

The generalized objective function is

$$M(X) = F(X) + K \sum_{j=1}^{m} g_j^2(X) u(g_j(X)) \tag{11.6.20}$$

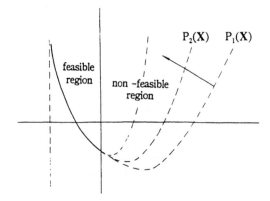

Fig. 11.6.4. Physical meaning of the penalty function method

11.6 Constrained optimization methods

where

$$u(g_j(X)) = \begin{cases} 0 & g_j(X) \geq 0 \\ 1 & g_j(X) < 0 \end{cases} \tag{11.6.21}$$

A similar sequence, in which the search is always in the feasible region is called the interior penalty function method or the barrier function method. A barrier function has to be added to the objective function, to construct a generalized objective function. This function insures that the search is always in the feasible region i.e.

$$M(X, K_j) = F(X) + K_j \beta_j(X) \tag{11.6.22}$$

$$\beta_j(X) = \sum_{j=1}^{m} \frac{1}{g_j(X)}. \tag{11.6.23}$$

$K_j \beta_j(X)$ is called a barrier term K_j is the jth barrier factor. $K_j > 0$ is monotonously decreased during the search in succession. When X lies on the boundary of the feasible region, at least one of the constrained functions $g_i(X) \to 0$, then the barrier term tends to ∞. In each subsequent search, the extrema lies within the feasible region, the boundary is an obstacle that is not crossed. The barrier function method is only suitable for inequality constraints.

The steps of the barrier function method are summarized as:

(1) Define an initial point in the feasible region, and the penalty factor K_1 (for instance $K_1 = 10$).
(2) Solve the unconstrained minimization X_j of $M(X, K_j)$.
(3) If the barrier term is less than a small value ε (for instance $\varepsilon = 10^{-6}$) X_j^* is the minimum of $F(X)$, otherwise let $K_{j+1} = cK_j$ (for instance $c = 0.1$–0.02). Then return to step 2.

If there are equality and inequality conditions, then the generalized objective function is:

$$M(X) = F(X) + K_j \alpha(X) \tag{11.6.24}$$

where

$$\alpha(X) = \sum h_i^2(X) + \sum_{j=1}^{m} g_j^2(X) u(g_j(X)) \tag{11.6.25}$$

$$\alpha(X) = \begin{cases} 0 & X \in D \\ 1 & X \notin D \end{cases} \quad (D \text{ represents the feasible region}) \tag{11.6.26}$$

Example 11.6.3

$$\begin{cases} \text{Minimize} & F(X) = x - 1 \\ \text{Subject to} & 0 \leq x \leq 1 \end{cases}$$

Solution.

$$M(X) = (x - 1) + \frac{K}{x} + \frac{K}{1-x}$$

Solve

$$\frac{\partial M}{\partial x} = 1 - \frac{K}{x^2} + \frac{K}{(1-x)^2} = 0$$

which yields

$$K = \frac{x^2(1-x)^2}{1-2x}$$

when

$$K \to 0, \quad x^* = 0 \quad \text{or} \quad x^* = 1.$$

In order to judge these two possible solutions, the behaviour of K in the vicinity of these two solutions must be examined.

Let

$$x = \varepsilon \ll 1, \quad K = \frac{\varepsilon^2(1-\varepsilon^2)}{1-2\varepsilon} > 0$$

if

$$x = 1 - \varepsilon, \varepsilon \ll 1, \quad K = \frac{(1-\varepsilon)^2 \varepsilon^2}{2\varepsilon - 1} < 0.$$

By definition, K must be a positive number, thus $x^* = 0$ is the minimizer.

11.6.4 Sequential unconstrained minimization technique

In real physical problems, the objective function is cumbersome. Therefore it is impossible to obtain the analytical solution as shown in Example 11.6.3. For a large value of K, the Hessian matrix of $M(X)$ is ill-conditioned. Consequently, the value of the penalty factor has to be increased step by step. The minimum of the original function is to be found by a series of unconstrained optima of the function $M_k(X)$:

$$M_k(X) = F(X) + K_k \sum_{i=1}^{m} \frac{W_i}{g_i(X)} \quad k = 1, \ldots, L \quad (11.6.27)$$

where $K_k \geq 0$ for all k and $W_i > 0$ for all i. W_i are weighting factors that remain fixed throughout the calculation, while K_k are parameters that decrease from one iteration to the next. The solution in each step will be the initial value of the next step. Thus the constrained optimization is transformed to solve a sequence of unconstrained optimization problems. This method is called a sequential unconstrained minization technique (SUMT). It is a highly developed form of the penalty function method in which function values are needed only at feasible points. Reference [19] gives a good example, in which uses the penalty function to deal with the equality constraint conditions and the barrier function to approach the inequality constraint conditions. The author used this method to design a permanent magnet synchronous machine.

11.7 Summary

In this chapter, most of the commonly used unconstrained optimization methods have been introduced. They are classified as analytical methods and direct methods. Both methods use iteration procedures to force the objective function to approach zero. In using the method of steepest descent and/or method of conjugate gradient, the search directions (p_k) are determined by the first derivative of the objective function. In other methods like DFP or BFGS and method of least squares, the direction of search is based on the second order derivative of the objective function. Thus the rate of convergence is high. In the DFP method, in order to avoid the calculation of the second order derivative of the objective function, it is replaced by an artificially constructed H matrix as shown in Eq. (11.4.30). The H matrix uses only the first order derivative of the objective function. This method is useful to solve problems with a large number of variables. Reference [12] give an example having 100 variables. Reference [13] uses this method to find the minimum of the objective function having 172 independent variables. The disadvantage of the DFP or BFGS method is that they require larger computer storage than the conjugate gradient method. In the method of least squares, the second order derivative of the objective function is approximated by $G(X) \cong 2JJ^T$. where J is the matrix of the first order derivative of the objective function. Hence, only the first order derivative of the objective function is calculated. For the conjugate gradient method, the least squares method or the BFGS method are preferred.

Finally, it should be mentioned that in using analytical methods, in the case where the analytical expression of the objective function does not exist or the derivative of the objective function is difficult to derive, the derivative of the objective function is replaced by the difference of the objective function.

The greatest advantage of the direct method is that only the function value is calculated during the search process. The disadvantage of the direct method is that the number of the independent variables are limited. The variables in using the polytope method is less than 10.

A linear search is usually incorporated to prevent divergence of the iteration. The quadratic and cubic interpolation methods and the golden section method are general methods in determining the optimum step during each iteration of the optimization of multivariable functions.

Finally, constrained optimization problems are usually solved by sequential unconstrained minimization techniques.

References

1. W. Murray (Ed), *Numerical Method for Unconstrained Optimization*, Academic Press, London, 1972
2. G.R. Walsh, *Methods of Optimization*. John Wiley & Sons, 1975
3. Philipe, E. Gill, W. Murray, M.H. Weight, *Practical Optimization*. Academic Press, 1981
4. David A. Wismer, R. Chattergy, *Introduction to Nonlinear Optimization (A Problem Solving Approach)*. Elsevier North-Holland, 1978
5. R. Fletcher, *Practical Methods of Optimization*, 2nd edn. John Wiley, 1987
6. J. Kuester, *Optimization Techniques with Fortran*, McGraw-Hill, 1973
7. William H. Press, Brian P. Flannery, et al, *Numerical Recipes, The Art of Scientific Computing*. Combridge University Press, 1986
8. H.H. Rosenbrock, An Automatic Method for Finding the Greatest or Least value of a Function, *Comp. J* 3 175–184, 1969
9. R. Fletcher, C.M. Reeves, Function Minimization by Conjugate Gradients, *Comp. J.*, 7(2), 149–153, 1964
10. W.C. Davidson, Variable Metric Method Minimization, *AEC Research and Development Report*, ANL-5990 (Rev)
11. R. Fletcher, M.J.D. Powell, A Rapidly Convergent Descent Method for Minimization, *Comp. J.*, 6, 163, 1963
12. R. Fletcher, A New Approach to Variable Metric Algorithm, *Comp. J.*, 13, 317–322, 1970
13. Fei Zeng-yao, Zhou Pei-bai, Optimum Design of the High Voltage Electrode, *5th ISH*, Braunschweig, 31.09, 1987
14. M.J. Hopper (Ed), Harwell Subroutine Library Catalogue, Theoretical Physics Division, A.E.R.E. Harwell, England, 1973
15. Wang Jing-guo, M.T. Ahmed, J.D. Lavers, Nonlinear Least Squares Optimization Applied to the Method of Fundamental Solutions for Eddy Current Problems, *IEEE Trans. on Mag.*, 1990
16. Sun Xiao-rui, Automatic Optimal Design of Electrode Figuration of SF_6 Circuit Breaker, *5th ISH*, Braunschweig, 31.08, 1987
17. P.E. Gill, W. Murray (Ed.), *Numerical Methods for Constrained Optimization* Academic Press, 1974
18. A.V. Fiacco, G.P. McCormick, *Nonlinear Programming: Sequential Unconstrained Minimization Techniques*. John Wiley, New York, 1968
19. S. Russenschuck, Mathematical Optimization Techniques for the Design of Permanent Magnet Synchronous Machines Based on Numerical Field Calculation, *IEEE Trans. on Mag.*, 26(2), 638–641, 1990

Chapter 12

Optimizing Electromagnetic Devices

12.1 Introduction

The problem of optimum design has been studied for a long time. One of the earliest example was to find a 2-D shape which occupied the maximum area with its circumference as a given constant. Before the 1950's classical mathematical methods such as the differential and variational methods were used to solve these problems. With digital computers and using numerical methods, the methods of optimal design developed rapidly.

The optimal shape design is to determine the dimensions of a device for a specific purpose and at the same time satisfying the limitation set by physical constraints. This is an inverse or a synthetic problem. Synthesis is always more complicated than the analysis. In general, the subject includes at least two aspects: numerical solution of partial differential equations and the optimization method. The optimal shape design of structural and mechanical devices are discussed in references [1-3]. With electromagnetic problems, the analysis methods of field probloems are well understood and the design of the devices has a long history but most of them are familiar with the treatment using the principle of equivalent circuit theory. Now it is possible to deal with design problems based on field analysis to obtain more accurate results. The inverse problem or the optimum design is important for the

(1) Design of devices to obtain desired field distribution and obtain a maximum economic efficiency.

(2) Determination of the source distribution to satisfy a required field distribution.

(3) To find discontinuities in materials. The method based on electromagnetic field analysis is the basis of non-destructive testing see reference [4].

(4) Determination of material constants. This is important both in electrical engineering and in biomedical science. For example, the aim of impedance tomography is to map the distribution of electrical conductivity within a body by applying a voltage to the body surface and measuring the injected current or vice versa.

These problems are more complicated than straight field analysis. Here two difficult problems are combined, one is the analysis of the field distribution, the other one is the determination of the design variables in such a way that the desired objectives are fulfilled. An optimm design is generally the result of a combination of mathematical results, empirical data and the experience of scientists and engineers. In this chapter two different kinds of the optimization methods are introduced. One is the combination of the numerical solution and the mathematical optimization procedures. The other is based on numerical solution using some physical principles. Detailed process is a combination of the solution method of a field analysis and the mathematical optimum. Hence in this chapter only the general methods of the optimal design are introduced, more information is given in references.

12.2 General concepts of optimum design

The process of optimal design is demonstrated in Fig. 12.2.1. The mathematical model includes both the differential and the integral equations which describe the boundary value problems and the objective function which describes the design puposes. The optimization procedure may be any one of the mathematical optimization methods or the method based on the physical principles. These methods are used to adjust the design variables to satisfy predetermined goals.

12.2.1 Objective functions

The objective function may have the following forms

$$F(\mathbf{X}) = \text{Const.} \qquad (12.2.1)$$

$$F(\mathbf{X}) = \max(\min) \qquad (12.2.2)$$

$$F(\mathbf{X}) = \alpha F_1(\mathbf{X}) + \beta F_2(\mathbf{X}) = \min(\max) \qquad (12.2.3)$$

$$F(\mathbf{X}) = \frac{F_1(\mathbf{X})}{F_2(\mathbf{X})} = \max(\min) \qquad (12.2.4)$$

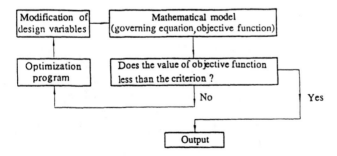

Fig. 12.2.1. Diagram of the optimum design

12.2 General concepts of optimum design

where **X** is a column matrix that represents the design variables. For instance, the distribution of the field strength along the surface of an electrode is required to be uniform and to have a predetermined value. The objective function is now expressed by Eq. (12.2.1). If the maximum field strength along the surface of an electrode is limited to a minimum value, Eq. (12.2.2) is the proper objective function. In designing a transformer, usually the combined price of copper and magnetic steel has to be minimal while the power losses must not exceed the desired value. Here the objective function will have the form of Eq. (12.2.3) where $F_1(X)$ is the function which relates the price of the material and the design variables, $F_2(X)$ is the function that relates the power losses and the design variables; α, β are weighting coefficients of the two functions $F_1(X)$ and $F_2(X)$. In designing electric machines, the maximum flux density in the air gap and the cost of magnetic materials must be minimized thus the proper function is given by Eq. (12.2.4).

Another problem is that the same results may be obtained by using different objective functions. For example, to obtain uniform distribution of field strength along the surface of an electrode, the objective function may be:

$$F(\mathbf{X}) = E_{max} - E_{min} = \min \quad (12.2.5)$$

or

$$F(\mathbf{X}) = E_{max} - E_{av} = \min \quad (12.2.6)$$

or

$$F(\mathbf{X}) = (E_{max} - E_{min})^2 = \min \quad (12.2.7)$$

where E_{av} is the average value of E along the contour of the electrode. The choice of the objective function determines the process of mathematical optimization. For example, if Eq. (12.2.7) is chosen as the objective function then the non-linear least square method is suitable to be chosen as the method of optimization. Thus the choice of the objective function is an important decision as it influences the efficiency even the success of the optimum design.

In most practical cases, the design variables are limited to have values within a given range. In this case, the problem is a constrained optimization problem i.e.

$$\begin{cases} \text{Minimize} & F(X) \\ \text{Subject to} & g_1(X) = 0 \\ & g_2(X) \geq 0 \end{cases} \quad (12.2.8)$$

where $g_1(X)$, $g_2(X)$ are constraints to be met by the design variables X.

The target of the objective function must consider the practical possibilities and usually needs to be adjusted during the process of optimization, as the initial target may be too limited or unreasonable.

12.2.2 Mathematical expressions of the boundary value problem

The solution method of a field problem depends on the equation which describes the physical problem. Either one of the domain methods may be used to solve differential equations which describe the problem, or the boundary methods are used to solve the corresponding integral equations which describe the problems. During the process of optimal design, the design variables are modified, e.g. dimensions, contours and boundaries. As the boundaries may change during the process, the boundary methods are more convenient than the domain methods used to analyze field distributions. Here the boundary element discretization is refined during each iteration. The drawback of domain methods is that it requires continuous use to refine the whole mesh during the search for an optimum design. Hence integral equation methods are preferred compared to the differential equation method in the process of optimization.

12.2.3 Optimization methods

Optimization methods are used to adjust the design variables automatically and to force the objective function to approach its desired value. There are two different methods to adjust the design variables. One depends on the mathematical programming which is discussed in Chap. 11. The choice of mathematical optimization methods depends on the number of the variables, the characteristic of the objective function and on the type of the computer available. The other method may be specially designed to suit the physical characteristics of the problem. For example the strength of the electric field is proportional to the curvature of the surface of the electrode. Therefore a uniform distribution of the field is obtained by adjusting the curvature of the electrode. Two of these types of methods will be introduced in discussing the contour optimization of the electrode.

During the process of optimization, all the design variables of the problem are considered together or the variables may be grouped and considered one by one. It is even possible to consider the change of variables one by one. It looks tedious but this simple strategy may save time and yield the best results.

The flow-chart of the optimization procedures is shown in Fig. 12.2.2.

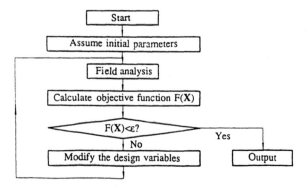

Fig. 12.2.2. The flow-chart of the optimization

12.2.4 Categories of optimization

The optimization problems of electromagnetics may be divided into two categories: domain optimization and contour optimization. The domain optimization finds a predetermined field distribution in a specific region. The design variables may be the source distribution or the shape of the devices. In the design of magnets, electron optics, mass spectrography or Tokomak devices, a specific field distribution within a region is usually be necessary. On the other hand, contour optimization determines the shape of the contour in such a way that the field distribution along the contour can satisfy specific requirements. In most high voltage devices, the field strength along the surface of the electrode or along the interface of dielectrics are parameters to meet the given conditions. In the next two subsections, both contour and domain optimization are discussed.

12.3 Contour optimization

In many electromagnetic devices, the field distribution must be homogeneous. One of the earliest paper discussing the problem of achieving an homogeneous field analysis for a capacitor was published in 1923 by W. Rogowski [5]. In this paper, the field calculation used conformal maps.

Advances in numerical methods for field analysis resulted in a number of methods to investigate the optimum design of the electrodes (see references [6–9]). The methods available are summarized in the following table.

Table 12.3.1. The methods for the contour optimization

Field calculation methods		Optimization methods
Domain method	Boundary methods	Curvature adjustment [6, 11]
FEM [10]	CSM [7, 11–13]	Optimization by charge redistribution [7, 12, 13]
	SSM [14–17]	Non-linear programming [9]
		Other physical methods [17]

12.3.1 Method of curvature adjustment

It is well known that the charge distribution along the surface of the electrode is dependent on the curvature of the contour. Figure 12.3.1 shows two general equipotential lines and two lines of the field strength. P_1 and P_2 are two points close to each other on equipotential lines. E_1 and E_2 are the field strengths at P_1 and P_2. Hence

$$E_1 l_1 = E_2 l_2 \qquad (12.3.1)$$

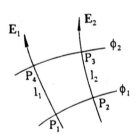 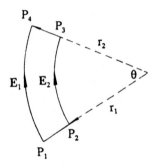

Fig. 12.3.1. A curvilinear rectangle composed of E, φ lines

Fig. 12.3.2. A curvilinear rectangle composed of two circular E lines

where l_1 and l_2 are distances between two equipotential lines φ_1 and φ_2, respectively. Through the simple manipulation of Eq. (12.3.1), one obtains

$$\frac{E_2 - E_1}{E_1 + E_2} = \frac{l_1 - l_2}{l_1 + l_2}. \tag{12.3.2}$$

Multiplying $1/2 \Delta n$ by the denominator of both sides of the last equation yields

$$\frac{E_2 - E_1}{1/2(E_1 + E_2)\Delta n} = \frac{l_1 - l_2}{1/2(l_1 + l_2)\Delta n} \tag{12.3.3}$$

where Δn is an incremental length perpendicular to the E line, i.e. $\Delta n = P_1 P_2$. If Δn is sufficiently small, the field strength E can be approximated by

$$E = \tfrac{1}{2}(E_1 + E_2). \tag{12.3.4}$$

Then the LHS of Eq. (12.3.3) becomes:

$$\frac{E_2 - E_1}{E \Delta n} = \frac{1}{E}\frac{\Delta E}{\Delta n}. \tag{12.3.5}$$

Assume that $P_1 P_2$ and $P_3 P_4$ as shown in Fig. 12.3.2, are two small sections of a circle.
  Let

$$\Delta S = S_{p_1 p_2 p_3 p_4} = (r_2 - r_1)\tfrac{1}{2}[\vartheta(r_1 + r_2)] \tag{12.3.6}$$

due to

$$\Delta l = \vartheta(r_2 - r_1) \tag{12.3.7}$$

then

$$\frac{\Delta l}{\Delta S} = \frac{1}{\tfrac{1}{2}(r_1 + r_2)} = \frac{1}{r_m} = C \tag{12.3.8}$$

where r_m is the geometric mean radius of a plane curve. By definition, C represents the mean curvature of a plane curve. A combination of Eq. (12.3.3), (12.3.5) and (12.3.8) shows that the relationship between the field strength and the

12.3 Contour optimization

curvature of the contour of an electrode is

$$\frac{1}{E}\frac{\partial E}{\partial n} = C. \qquad (12.3.9)$$

Equation (12.3.9) is the one used to modify the curvature so as to achieve any desired change of the field strength. The deviation of the field strength ΔE from the desired value is calculated by the differentiating Eq. (12.3.9)

$$\Delta C \cong \frac{1}{E^2}\frac{\partial E}{\partial n}\Delta E. \qquad (12.3.10)$$

Applications of this method are given in references [6] and [11].

In a 3-D case, the curvature of a curved surface is defined as in reference [18]

$$C = \frac{1}{R_1} + \frac{1}{R_2} \qquad (12.3.11)$$

where $1/R_1$, $1/R_2$ are two main curvatures orthogonal to each other.

12.3.2 Method of charge redistribution

This method is an old one, having been already suggested in 1979 in reference [7]. It was further developed in references [12] and [13]. The method was based on the advantages of the charge simulation method. When the simulated charge is determined, the equi-potential lines are calculated and one of these lines is then the first approximation to the desired contour of the electrode. Next, the simulated charges are altered to obtain a new contour. Repeating this procedure several times yields the desired contour of the electrode.

Figure 12.3.3(a) shows a pair of axisymmetric electrodes, the figure of the top of the electrode is to be optimized. In order to simplify the calculation, the contour of the electrode is subdivided into two parts: a fixed part AB and a part BC which is to be optimized.

Charges 1, 2 and 3 simulated the part AB of the contour, these simulated charges are denoted as Q_j. The position of these charges are fixed while the contour BC is changing. Charges 4, 5 and 6 corresponding to the part BC of the contour are the charges to be optimized and are denoted as $\tilde{Q}_j$. These are adjusted to fit the optimization of the electrode contour. The equation of CSM is

$$\sum_{j=1}^{m} P_{ij}Q_j + \sum_{j=m+1}^{n} P_{ij}\tilde{Q}_j = \varphi_i \qquad (12.3.12)$$

where m is the number of the simulated charges and $n - m$ is the number of the charges to be optimized. φ_i are the known potentials on the part AB of the electrode. In Eq. (12.3.12), both the positions and amplitudes of charges $\tilde{Q}_j$ are known. The unknown chargwes Q_j are determined by Eq. (12.3.12) at the

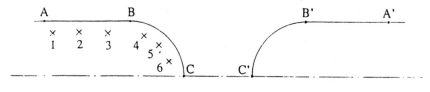

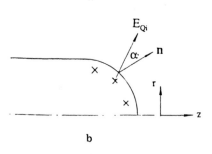

Fig. 12.3.3a, b. Method of charge redistribution

assumed position of Q_j. The field strength on the optimized part is calculated by

$$[f_r]\{Q\} = \{E_r\}$$
$$[f_z]\{Q\} = \{E_z\}$$
$$\{Q\} = \{Q_1 \dots Q_m \dots Q_n\} \tag{12.3.13}$$

where E_r, E_z are the components of E while $[f_r]$ and $[f_z]$ are components of the coefficient matrices of field strength. The change of magnitude of the field strength is dependent on the change of the charge $\{Q + \tilde{Q}\}$, i.e.,

$$[f]\Delta\{Q + \tilde{Q}\} = \{\Delta E\} \tag{12.3.14}$$

where $\{\Delta E\}$ is the difference between the desired value $\{E_0\}$ and the calculated value $\{E\}$, i.e.,

$$\{\Delta E\} = \{E\} - \{E_0\}. \tag{12.3.15}$$

Equation (12.3.14) illustrates that the increment $\{\Delta E\}$ is proportional to the increment of $\Delta\{Q + \tilde{Q}\}$. The relative coefficient matrix $[f]$ can be obtained by using different methods as shown in references [7, 12, 13], respectively. Reference [7] considers that the optimized charge $\tilde{Q}_j$ is the main factor to influence the field strength along the contour BC, hence

$$[\tilde{f}]\{\Delta\tilde{Q}\} = \{\Delta E\} \tag{12.3.16}$$

where

$$\tilde{f}_{ij} = (\tilde{f}_{r_{i,j}}^2 + \tilde{f}_{z_{i,j}}^2)^{1/2} \cos\alpha_i \tag{12.3.17}$$

and α_i is an angle shown in Fig. 12.3.3(b). $[\tilde{f}]$ is the coefficient of field strength corresponding to $\tilde{Q}$. Reference [12] derives a corrective coefficient matrix $[C]$,

12.3 Contour optimization

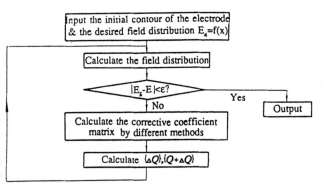

Fig. 12.3.4. The flow-chart of charge redistribution method

it modifies the values of the optimization charges $\{Q\}$ while the position of the charges $\{\tilde{Q}\}$ is fixed. Reference [13] changes both the position and the magnitudes of $\{\tilde{Q}\}$. The flow-chart of this method is shown in Fig. 12.3.4.

12.3.3 Contour optimization by using non-linear programming

This method uses mathematical optimization to find the extremum point of the objective function directly.

The method of surface charge simulation states

$$\mathbf{P}\{\sigma\} = \{\varphi\} \qquad (12.3.18)$$

that the field strength on the electrode surface or on the interface of the dielectrics is directly proportional to the surface charge $\{\sigma\}$. The objective function is chosen as

$$F(X) = \sum_{i=1}^{m} (E_i - E_{i0})^2 = \min \qquad (12.3.19)$$

where E_i are the calculated values and E_{i0} are the desired values along the contour of the electrode. The total number of the discretization points on the contour is m. In order to reduce the number of design variables, assume that any surface to be optimized is moved only in a normal direction to the contour. Any method of nonlinear optimization to minimize the quadratic function introduced in Chap. 11 may be used to solve Eq. (12.3.19). The derivative of the objective function is obtained and used whenever possible in analytical form. Whether this is possible depends on the form of the approximating function of the charge distribution $\{\sigma\}$. If no analytical function exists, the derivative of the objective function is replaced by the differences of the objective function. Reference [9] compares different methods of optimization such as the method of steepest descent, quasi-Newton's methods and the conjugate gradient method to solve practical problems.

12.4 Problems of domain optimization

The field distribution that has to satisfy a specific requirement within a region is defined as a domain optimization. For example the design of the magnets may requires

(a) a constant air gap flux density
(b) uniform air gap field
(c) prescribed field distribution in the air gap
(d) prescribed field distribution in a specific region

These problems are regarded as the domain optimization. The predetermined targets are met with suitable dimensions of the magnet and/or by using a suitable exciting coil. In the past, these problems were treated by circuit analysis. Recently, optimum design of magnets is obtained by field analysis [19, 20]. Several methods are shown in the following sub-sections.

12.4.1 Field synthesis by using Fredholm's integral equation

A predetermined distribution of the magnetic field strength along the axis of a solenoid [21] or in a plane perpendicular to the axis of the solenoid [22] or in a volume within the solenoid [23] are commonly required. All these problems are solvable by using Fredholm's integral equation of the first kind.

Assume that the solenoid is composed of a number of similar sections shown in Fig. 12.4.1(a). Each of these sections may have different current densities. The vector potential of a filamentary circular loop with radius r' (Fig. 12.4.1(b)) is

$$A = A(r, z)n_0 \qquad (12.4.1)$$

where

$$A(r, z) = \frac{\mu_0 I}{2\pi}\left(\frac{r}{r'}\right)^{1/2} f(k) \qquad (12.4.2)$$

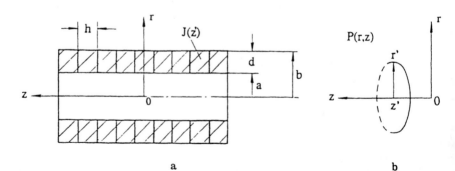

Fig. 12.4.1a, b. A solenoid composed of segments

12.4 Problems of domain optimization

(r, z) is the location of the observation point and I is the current in the loop. In Eq. (12.4.2),

$$f(k) = \left(\frac{2}{k} - k\right)K(k) - \frac{2}{k}E(k) \tag{12.4.3}$$

where

$$k^2 = \frac{4rr'}{(r + r')^2 + (z - z')^2} \tag{12.4.4}$$

$K(k)$, $E(k)$ are complete elliptic integrals of the first and second kind. Take the curl of vector potential A to obtain the magnetic field strength $\mathbf{H}$

$$\mathbf{H} = \frac{1}{\mu_0} \nabla \times A \ . \tag{12.4.5}$$

The component of the magnetic field strength along z direction is

$$H_z(r, z) = \frac{1}{\mu_0 r} \frac{\partial}{\partial r}[rA(r, z)] = \frac{I}{2\pi} g(r, r'\, z, z') \tag{12.4.6}$$

where

$$g(r, r'\, z, z') = \frac{1}{[(r + r')^2 + (z - z')^2]^{1/2}}$$

$$\times \left[K(k) + \frac{r'^2 - r^2 + (z - z')^2}{(r - r')^2 + (z - z')^2} E(k)\right] . \tag{12.4.7}$$

Assuming that the current density is uniform in each small section of length h and width d, hence the field strength is

$$H(r, z) = \frac{J_i}{2\pi} \int_a^b \int_{z'-h/2}^{z'+h/2} g(r, r', z, z') dr'\, dz' \ . \tag{12.4.8}$$

Consider the symmetry of the solenoid, the magnetic field strength produced by whole solenoid is

$$H(r, z) = \frac{1}{2\pi} \sum_{i=1}^{N} J_i \int_a^b \int_{z'-h/2}^{z'+h/2} [g(r, r', z, z') + g(r, r', z, -z')] dr'\, dz' \tag{12.4.9}$$

where N is half of the total number of sections. If $r = 0$, the integrand of Eq. (12.4.9) is

$$g(r, r', z, z') = r'^2 [r'^2 + (z - z')^2]^{-3/2} \ . \tag{12.4.10}$$

The discretized form of Eq. (12.4.9) is

$$H(r_j, z_j) = \frac{1}{2\pi} \sum_{i=1}^{N} a_{ij} J_i \ . \tag{12.4.11}$$

This is a set of algebraic equations. It can be written in a general form

$$\mathbf{AX} = \mathbf{B} \tag{12.4.12}$$

The **X** is the unknown current density {**J**} in each section of the solenoid, **B** is the desired field distribution. The elements of matrix **A** are

$$a_{ij} = \int_a^b \int_{z'-h/2}^{z'+h/2} [g(r, r', z, z') + g(r, r' z, -z')] dr' dz' . \qquad (12.4.13)$$

The solution of Eq. (12.4.12) yields the current distribution. Unfortunately, the solution of Eq. (12.4.12) is unstable due to the fact that Fredholm's integral equation of the first kind (Eq. (12.4.8)) is ill-conditioned[†]. A small error in the data yields great errors in the results. To overcome this problem, one method is to transform Eq. (12.4.12) as

$$\mathbf{AX} + \alpha \mathbf{IX} = \mathbf{B} \qquad (12.4.14)$$

In Eq. (12.4.14) **I** is a unit matrix, α is a regularization parameter. α is hard to estimate. Initially it can be assumed as an arbitrary constant until the final results satisfy Eq. (12.4.12). References [21, 23] describe the method of this solution. The influence of the value of α on the results is also given in these papers.

12.4.2 Domain optimization by using non-linear programming

In designing various magnetic devices, the design variables may be the contour of the iron yoke, the profile of the coil [19] or the dimensions of the winding [24, 25]. All these are shown in Fig. 12.4.2. The method using non-linear programming for domain optimization is explained as follows.

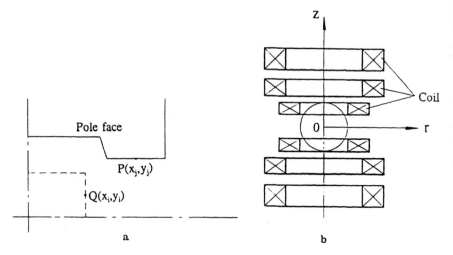

Fig. 12.4.2a, b. Different objects to be optimized

[†] A matrix **A** whose inverse $\mathbf{A}^{-1}$ contains very large elements comparison with those of matrix **A** is a set of ill-posed equations.

12.4 Problems of domain optimization

Assume that the contour of the iron boundary shown in Fig. 12.4.2(a) which is to be modified in order to find the magnetic field over a prescribed region (e.g. the region enclosed by the line of dashes) approaches a predetermined distribution. The points $P(x_j, y_j)$ on the pole face are varied, the points $Q(x_i, y_i)$ within the space are fixed. Then the flux density within the space can be expressed by:

$$B_i = f(x_i, y_i; x_j, y_j) \quad i = 1, \ldots, n \quad j = 1, \ldots, m. \tag{12.4.15}$$

Assume now that the required value of the flux density inside the region enclosed by the line of dashes is B_0. Then the objective function $F(X)$ is:

$$F(X) = \sum_{j=1}^{m} (\Delta B_j / B_0)^2 = \min \tag{12.4.16}$$

where

$$\Delta B_j = B_j - B_0. \tag{12.4.17}$$

The variables x_j and y_j are obtained by minimizing the function $F(X)$.

No matter whether the objective function can be expressed analytically or not, non-linear programming can adjust the design variables automatically so as to force the objective function to tend to a minimum. Reference [25] introduces different mathematical programming to obtain a uniform field distribution within a sphere centred at the origin of the coordinates shown in Fig. 12.4.2(b) where the dimensions of the six coils are design variables.

One example of the constraint optimization design is given in reference [26]. The object is to obtain the maximum flux density in the air gap of a permanent magnet machine shown in Fig. 12.4.3.

Here the objective function is defined as:

$$\text{Max } F(X) = \frac{B_\delta(X)}{V(X)} \tag{12.4.18}$$

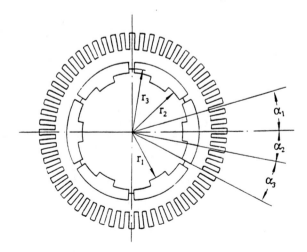

Fig. 12.4.3. The optimum design of permanent magnet machine

where $V(X)$ is volume of the rotor of the permanent magnet machine. The design variables $X(\alpha_i, r_i)$ are the sizes and the geometry of the magnet pole. These variables r_i, α_i ($i = 1, 2, 3$) have to satisfy the following constraints, e.g.

$$a_i \leq r_i \leq b_i \quad \text{and} \quad \alpha_i \leq \vartheta_i \leq \beta_i \,. \tag{12.4.19}$$

Using the barrier function method, the constrained optimization problem is transformed to a sequence of unconstrained optimization problems, i.e. the equivalent objective function is

$$\max F(X) = \max \left\{ \frac{B_\delta(X)}{V(X)} - \sum_{i=1}^{6} \frac{1}{p} \left[\left(\frac{x_i}{x_{iu} - x_i} \right) + \left(\frac{x_i}{x_i - x_{il}} \right) \right] \right\} \tag{12.4.20}$$

where p is the penalty parameter and the subscripts u and l indicate the upper and the lower bound of the design variables. Any of the unconstrained optimization methods which were introduced in Chap. 11 can be used to solve Eq. (12.4.20). Results using different optimization methods are given in reference [26].

12.5 Summary

The optimum design of electromagnetic devices is achieved by combining field analysis and mathematical optimization. The general idea of optimum design has been introduced in this chapter. The methods of solution are problem dependent and the results are not unique. Except in a few well defined problem it is unlikely that the first automatically-optimized solution would be entirely satisfactory. The user's intervention in redefining the requirements and constraints is needed during the process of optimum design.

References

1. James N. Siddall, *Optimal Engineering Design*, Marcel Dekker, Inc. 1982
2. O. Pironnean, *Optimal Shape Design for Elliptic Systems*, Springer-Verlag, 1984
3. C.H. Tseng, J.S. Arora, On Implementation of Computational Algorithms for Optimal Design, *Int. J. Num.* (Ms in Eng), 26(6), 1365–1402, 1988
4. Y. Iwamura, K. Miya, *Characterization of a Crack Shape by Potential Method, Electromagnetic Fields in Electrical Engineering*. International Academic Publishers (Proc. of BISEF '88), pp. 300–303, 1989
5. W. Rogowski, Die Elektrische Festigkeit am Rande des Platten-kondensators, *Arch, Electrotech*, 12, 1–14, 1923
6. H. Singer, P. Grafoner, Optimization of Electrode and Insulator Contours, *2nd Int. Symp. on High Voltage Engn*, Zurich, pp. 11–116, 1975
7. D. Metz, Optimization of High Voltage Fields, *3rd Int. Symp. on High Voltage Engines.*, Milan, 11.12, 1979
8. H.H. Daumling, H. Singer, Investigation on Field Optimization of Insulator Geometries, *IEEE*, SM 568-8, 1988

References

9. H. Tsusor; T. Misaki, The Optimum Design of Electrode and Insulator Contours By Nonlinear Programming Using the Surface Charge Simulation Method, *IEEE Trans. on Mag*, 24(1), 35-38, 1988
10. Sun Xiao-rui, Automatic Optimal Design of Electrode Figuration of SF_6 Circuit Breaker, *5th Int. Symp. on High Volt. Engines*, Braunschweig, 31.08, 1987
11. H. Klement, B. Bachmann, H. Gronewald, H. Singer, Layout of HV Components by Computer Design, *CIGRE Paper*, 2, 33-04, 1980
12. Fei Zengyao, Zhou Peibai, Optimum Design of the High Voltage Electrode, *5th Int. Symp. on High Volt. Engines*, Braunschweig, 31.09, 1987
13. Liu Jin, Sheng Jianni, The Optimization of the High Voltage Axisymmetrical Electrode Contour, *IEEE Trans. on Mag.*, 24(1), 39-42, 1988
14. T. Misaki, H. Tsuboi, K. Itaka, T. Hara, Computation of Three-Dimensional Electric Field Problems by a Surface Charge Method and its Application to Optimum Insulator Design, *IEEE Trans. on PAS*, 101(3), 627-634, 1982
15. T. Misaki, Optimization of 3-D electrode Contour Based on Surface Charge Method and its Application to Insulation Design, *IEEE Trans. PAS*, 102(6), 1687-1792, 1983
16. H. Singer, H.H. Daumling, *Computer-Aided Design of Insulators*, ICPADAM-88, H-09, Beijing, 1988
17. Z. Andjelic, B. Krstajic et al, A Procedure for Automatic Optimal Shape Investigation of Interfaces Between Different Media, *IEEE Trans. on Mag.*, 24(1), 415-418, 1988
18. Von J. Spielrein, Geometrisches zur Elektrischen Festigkeitsrechnung II, *Archiv für Elektrotechnik*, V. 7, 244-254, 1917
19. T. Nakata, N. Takahashi, Direct Finite Element Analysis of Flux and Current Distribution Under Specified Conditions, *IEEE Trans. on Mag.*, vol. 18(2), 325-330, 1982
20. A.G.A.M. Armstrong, M.W. Fan, J. Simkin, C.W. Trowbridge, Automated Optimization of Magnet Design Using the Boundary Integral Method, *IEEE Trans. on Mag.* 18(2), 620-623, 1982
21. R. Sikora, J. Purczynski, K. Adamiake, Szczecin, The Magnetic Field Synthesis on a Cylinder Solenoid Axis by Means of Tichonov's Regularization Method, *Archiv für Elek.*, 60, 83-S6, 1978
22. R. Sikora, P. Krason, M. Gramz, Magnetic Field Synthesis at the Plane Perpendicular to the Axis of A Solenoid, *Arch. Elek.* 62, 153-156, 1980
23. K. Adamik, Kielce, Synthesis of Homogeneous Magnetic Field in Internal Region of Cylindrical Solenoid, *Arch. für Elek.* 62, 75-79, 1980
24. Massimo Guarnieri, Andrea Stella, A Procedure for Axisymmetric Winding Design Under Parametric Constraints: An application to the RFX Poloidal Transformer, *IEEE Trans. on Mag.*, 21(6), 2448-2452, 1985
25. Ales Gottrald, Comparative Analysis of Optimization Methods for Magnetostatic, *IEEE Trans. on Mag.*, 24, (1) 411-414, 1988
26. R. Russenschuck, Mathematical Optimization Techniques for the Design of Permanent Synchronous Machines Based on Numerical Field Calculation, *IEEE Trans. on Mag.*, 26(2), 638-641, 1990

Subject Index

Bandwidth 108
Basis 42
  global 330, 345
  sub-region 334
  sub-sectional 337, 342
Boundary condition
  Dirichlet 10, 111, 145, 216, 241
  essential 10
  inhomogeneous 21, 104, 144, 161, 162
  interfacial 9, 83, 131, 148, 221, 254, 271
  mixed 47, 125, 218
  natural 131, 146
  Neumann 10, 21, 145, 218
  of the first kind 10, 143, 161
  of the second kind 10, 104, 146, 161
  of the third kind 10, 143, 146
  Robin 10, 145
  surface impedence 316
Boundary value problems 10, 21, 46, 47, 111, 145, 150, 389
Brebbia C. A. 287

Charge
  double layer 22, 24, 25
  line 223, 238
  magnetic surface 270, 280
  point 222
  ring 223, 238
  simulated 215, 217, 221, 370
  single layer 20
  surface 8, 20, 252
Condition
  constrained 146, 236, 352
  extremum 136, 137, 159
  Kuhn-Tucker 378
  necessary 136, 352, 380
  sufficient 353, 380
Condition number 42, 119, 358
Constraints
  equality 377
  inequality 379, 381, 382
Coordinates
  area 165, 178
  Catisian 64, 199, 225, 241
  cylindrical 161, 223, 306
  elliptic cylindircal 225
  generalized 173
  global 176, 179
  local 176, 183, 263, 268, 296
  natural 176, 177, 180, 182
  polar 69
Convergence 79, 363, 367, 368
  global 357
  quadratic 364
  superlinear 358
Criterion 236
  convergent 80, 118, 357
  for slow and rapid 6
  of quasi-static 8
Current
  eddy 7, 310, 341
  impressed 279, 311
  magnetization 28
  surface 8, 21

Decomposition
  Cholesky's 115, 241, 346, 371
  triangular 115
Difference quotients 64, 67, 89
Dipole 22
  magnetic 25
Discretization
  of the boundary 313
  domain 97, 155

Eigen value 42, 43, 165
Element 36, 95, 173
  arced 262
  edge 205
  constant 296
  hexahedral 188
  high order 197, 268
  isoparametric 178, 190
  linear 176, 178, 181, 182, 296, 302
  planar 259
  quadratic 205, 296
  quadrilateral 186, 205
  rectangular 186

Subject Index

Element
  ring 258, 264
  spline function 269
  sub-parametric 190, 205
  sup-parametric 190
  tetragonal 267, 278
  tetrahedral 181, 188
  triangular 98, 155, 165, 179, 184, 267
Ellipse 354
Ellipsoid 354, 363
Equation
  boundary element 56, 288, 299
  boundary integral 287, 292, 294, 295
  charge simulation 216, 228
  Crank-Nicolson 75
  difference 64, 90
  differential integral 35, 314
  diffusion 19, 45, 73, 310, 320
  Euler-Lagrange 138
  Euler's 130, 137, 139, 145, 150
  finite element 113, 156, 161, 163, 164
  Fredholm integral 45, 55, 252, 255, 257, 271, 337, 396
  integral 14, 254, 343
  Hallen's 344
  Helmholtz 19, 45, 148, 164, 295, 311, 315
  Laplace's 8, 19
  matrix 36, 57, 96, 113, 217, 218, 219, 220, 305, 306
    element 101, 191
  Maxwell's 3, 6, 8, 35, 148, 310
  operator 15, 16, 37, 56, 141
  Pocklington's 344
  Poisson's 8, 28, 97, 131, 143, 147, 159, 219, 291, 304
  superparametric 235
  wave 4, 19
Error
  minimization 235, 241, 242
  round off 371, 374
  truncation 89

Fields
  axisymmetric 72
  electromagnetic 142
  magnetic 73, 163, 150, 161, 246
  quasi static 7, 8, 43, 241
  static 7, 8, 43, 241, 336
  steady-state 6
  translational symmetric 341
Five-point star 66
  asymmetric 69, 90
  symmetric 69, 90, 91
Formulation
  explicit 74

  implicit 74, 87
  weak 55, 174, 175, 290
Function
  approximating 173, 174, 290, 327
  Bessel 312, 323, 325
  basis 54, 328, 330, 337, 340, 343, 346
  Dirac-delta 15, 46, 51, 55, 292, 332, 333, 337
  expansion 327
  extremum 130, 136, 153
  Green's 14, 16, 18, 28, 31
  Hankel 19, 315, 324
  interpolation 56, 96, 173, 191, 296, 331
    cubic 362
    linear 178, 189
    quadratic 189, 331, 360
    spline 331
  Kronecker 100
  multivariable 138, 353, 358
  objective 236, 351, 367, 372, 388, 395, 399
  pulse 46, 330, 334, 340, 342
  quadratic 355
  response 15
  shape 91, 97, 99, 165, 177, 180, 191, 296, 315
    exponential 205
    Hermite 173, 199, 200
    Lagrange 173, 183
    spline 205
  sinusoidal 331, 345
  target 351
  transformation 240
  trial 36, 52, 54, 98, 153, 155
  weighting 3, 54, 55, 100, 290, 327, 328, 330, 333, 342
Functional 52, 56, 132, 291
  constrained 131, 145, 149
  equivalent 56, 130, 141, 144, 159

Galerkin B. 54, 100
Gauge 9
  Coulomb's 7, 8, 149, 311
  Lorentz's 5, 344
Gaussian quadrature 273, 275, 307, 340, 346
Grid 64
  regular 36

Harrington R. F. 53, 327

Inner product 39, 327, 339, 347
Integral
  elliptic 224, 257, 264, 276, 397
  singular 251, 272, 339
Iteration
  block 79

Subject Index

Gauss-Seidel 78, 80
Jacobin 78
over-relaxation 78, 118
under-relaxation 73

Kernal 16, 41, 44, 272, 330
  separable 43

Magnetization vector 4, 270
Matrix
  banded sparse 77, 97, 159
  element 101, 105, 157
    Hermite 202
    Lagrangian 193
  global 106
  Hessian 353, 358, 367
  ill-conditioned 226, 235, 358, 363
  Jacobian 192
  positive definite 77, 97, 108, 159, 354
  positive semi-definite 354
  stiffness 97, 107
  symmetric 97, 159
  system 97, 109, 159
  transformation 233, 234
  universal 195, 197
  well-conditioned 328
Mesh generation 121
  adaptive 123
  automatic 123
Method
  approximate 130
  ballooning 168
  barrier function 383, 400
  boundary 213
  boundary element (BEM) 37, 55, 251, 287, 327
  charge simulation (CSM) 14, 215, 220, 251, 327, 333, 393
  collocation 48, 55
  conjugate gradient 119, 364
  DFP 238, 368, 373
  domain 288, 309
  finite difference (FDM) 36, 63
  finite element (FEM) 36, 96, 155, 166, 327
  function comparison 374
  Galerkin's 48, 54, 332
  Gaussian elimination 77, 109, 113, 226
  Gauss-Newton's 372
  golden section 358
  integral equation 309
  iterative 117
  Lagrange multiplier 378
  Levenberg-Marquardt 373
  magnetic surface charge simulation 270
  Newton's 367
  optimization 358, 390
  optimized charge simulation (OCSM) 235, 244, 344
  penalty function 383
  polytype 374
  quasi-Newton's 367
  of charge redistribution 393
  of curvature adjustment 391
  of equivalent source 251
  of least square 372
  of moment 48, 53, 327, 336, 343
  of polynomial interpolation 360
  of steepest descent 363, 367
  of weighted residuals 53
  orthogonal projection 51
  point matching 55, 90, 215, 332, 344, 345
  Powell's 375
  preconditioned conjugate gradient (PCGM) 119
  quasi-Newton's 367
  Ritz 153, 154, 155
  simplex 375
  sub-domain 48
  surface charge simulation (SSM) 251, 327
  variable matrix 368
  variational 52
Minimizer 235, 354, 363, 367
  local 354
  strong 354

Nodes 36, 64, 173

Optimization
  constrained 236, 352, 377, 389, 399
  contour 391
  domain 396, 398
  unconstrained 237, 356, 362
Operator 133
  bounded 41
  continuous 41
  completely continuous 41, 43
  differential 16
  integral 16
  inverse 42, 43
  linear 40, 317
  non-linear 317
  non-self-adjoint 45, 46, 58
  positive definite 41, 52, 142, 288
  positive semi-definite 41
  projection 49
  self-adjoint 41, 43, 143, 288
  symmetric 40, 52
  symmetric positive definite 43
Orthogonal 39
Orthonormal 39

Point
  collocation 217
  matching 217, 220, 229, 235, 242, 244
Polarization vector 4, 26, 220
Polynomial 46, 98, 176
  Chebyshev 330
  complete 191
  Hermite 199
  Lagrange 183
  Legendre 330
Powell M. J. D. 360, 368
Potential
  floating 166
  Scalar 5, 6, 8
  Vector 5, 6, 8, 14
Principal value 134, 135
Principle
  charge conservation 3
  extremum 57, 159
  of error minimum 36, 46, 58, 287
  of superposition 37, 216, 317
  of weighted residuals 36, 43, 47, 58, 90, 96, 287
  variational 36, 51, 58, 91, 96, 336
Problem
  axisymmetric 149, 220, 258, 306
  eddy current 205, 310, 312, 313, 341
  Laplacian 69, 156
  non-linear 96, 317
  open boundary 167
  time-dependent 73, 96, 320
  variational 130, 138, 141
Projection
  orthogonal 49, 58

Schwarz inequality 39
Search
  direction 362, 367
    conjugate 364
    orthogonal 363
    steepest descent 363
  linear 358, 362, 369
Series
  orthogonal 153
  power 53, 328, 340
  Taylor's 65, 74, 134, 353, 354

Silvester P. P. 95, 184
Singer H. 251
Singular, Singularity 20, 58, 273, 293, 367, 371, 375
  weak 272
  strong 272
Solution
  approximate 36, 46, 53, 342
  fundamental 16, 18, 20, 31, 55, 241, 287, 292, 299, 306, 315
Source
  dipole layer 24
  equivalent 14, 19, 215
  equivalent current 246
  double layer 23, 28, 251, 254, 287
  impressed 256, 315
  single layer 21, 28, 220, 251, 254, 287
Space
  complete inner product 40
  Hilbert 38
  inner product 39
  linear 38, 48
  metric 38
  normed linear 39
  unitary 39
Stability 227, 355
SUMT (sequential unconstrained minization technique) 384

theorem
  Green's 12, 14, 28, 139, 144, 148, 288, 291
  Poynting 315
  uniqueness 11, 215
Transformation
  coordinate 192, 225, 232, 259, 262, 274, 278, 280
  of constrained condition 237

Variation 56, 143, 335
  of function 134
  of functional 135
  quadratic 136

Wave length 8, 343

Zienkiewiecz O. C. 95

CPSIA information can be obtained at www.ICGtesting.com
Printed in the USA
LVOW121107280413

331269LV00017B/327/P